Vascular Endothelium
Physiological Basis
of Clinical Problems

NATO ASI Series

Advanced Science Institutes Series

A series presenting the results of activities sponsored by the NATO Science Committee, which aims at the dissemination of advanced scientific and technological knowledge, with a view to strengthening links between scientific communities.

The series is published by an international board of publishers in conjunction with the NATO Scientific Affairs Division

A	Life Sciences	Plenum Publishing Corporation
B	Physics	New York and London
C	Mathematical and Physical Sciences	Kluwer Academic Publishers
D	Behavioral and Social Sciences	Dordrecht, Boston, and London
E	Applied Sciences	
F	Computer and Systems Sciences	Springer-Verlag
G	Ecological Sciences	Berlin, Heidelberg, New York, London,
H	Cell Biology	Paris, Tokyo, Hong Kong, and Barcelona
I	Global Environmental Change	

Recent Volumes in this Series

Volume 202—Molecular Aspects of Monooxygenases and Bioactivation of Toxic Compounds
edited by Emel Arınç, John B. Schenkman, and Ernest Hodgson

Volume 203—From Pigments to Perception: Advances in Understanding Visual Processes
edited by Arne Valberg and Barry B. Lee

Volume 204—Role of Melatonin and Pineal Peptides in Neuroimmunomodulation
edited by Franco Fraschini and Russel J. Reiter

Volume 205—Developmental Patterning of the Vertebrate Limb
edited by J. Richard Hinchliffe, Juan M. Hurle,
and Dennis Summerbell

Volume 206—Alcoholism: A Molecular Perspective
edited by T. Norman Palmer

Volume 207—Bioorganic Chemistry in Healthcare and Technology
edited by U. K. Pandit and F. C. Alderweireldt

Volume 208—Vascular Endothelium: Physiological Basis of Clinical Problems
edited by John D. Catravas, Allan D. Callow,
C. Norman Gillis, and Una S. Ryan

Series A: Life Sciences

Vascular Endothelium
Physiological Basis of Clinical Problems

Edited by
John D. Catravas
Medical College of Georgia
Augusta, Georgia

Allan D. Callow
Washington University Medical Center
St. Louis, Missouri

C. Norman Gillis
Yale University
New Haven, Connecticut

and
Una S. Ryan
Monsanto Company
St. Louis, Missouri

Springer Science+Business Media, LLC

Proceedings of a NATO Advanced Study Institute on
Vascular Endothelium: Physiological Basis of Clinical Problems,
held June 18-28, 1990,
in Corfu, Greece

Library of Congress Cataloging in Publication Data

NATO Advanced Study Institute on Vascular Endothelium: Physiological Basis of Clinical Problems (1990: Kerkyra, Greece)
  Vascular endothelium: physiological basis of clinical problems / edited by John D. Catravas ... [et al.].
    p.   cm.—(NATO ASI series. Series A, Life sciences; v. 208)
  "Proceedings of a NATO Advanced Study Institute on Vascular Endothelium: Physiological Basis of Clinical Problems, held June 18-28, 1990, in Corfu, Greece"—T.p. verso.
  "Published in cooperation with NATO Scientific Affairs Division."
  Includes bibliographical references and index.
  ISBN 978-1-4613-6663-8   ISBN 978-1-4615-3736-6 (eBook)
  DOI 10.1007/978-1-4615-3736-6
  1. Vascular endothelium—Pathophysiology—Congresses. I. Catravas, John D. II. North Atlantic Treaty Organization, Scientific Affairs Division. III. Title. IV. Series.
  [DNLM: 1. Endothelium, Vascular—physiology—congresses. QS 532.5.E7 N279va 1990]
RC691.4N38   1990
616.1'3-dc20
DNLM/DLC   91-24022
for Library of Congress   CIP

ISBN 978-1-4613-6663-8

© 1991 Springer Science+Business Media New York
Originally published by Plenum Press, New York in 1991
Softcover reprint of the hardcover 1st edition 1991

All rights reserved

No part of this book may be reproduced, stored in a retrieval system, or transmitted in any form or by any means, electronic, mechanical, photocopying, microfilming, recording, or otherwise, without written permission from the Publisher

PREFACE

This monograph contains the proceedings from the Advanced Study Institute on *"Vascular Endothelium: Physiological Basis of Clinical Problems"* which took place in Corfu, Greece in June 1990. The meeting consisted of twenty-eight lectures, most of them adapted as full length papers in this volume, as well as numerous short oral and poster communications which are abstracted and also included in alphabetical order (pages 239-302). There were ninety-six participants from ten NATO and four other European countries. The meeting was the second in as many years dealing with a specific subject in Endothelial Cell biology. Following the 1988 discussion on *"Receptors and Transduction Mechanisms"*, the present ASI recognized and tried to deal with the increasing overlap in interest between basic scientists studying endothelial cell functions and clinicians facing problems of known or suspected endothelial pathological involvement. As with any similar effort, we opted to be selective, rather than fail by trying to be inclusive, in the subjects covered. We chose to discuss diseases, such as atherosclerosis, sepsis, ARDS and stroke, based on their relevance to endothelial cell function and urgent need for new insights into their pathogenesis and treatment. Similarly, we examined endothelial cell functions by considering their relevance to disease and their potential for elucidating important pathologies. Obviously, some areas were covered superficially or not at all; this should not distract from their importance, but rather reflect on the constraints of time and - not at all negligibly - the bias of the organizing committee.

As a product of the Advanced Study Institute, this book is the result of efforts by many individuals. The International Organizing Committee, composed of the Co-Directors, Dr. Alberto Mantovani (Italy), Dr. Salvador Moncada (U.K.) and Dr. Magdi Yacoub (U.K.), put together the scientific program and selected the Lecturers, each one of whom was kind enough to generously offer time and expertise. The Local Organizing Committee, composed of Drs. Orfanos and Maragoudakis and Ms. Lydia Argyropoulos, assumed numerous important responsibilities, relating to both the scientific aspects as well as the arrangement of local facilities, necessary for the smooth execution of the conference. Dianne Rosenquist was again the Co-ordinator of the ASI who maintained order ably and cheerfully and seemed unflappable at the sight of chaos. As in 1988, Mary Ann Roupp diligently and enthusiastically processed all manuscripts and abstracts, provided original contribution towards the style of the book and, with the help of Jim Parkerson, Connie Snead, Adrienne Grzeskiewicz, Stavros Topouzis, Nandy Marczin and Stelios Orfanos, proof read the entire text. To all these individuals, we are deeply grateful. Lastly, and perhaps most importantly, we express our sincere appreciation to all participants, whose interest in the field of Endothelial Biology and active participation assured the success of this meeting and who were gracious enough to write or call with flattering comments.

John D. Catravas (Augusta)
Allan D. Callow (St.Louis)
C. Norman Gillis (New Haven)
Una S. Ryan (St. Louis)

CONTENTS

Abbreviations ix

I. CLINICAL TOPICS

    The Clinical Profile of Sepsis and the Adult Respiratory Distress
        Syndrome 3
    Kenneth L. Brigham

    The Participation of the Complement System in Atherosclerotic Vascular
        Disease 13
    Paul S. Seifert, S. Bhakdi and M.D. Kazatchkine

    The Spectrum of Atherosclerosis in the Human 27
    Allan D. Callow

    Cerebral Endothelial Injury in Stroke, Brain Trauma and Hypertension 47
    Hermes A. Kontos

II. GENERAL CONCEPTS IN ENDOTHELIAL CELL PATHOPHYSIOLOGY

    Free-Radical Mediated Actions on Endothelial Cells of the Intact Lung 55
    C. Norman Gillis, X. Chen and M. Merker

    Regulation of Vascular Function by Vascular Permeability Factor 69
    Daniel T. Connolly

III. CHEMICAL MEDIATORS OF ENDOTHELIAL CELL INJURY

    Endothelial Cells as Targets for and Producers of Cytokines 79
    E. Dejana, G. Bazzoni, I. Martin-Padura, S. Walter
    and Alberto Mantovani

    Roles of Vascular Cells in Inflammation and Immunopathology 87
    Peter Libby

    The Biochemistry, Cell and Molecular Biology of Type 1 Plasminogen
        Activator Inhibitor 97
    David J. Loskutoff

IV. ENDOTHELIAL CELL INTERACTION WITH BLOOD COMPONENTS

    Purine Regulation of Endothelial Cells: Relevance to Pathophysiology 111
    John L. Gordon

Autoantibodies to Endothelial Cells 117
    Jeremy D. Pearson

Leukocyte-Endothelial Cell Interactions 127
    James Varani

V. VASCULAR RESPONSES TO ENDOTHELIAL CELL INJURY

Endothelial Function in Human Coronary Bypass Grafts 139
    Thomas F. Lüscher and Z. Yang

Altered Renovascular Endothelial Functions During Nephrotoxicity 157
    N. Perico, C. Zoja and Giuseppe Remuzzi

Cerebral Endothelial Function: Physiology and Pathophysiology 167
    Donald D. Heistad, F. Faraci, and G. Baumbach

Mechanisms of Altered Reactivity in the Cerebral Microcirculation 175
    Hermes A. Kontos

VI. MECHANISMS OF ENDOTHELIAL CELL DYSFUNCTION

The Role of Apolipoprotein E and Apolipoprotein B in
        Atherosclerosis 183
    Thomas L. Innerarity

Endothelial Cell-Matrix Interactions in Health and Disease 197
    Elisabetta Dejana, A. Zanetti, C. Dominguez-Jimenez
    and G. Conforti

Biosynthesis and Assembly of von Willebrand Factor by Vascular Endothelial
        Cells: Relevance to Pathophysiology 203
    Jan A. van Mourik

VII. MARKERS OF ENDOTHELIAL CELL INJURY AND REPAIR

Monitoring of Endothelial Plasmalemmal Ectoenzyme Function as an
        Index of Endothelial Injury and Repair 213
    John D. Catravas

Pathophysiological Significance of Endothelial Cell Integrins 225
    Jan A. van Mourik, Jacques G. Giltay and Albert E.G. Kr. von
    dem Borne

VIII. EPILOGUE

Vascular Endothelium: Physiological Basis of Clinical Problems 233
    Kenneth L. Brigham

IX. ABSTRACTS OF ORAL AND POSTER PRESENTATIONS 239

X. LIST OF PARTICIPANTS 303

XI. INDEX 313

ABBREVIATIONS

β-VLDL	beta-very low density lipoprotein
15-HPETE	15-hydroperoxyeicositetraeinoic acid
5-HIAA	5-hydroxyindoleacetic acid
5-HT	serotonin, 5-hydroxytryptamine
ACE	angiotensin converting enzyme
ACh	acetylcholine
ADP	adenosine 5´-diphosphate
AECA	anti-endothelial cell antibodies
AI	angiotensin I
AII	angiotensin II
apo-B100,B48,-E,-E3,-E4	apolipoprotein-B100,B48,-E,-E3,E4
BAE	bovine aortic endothelial
BBB	blood-brain barrier
bFGF	basic fibroblast growth factor
BK	bradykinin
BSA	bovine serum albumin
CHD	coronary heart disease
CR	complement receptors
CVD	cardiovascular disease
CyA	cyclosporine A
DAF	decay-accelerating factor
DMEM	defined minimum essential medium
EC	endothelial cells
ECM	extracellular matrix
EDCF	endothelium-derived constricting factor
EDRF	endothelium-derived relaxing factor
ELAM	endothelial-leukocyte adhesion molecule
ELISA	enzyme linked immunosorbent assay
FDB	familial defective apo-B100
FH	familial hypercholesterolemia
G-CSF	granulocyte-colony stimulating factor
GFR	glomerular filtration rate
GLUT-1,-4	glucose transporter isotype
GM-CSF	granulocyte-macrophage-colony stimulating factor
HPLC	high pressure liquid chromatography
HR	heart rate
HRF	homologous restriction factors
HUE	human umbilical vein endothelial cells
HUS	hemolytic-uremic syndrome

ICAM	intercellular adhesion molecule
IDL	intermediate density lipoprotein
IFN	interferon
IgE,G,M	immunoglobulin E,G,M
IL-1,-6	interleukin-1,-6
L-NMMA	L-N-mono-methylarginine
LDL	low density lipoprotein
LFA	leukocyte function associated molecule
LPS	lipopolysaccharide
MAb	monoclonal antibodies
MAO	monoaminooxidase
MCP	membrane cofactor protein
MEL	mouse endothelial lectin
NAP	neutrophil activating protein
NO	nitric oxide
PA	plasminogen activator
PAF	platelet activating factor
PAI-1	type 1 plasminogen activator inhibitor
PAI-2	type 2 plasminogen activator inhibitor
PCR	polymerase chain reaction
PDGF	platelet-derived growth factor
PGH_2	prostaglandin H_2
PGI_2	prostaglandin I_2, prostacyclin
PMA	phorbol myristate acetate
PR	pulse rate
RR	respiratory rate
SDS-PAGE	sodium dodecyl sulfate-polyacrylamide gel electrophoresis
SHRSP	stroke-prone spontaneously hypersensitive rats
SLE	systemic lupus erythematosus
SNP	sodium nitroprusside
SOD	superoxide dismutase
t-PA	tissue type-PA
TG	triglycerides
TGF	transforming growth factor
TIAs	transient ischemic attacks
TNF	tumor necrosis factor
TTP	thrombotic thrombocytic purpura
TxA_2,B_2	thromboxane A_2,B_2
u-PA	urokinase-type-PA
VCAM	vascular cell adhesion molecule
VEGF	vascular endothelial growth factor
VLDL	very low density lipoprotein
Vn	vitronectin
VPF	vascular permeability factor
VWF	von Willebrand factor
WHHL	Watanabe heritable hyperlidemic
WKY	Wistar-Kyoto

I.   CLINICAL TOPICS

THE CLINICAL PROFILE OF SEPSIS AND THE ADULT RESPIRATORY DISTRESS SYNDROME

Kenneth L. Brigham

Center for Lung Research and Division
of Pulmonary and Critical Care Medicine
Vanderbilt University
Nashville, Tennessee 37232, U.S.A.

INTRODUCTION

Both sepsis syndrome and the adult respiratory distress syndrome (ARDS) are empirically defined clinical entities *(Brigham, 1983; Bernard and Brigham, 1984)*. Definition of the syndromes, that is, the specific constellation of abnormalities which qualifies for the diagnosis, has evolved from clinical experience with little regard for etiology or for the spectrum of pathophysiology which may be present. Most definitions of sepsis syndrome include fever, tachycardia, tachypnea and evidence of failure of at least one organ (usually brain, lungs, liver or kidneys). ARDS is usually defined as non-cardiogenic pulmonary edema with refractory hypoxemia. The actual numbers for each of the variables used to define the syndromes vary. Although such empirical definitions might be expected to be nonspecific, including very heterogenous groups of patients, in practice that is not the case. For example, using several criteria for selecting patients with ARDS, different series from different parts of the United States appear to identify groups of patients with similar distributions of etiologies, similar mortality rates and similar clinical behavior *(Brigham, 1985; Bernard and Brigham, 1985)*. The same appears to be true of sepsis syndrome.

Simple clinical definitions based on arbitrary degrees of dysfunction appear to be relatively specific in selecting similar patients but are almost certainly insensitive. There is a great deal of experimental data to indicate that ARDS and sepsis syndrome are consequent to injuries to the lungs or other organs, which result in organ failure only at the most severe end of the spectrum. The usual definition of "at risk" categories of patients probably selects patients who already have a mild form of organ injury which is pathogenically identical to the full-blown syndrome and which may or may not progress to dysfunction of adequate severity to meet diagnostic criteria *(Bernard and Brigham, 1986)*. That fact could be important in evaluating interventions in the clinical setting where distinctions between early treatment and prophylaxis may not be easily established on the basis of timing of intervention relative to the available clinical data.

Both sepsis syndrome and ARDS are grave clinical conditions, often with fatal outcome. The two syndromes share pathophysiology, and there is good reason to think that pathogenetic mechanisms, which apply to acute lung injury, may also be at play in injury of other organs which occurs in sepsis syndrome. Thus, the two syndromes are interrelated both conceptually and, potentially, therapeutically as well.

THE ADULT RESPIRATORY DISTRESS SYNDROME (ARDS)

ARDS was defined about 20 years ago as non-cardiogenic pulmonary edema and refractory hypoxemia with decreased lung compliance. It has been said repeatedly in the

TABLE I: Criteria for Diagnosis of ARDS

- $PaO_2 < 70$ torr with $F_IO_2 < 0.4$
- $PaO_2/PAO_2 < 0.3$
- Chest X-ray - bilateral diffuse infiltrates consistent with pulmonary edema
- $Ppaw < 18$ torr

literature that there are approximately 150,000 cases of ARDS in the United States each year. Mortality in reported series of patients has continued to be in excess of 50% over the 20 years since the original description of the syndrome. There is some question whether persistently high mortality rates in series reported in recent years are a consequence of selection of more seriously ill patients than was the case in earlier reports, but at least in many patient groups, which meet the clinical criteria for diagnosis of ARDS, the mortality rate remains high *(Bernard and Brigham, 1989)*.

The clinical diagnosis of ARDS is usually made on the basis of a set of arbitrary criteria for several clinically measurable variables. Those criteria are enumerated in Table I. The actual numbers assigned to the gas exchange variables and wedge pressure differ somewhat from different series and, in any case, are arbitrary; the criteria shown in Table I are those which we have used in selecting groups of patients with an unambiguous diagnosis of ARDS. It has been argued that etiology should also be considered in selecting patients with ARDS, since the pathogenesis of respiratory failure may differ with different etiologies. However, it is not yet clear that patients with the clinical syndrome, that is, the constellation of findings listed in Table I, compose pathogenically distinct subsets which influence the clinical course of the disease.

When groups of patients are selected prospectively by the criteria shown in Table I, the distribution of etiologies appears to be quite similar in different series from different institutions. The most common etiology is sepsis which may be present with or without other risk factors for ARDS *(Bernard et al, 1987)*. In our experience, almost half of the patients with ARDS have sepsis as a predisposing factor. The second most common etiology is usually aspiration, and other causes account for much smaller numbers of patients. The fact that the selection criteria commonly used appear to select similar groups of patients even though the criteria are quite simple is consistent with the notion that pathophysiology of ARDS is similar regardless of the etiology.

Some of the clinical features of ARDS can be illustrated with data taken from a prospective placebo-controlled clinical trial of the effects of corticosteroids in ARDS published several years ago *(Bernard et al, 1987)*. The presentation of these data should be prefaced by the conclusion that we could see no effects of high dose methylprednisolone administered over 24 hours on mortality, time to resolution or any other variable measured.

Table II lists average numbers for several variables in the 99 patients chosen prospectively according to the criteria in Table I *(Bernard et al, 1987)*. It is obvious from those data that the patients had moderately severe respiratory failure with a shunt fraction of 0.28 and a chest X-ray severity score of 2.4 (on a 0-3 severity scale). On average, there was a modest elevation of pulmonary artery pressure, and lung compliance was decreased. If reversal of ARDS is defined as reversal of the blood gas criteria for the original definition of the syndrome, then over the 45 days in which these patients were studied, 60% of them reversed their respiratory failure. Virtually all of these patients had reversed their respiratory failure in the first 5 days following admission into the study. These data indicate that most patients who will recover normal oxygenation following ARDS will do so within the first week following diagnosis. The time course for mortality does not parallel the time course for reversal of ARDS. In this series of patients, there was a 60% mortality at 45 days following entry into the study. About half of those patients died in the first week, and the other half over the ensuing weeks. Table II shows the time course for improvement in chest radiograph, static lung compliance and oxygenation over the first 5 days in this group of patients. Both

TABLE II: Physiologic Variables in Patients with ARDS (N=99)

Age (yrs)	56
Q_s/Q_t	0.28
Chest X-ray score (0-3)	2.4
Total thoracic compliance (ml/cmH$_2$O)	38
Mean P_{pa} (torr)	25
P_{paw} (torr)	12
Cardiac Index (L/min/m^2)	3.67
Serum bilirubin (mg/dl)	4.1
Serum creatinine (mg/dl)	2.0

chest X-ray and oxygenation showed prompt and continued improvement over the first 5 days following diagnosis indicating the resolution of the lung injury. Static compliance changed little over those first 5 days.

These data demonstrate that ARDS is a highly lethal clinical syndrome occurring often in the setting of sepsis. Patients with this syndrome who recover are likely to do so within the first week, and patients whose respiratory failure persists past the first week may have prolonged respiratory failure with a late fatal outcome.

TABLE III: Time Course of Lung Dysfunction in ARDS

	Day				
Variable	1	2	3	4	5
Chest X-ray Score (0-3)	2.4	2.2	2.0	1.8	1.7
Static Compliance (L/cmH$_2$O)	0.039	0.041	0.040	0.041	0.042
PaO_2/PAO_2	0.24	0.26	0.29	0.31	0.31

TABLE IV: Criteria for Diagnosis of Sepsis Syndrome

- **Abnormal Vital Signs**

    T ≥ 101°F or T ≤ 96°F (rectal)

    PR ≥ 90 beats/min

    RR > 20 breaths/min or
    minute ventilation > 10 L/min

- **Organ System Failure**

    Altered mental status

    PaO_2 < 75 torr on room air (PaO_2/PAO_2 < 0.75)

    ↑ plasma lactate

    urine output < 30 ml/hr

SEPSIS SYNDROME

As with ARDS, it is possible, using fairly simple clinical criteria, to identify groups of patients which behave similarly in both pathophysiological changes and clinical course where the primary insult appears to be infection. Most clinicians would now agree that demonstration of viable bacteria in the bloodstream is not a *sine qua non* for the clinical syndrome of sepsis. The presumption is that bacterial products, the most thoroughly studied of which is gram negative bacterial endotoxin, may access the bloodstream and therefore all the organs in the body with the consequence that multiple organ dysfunction occurs. That pathogenetic sequence must remain somewhat speculative at the present time *(Brigham and Bernard, 1986; Brigham and Meyrick, 1986)*.

Table IV lists criteria for diagnosis of sepsis syndrome in the clinical setting which have been employed in one or another variation by many clinical investigators. The usual triad includes some apparent clinical source of infection, abnormal vital signs, and evidence of failure of at least one end organ. Temperature may be either abnormally high or abnormally low, and tachypnea is common. Either subjective evidence of central nervous system dysfunction or objective evidence of kidney or lung dysfunction is usually accepted as sufficient clinical evidence of failure of an end organ. Acidosis is a common finding, especially in patients in shock.

Table V shows the sources of sepsis in patients selected prospectively by the criteria in Table IV. These data are from a fairly small study of about 30 patients conducted by Dr. Gordon Bernard at the Vanderbilt Center for Lung Research. In this group of patients, only about a third of blood cultures were positive for pathogenic bacteria. Almost two-thirds of the patients had the lungs as a source of the sepsis, with the second most frequent site as the peritoneum. Virtually all of the patients with a clinically suspected site of infection proved to have a documented pathogen cultured from some site. Physiological variables from the same group of patients, from which the data in Table V were taken, are shown in Table VI. Of particular note is the fact that even though these patients were not selected exclusively for respiratory dysfunction, on average they had a shunt fraction of 26% and an arterial to alveolar partial pressure of oxygen ratio of 0.36. The chest X-ray score also indicated that on average there were early abnormalities on the chest X-ray. As indicated in Table VII, half the patients selected by the criteria in Table IV prospectively were in clinical shock at the time at which they met those criteria. Of the patients presenting in shock, approximately half reversed their shock, and this occurred within the first few days following diagnosis. The time

TABLE V: Sources of Sepsis in Sepsis Syndrome

Lung	60%
Peritoneum	17%
Urinary tract	3%
Skin/soft tissue	7%
Other	13%

(Blood cultures positive in 37%)

course of patients with sepsis syndrome was different from that in patients with ARDS. By day 30, mortality was 43%, but only 1% mortality was present at day 5. Thus, unlike ARDS, most of the patients who die do so after the first week following diagnosis rather than during that period.

Since ARDS occurs commonly as a complication of sepsis, one perspective could be to view ARDS as simply a subset of patients with sepsis syndrome. The physiologic criteria for making the diagnosis are arbitrary, and many patients with sepsis syndrome have abnormalities in gas exchange which may not be severe enough to meet the criteria for ARDS. It may well be that the pathogenesis of organ injury in all patients with sepsis is similar and that the pathogenesis of lung injury in patients with ARDS reflects exactly the same process as that which occurs in other organs in the setting of sepsis. If that is true, then pharmacological interventions developed from rationales emerging from investigations of acute lung injury may be more broadly applicable in the setting of sepsis syndrome.

TABLE VI: Physiologic Variables in Patients with Sepsis Syndrome (n=30)

Age (yrs)	54
T (°F)	100.5
HR (bpm)	110
RR (b/min)	19
Q_s/Q_t	0.26
PaO_2/P_AO_2	0.36
Static C_L (ml/cmH$_2$O)	50
Chest X-ray score	1.2
P_{paw}	15

TABLE VII: Shock Reversal and Mortality in Sepsis Syndrome

	Day 0	Day 5	Day 30
Shock	50%	25%	---
Mortality	---	1%	43%

PATHOGENESIS OF ENDOTOXIN-INDUCED INJURY

A large amount of research in a broad spectrum of preparations over the last 20 years has resulted in two general theories about the pathogenesis of tissue injury caused by gram-negative bacterial endotoxin. The two theories are not mutually exclusive but rather are interrelated.

What could be called a cell theory emphasizes inflammation as the central process and focuses on the polymorphonuclear leukocyte as the main player in the pathogenetic scenario *(Brigham and Meyrick, 1986; Snapper and Brigham, 1986; Bernard and Brigham, 1986a; Brigham, 1986; Brigham, 1988)*. This theory proposes that neutrophils are activated, that they sequester in the microcirculation adhering closely to endothelial cells and generate noxious substances including proteolytic enzymes and aggressive oxygen species. These noxious substances then injure the cells in close proximity, probably endothelial cells, early in the pathogenesis of injury. This theory would provide a number of sites in the pathogenetic sequence at which pharmacologic therapy might be aimed. One of those would be antioxidants which would be capable of preventing the effects of oxidant stress resulting from activated neutrophils. A second possible site would be to interfere with the association of activated neutrophils with endothelium. Endothelium, particularly microvascular endothelium, generates prostanoids which have some anti-inflammatory effects, principally prostacyclin and prostaglandin E_2. Other pro-inflammatory prostanoids are also produced in many forms of injury including that resulting from endotoxin. Thus, manipulation of prostanoids might also be a site at which intervention could be directed.

A second theory could be called generally, the mediator theory. Although obviously not exclusive of the inflammation, this hypothesis focuses on the generation of mediators by cells in the lungs as well as, perhaps, inflammatory cells *(Brigham et al., 1983; Niedermeyer et al., 1984; Loyd et al., 1984; Brigham and Meyrick, 1984; Hinson et al., 1984)*. These mediators include aggressive oxygen species which may be made by lung cells creating a sort of positive feedback loop in which an insult to lung cells induces the lung cells to produce substances which are then injurious to the cells. If this were true, then antioxidants, especially those which access the cell interior, might be efficacious in interrupting the pathogenetic sequence *(Brigham, 1987; Brigham, 1987a; Brigham et al., 1987)*. The generation of prostanoids by lung cells might also be important in the pathogenesis of lung dysfunction. The same reasoning which holds for the lung could apply to other organs which are injured by endotoxin.

THERAPEUTIC IMPLICATIONS

The pathogenetic theories alluded to above provide rationales for pharmacologic intervention in the clinical setting. In addition, a large body of animal experimentation suggests that intervention, particularly with antioxidants *(Bernard et al., 1984)* and antiinflammatory agents which appear to act by inhibiting prostanoid synthesis *(Snapper et al., 1983; Niedermeyer and Brigham, 1984)*, show some promise in preventing organ dysfunction and even death following gram negative septicemia.

The prospective clinical study with high dose corticosteroids alluded to earlier showed conclusively that patients with ARDS of moderate severity do not respond to administration of corticosteroids *(Bernard et al, 1987)*; however, other clinical trials are now beginning. For example, based upon both *in vitro* and animal experimentation data, the antioxidant N-acetly-

TABLE VIII: Expression of CAT Gene in Organs of Mice* Transfected *In Vivo**

Route	% chloramphenicol acetylation per hr per mg protein x 10^{-2}		
	Lungs	Liver	Kidneys
Intravenous			
30 mg DNA	24.5	0	0
15 mg DNA	10.1	0	0
Intratracheal			
30 mg DNA	35.7	0	0
Intraperitoneal			
30 mg DNA	0	0	0

* Reprinted with permission from *Brigham et al., Am. J. Med. Sci. 1989a.*

cysteine has undergone preliminary studies in patients with ARDS. N-acetlycysteine can be administered safely in large doses to humans. The early studies suggest that even in patients with established ARDS there may be a more rapid resolution of the pulmonary edema and respiratory failure in these patients. Currently, a larger prospective trial is being planned. To my knowledge this is the first pharmacologic intervention in ARDS patients which looks promising.

Also currently in the preliminary stages is a prospective study of the non-steroidal antiinflammatory agent, ibuprofen, in patients with sepsis syndrome. As with N-acetlycysteine and ARDS, this study is based on a large amount of *in vitro* and animal data suggesting that non-steroidal antiinflammatory agents may be beneficial in the treatment of responses to endotoxin. Preliminary data from Dr. Bernard and his co-workers also suggest the possibility that this drug may be efficacious in sepsis syndrome.

TRANSIENT GENE THERAPY

The veritable explosion of technology dealing with the manipulation of DNA in recent years has raised the possibility of gene therapy in a variety of diseases. Although largely these efforts have been focused on the treatment of diseases which are consequences of genetic dysfunction, it might be logical to expand that concept to include diseases which are acute and limited in duration. ARDS and sepsis syndrome might fall into this category.

Several proteins normally made in mammalian cells could prove beneficial in the treatment of endotoxin-induced injury. For example, the antioxidant enzymes superoxide dismutase and catalase, when present in high quantities, enhance resistance of cells to oxidant stress. The genes for these proteins have been cloned and at least some of them expressed in mammalian cells. Endogenous antiproteases, particularly alpha-1 antitrypsin, might fall into the same category.

The rationale for this "transient gene therapy" would be that if functioning foreign DNA could be introduced into the cells of the host-DNA designed in a manner which would generate increased quantities of a desired protein over a specified and transient period of time-this intervention might render the host cells resistant to injury or enhance their rate of repair. The trick, of course, is getting the DNA into the host cells.

We have demonstrated recently that functioning foreign genes can be introduced into the lungs of animals by the simple intravenous or intratracheal injection of plasmid constructs complexed to small cationic liposomes *(Brigham et al, 1989; Brigham et al., 1989a)*. Table VIII shows data from our initial studies published a few months ago. In mice we injected a plasmid construct containing chloramphenicol acetyl transferase (CAT) as a reporter gene. The plasmid was complexed to small cationic liposomes and injected either intravenously, intraperitoneally or intratracheally. Three days following injection, animals were killed, the lungs, liver and kidneys removed, homogenized, and CAT activity measured as acetylation of radio labeled chloramphenicol. We found expression of the CAT gene exclusively in the lungs. Expression of the gene was related to the dose of DNA administered when given intravenously and expression for a given dose was higher in the lungs when the DNA was administered intratracheally than when given intravenously. Intraperitoneal administration resulted in no expression of the gene in any organ.

We have subsequently investigated the time course of this response and shown peak expression of the foreign gene at 3 days following injection with decreasing expression by one week. In addition, we have shown that other fusion genes, including those containing human growth hormone as a reporter gene and those containing beta galactosidase as a reporter gene, showed similar results. Promoters that appear to function well in this system include the SV40 promoter, the metallothionein promoter and the CMV promoter.

There is a rapidly emerging technology which will make the administration of functioning foreign DNA into host cells a feasible goal. The application of gene therapy to the large number of diseases which are not a direct consequence of genetic dysfunction will dramatically expand the usefulness of this tool in a clinical setting. Administration of foreign DNA in a manner which assures only transient expression may be safer to apply to humans than methods which result in permanent alteration of the host genome.

ACKNOWLEDGEMENTS:

This work was supported by NIH, National Heart, Lung and Blood Institute, Grants: HL 19153, SCOR in Pulmonary Vascular Diseases); HL 07123 (NRSA Institutional Training Grant in Multidisciplinary Respiratory Diseases); RO1 HL 34208. Also, grants from the John W. Cooke and Laura W. Cook Foundation; the Bernard Werthan, Sr. Foundation; the Harry H. and Martha W. Straus Foundation; and The Upjohn Company. Dr. Brigham holds the Joe and Morris Werthan Chair for Investigative Medicine at Vanderbilt University School of Medicine.

REFERENCES

Bernard, G.R., and Brigham, K.L. Adult respiratory distress syndrome. *Baylor College of Medicine Cardiology Series* H.D. McIntosh, Editor, 7(5):5-22, 1984.

Bernard, G.R., Lucht, W.D., Niedermeyer, M.E.,Snapper, J.R., Ogletree, M.L., and Brigham, K.L. Effect of n-acetylcysteine on the pulmonary response to endotoxin in the awake sheep and upon *in vitro* granulocyte function. *J. Clin. Invest.* 73:1772-1784, 1984.

Bernard, G.R., and Brigham, K.L. The adult respiratory distress syndrome. *Ann. Rev. Med.* 36:195-205, 1985.

Bernard, G.R., and Brigham, K.L. Non-cardiac pulmonary edema. *Comprehensive Therapy* 12(7):64-69, 1986.

Bernard, G.R., and Brigham, K.L. Pulmonary edema: Pathophysiologic mechanisms and new approaches to therapy. *Chest* 89(4):594-600, 1986a.

Bernard, G.R., Harris, T., Luce, J.E., Rinaldo, J., Sibbald, W.J., Sprung, C., Tate, R.M., Higgins, S., Kariman, K., Bradley, R., and Brigham, K.L. High dose corticosteroids in patients with the adult respiratory distress syndrome: A randomized double-blind trial. *N. Engl. J. Med.* 317:1565-1570, 1987.

Bernard, G.R., and Brigham, K.L. Increased lung vascular permeability: Mediators and therapies. In: *Textbook of Critical Care* Eds: W.C. Shoemaker, et al. W.B. Saunders, Co., Inc. Second Edition, Chapter 118, pp. 1049-1055, 1989.

Brigham, K.L. Primary (high permeability) pulmonary edema. *Seminars in Respiratory Medicine* 4(4):285-288, 1983.

Brigham, K.L., Begley, C.J., Bernard, G.R., Hutchison, A.A., Loyd, J.E., Lucht, W.D., Meyrick, B., Newman, J.H., Niedermeyer, M.E., Ogletree, M.L., Sheller, J.R., and Snapper, J.R. Septicemia and lung injury. *Clin. Lab. Med.* 3(4):719-744, 1983.

Brigham, K.L., and Meyrick, B. Granulocyte dependent injury of pulmonary endothelium: A case of miscommunication? *Tissue and Cell* 16(2):137-155, 1984.

Brigham, K.L. Fluid and solute transport in the acutely injured lung. *Acute Respiratory Failure* Zapol, W.M., Falke, K.J., Eds., New York: Marcel Dekker, pp. 209-226, 1985.

Brigham, K.L. Interactions of neutrophils in the pulmonary vascular bed in the adult respiratory distress syndrome (ARDS). *Acute Lung Injury Pathogenesis of Adult Respiratory Distress Syndrome* PSG Publishing Co., Littleton, Massachusetts. 11:139-143, 1986.

Brigham, K.L., and Meyrick, B.O. Endotoxin and lung injury: State of the art review. *Am. Rev. Resp. Dis.* 133(5):913-927, 1986.

Brigham, K.L., and Bernard, G.R. Pulmonary edema. *Internal Medicine*, 2nd edition, J.H. Stein, Ed.: Little, Brown and Company, Boston. pp. 621-627, 1986.

Brigham, K.L. Mechanisms of endothelial injury. In: *Pulmonary Endothelium in Health and Disease* Una S. Ryan, Eds.: Marcel Dekker, Inc., Chapter 9, pp. 207-236, 1987a.

Brigham, K.L. Mechanisms of lung endothelial injury. *Pulmonary Circulation in Health and Disease* J.A. Will, C.A. Dawson, E.K. Weir, C.K. Buckner, Eds.: Academic Press, Inc. Florida, pp. 363-369, 1987.

Brigham, K.L., Meyrick, B.O., Berry, L.C., and Repine, J.E. Antioxidants protect cultured bovine lung endothelial cells from injury by endotoxin. *J. Appl. Physiol.* 63(2):840-850, 1987.

Brigham, K.L. Diffuse lung injury. In: *High-Altitude Medical Science* G. Ueda, S. Kusama and N.F. Voelkel, eds. Shinshu University, Matsumoto, Japan. pp. 151-155, 1988.

Brigham, K.L., Meyrick, B., Christman, B., Magnuson, M., King, G., and Berry, L.C., Jr. *In vivo* transfection of murine lungs with a functioning prokaryotic gene using a liposome vehicle. *Am. J. Med. Sci.* 298(4): 278-281, 1989a.

Brigham, K.L., Meyrick, B., Christman, B., Berry, L.C., Jr., and King, G. Expression of a prokaryotic gene in cultured lung endothelial cells following lipofection with a plasmid vector. *Am. J. Resp. Cell Molec. Biol.* 1:95-100, 1989.

Hinson, J.M., Jr., Hutchison, A.A., Brigham, K.L., Meyrick, B.O., and Snapper, J.R. Effects of granulocyte depletion on pulmonary responsiveness to aerosol histamine. *J. Appl. Physiol.: Respirat. Environ. Exercise Physiol.* 56(2):411-417, 1984.

Loyd, J.E., Newman, J.H., and Brigham, K.L. Permeability pulmonary edema: Diagnosis and management. *Arch. Intern. Med.* 144:143-147, 1984.

Niedermeyer, M.E., and Brigham, K.L. Prospects for therapeutic interventions in acute respiratory failure. *Resp. Therapy* 14(6):15-26, 1984.

Niedermeyer, M.E., Sheller, J.R., and Brigham, K.L. Pathogenesis of the adult respiratory distress syndrome. *Current Pulmonology* 5:229-254, 1984.

Snapper, J.R., and Brigham, K.L. Pulmonary edema. *Hospital Practice* 5:87-101, 1986.

Snapper, J.R., Hutchison, A.A., Ogletree, M.L., and Brigham, K.L. Effects of cyclooxygenase inhibitors on the alterations in lung mechanics caused by endotoxemia in the unanesthetized sheep. *J. Clin. Invest.* 72:63-76, 1983.

THE PARTICIPATION OF THE COMPLEMENT SYSTEM IN ATHEROSCLEROTIC VASCULAR DISEASE

*Paul S. Seifert, **S. Bhakdi and *M.D. Kazatchkine

*Laboratoire d'Immunologie
Hopital Broussais
Paris, France

**The Institute for Medical Microbiology
Johannes-Gutenberg University
Mainz, Federal Republic of Germany

INTRODUCTION

Atherosclerosis is a vascular disease of large and medium-sized arteries wherein the tunica intima becomes thickened due to lipid accumulation, mostly cholesterol and its esters, smooth muscle cell proliferation, and increased deposition of connective tissue matrix. A major risk factor in the development of this disease is hypercholesterolemia arising from elevated levels of low density lipoproteins (LDL). The earliest recognizable lesion, which may be a precursor to the fibrofatty plaque, is the fatty streak. It is predominantly composed of monocyte-derived macrophage foam cells, i.e. cells ladened with intracellular lipid droplets. Hence, a fundamental aspect of atherogenesis is the insudation and accumulation of LDL-derived cholesterol and the attempt by monocyte/macrophages to clear it from the arterial wall. As lesion development progresses, the lipid component becomes surrounded by a fibrotic cap laid down by vascular smooth muscle cells. The proliferation of smooth muscle cells and induction of connective tissue synthesis is probably driven, at least in part, by the products released from resident macrophages. This chapter outlines the participation of the complement system in atherosclerosis. The ability of complement activation products to mediate proinflammatory functions of macrophages implicates this system in the pathogenesis of atherosclerotic lesion development.

THE COMPLEMENT SYSTEM

A brief outline of the complement system is provided here for the reader unfamiliar with this system (for more extensive descriptions see *Kazatchkine and Nydegger, 1982; Vogt, 1986; Muller-Eberhard, 1988*). The core of the complement system is composed of 9 plasma proteins designated C1 through C9 (Figure 1). Classical pathway complement activation takes place through the autocatalytic activation of component Cl when it encounters immune-complexes containing IgG or IgM. Subsequent binding and cleavage of C2 and C4 components results in the formation of a C3 convertase, i.e. a C3 cleaving enzyme. In the presence of an activating surface, C3 can form C3 convertases itself, i.e. in the absence of immunoglobulin, by binding factors B and D. This constitutes the alternative pathway activation. C3 cleavage by C3 convertases results in two cleavage fragments, C3a and C3b. C3a is an anaphylatoxin and is released into the fluid-phase following C3 activation. Specific receptors for C3a are present on granulocytes and monocytes/macrophages (Table 1). Engagement of the C3a receptor modifies the functional behavior of these cell types (Table 2).

The C3b fragment can covalently bind to activating surfaces where it serves to promote further C3 convertase formation, and whereby, it facilitates ingestion of the foreign surface by phagocytes bearing complement receptors. The ligands, cellular distribution and numerical designation of the complement receptors are outlined in Table 1.

When C3b molecules are deposited on the activating surface in close proximity to a C3 convertase, they form a C5 convertase. C5 convertases cleave C5 into C5a and C5b. Like C3a, C5a is an anaphylatoxin; it is released into the fluid phase during activation and it binds to specific receptors located on leukocytes thereby altering their function (Table 2).

C5b exhibits a binding site for C6 and C7 which, in turn, form a C5b,6,7 complex that exhibits binding affinity for cell surface phospholipids. Subsequent binding of C8 and C9 (with C9 polymerization) results in a C5b-9 complex that can form a porous channel through the lipid bilayer. A C5b-9 complex which forms in a lipid bilayer is termed C5b-9(m). The result of C5b-9(m) formation on a cell can be irreversible cell injury. In nucleated cells, lethal injury requires the formation of many C5b-9(m) complexes (probably one thousand or more). Sublytic C5b-9(m) formation stimulates calcium-dependent, protein kinase C-mediated physiologic changes in cells, e.g. leukotriene and prostaglandin production (*Imagawa et al., 1986; Suttorp et al., 1987; Bhakdi, 1988*), probably by acting like a calcium ionophore. Some nucleated cell types are known to protect themselves from C5b-9(m)-mediated damage by either shedding complexes or by actively internalizing them (*Ram et al., 1983; Morgan et al., 1987*). In addition, many cell types possess plasma membrane proteins that inhibit C5b-9(m) formation through the binding of C8 and/or C9. These proteins have been termed homologous restriction factors (HRF) since they have been shown to be responsible for inhibiting complement-mediated cell lysis in homologous systems (*Ojcius et al., 1990*). Two such proteins have so far been described: HRF 20 (CD59) and HRF 65 (the numbers derive from their molecular weights). Additionally, if C5b-9 fails to encounter a lipid bilayer it can bind the ubiquitously present S protein, also known as vitronectin, which renders it water-soluble and therefore cytolytically inactive. Thus, fluid phase C5b-9 generation will result in SC5b-9.

As mentioned above, regulatory proteins exist for the control of complement activation and C5b-9 formation. They occur as plasma proteins and cell surface molecules. Further discussion of these regulatory proteins will be considered later in specific reference to their participation in atherosclerosis.

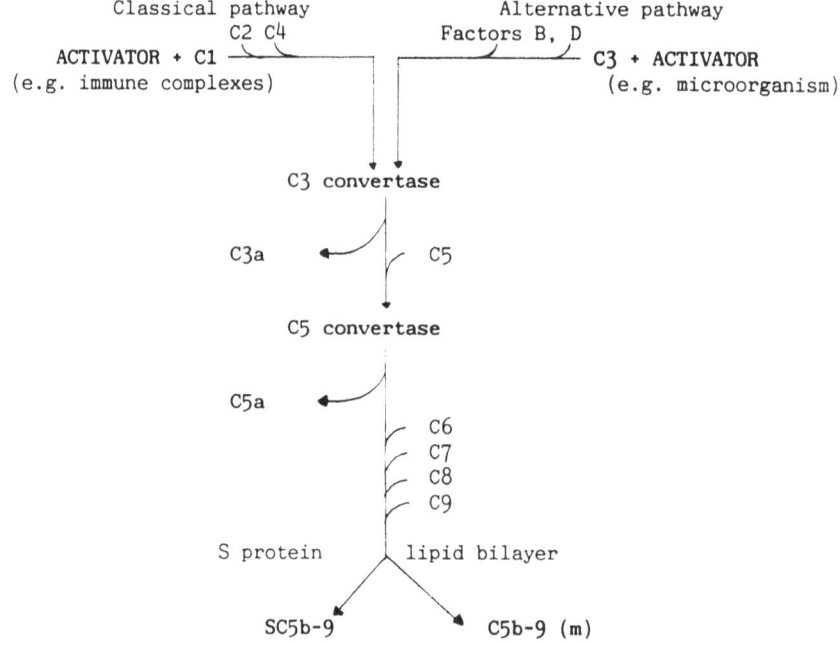

FIGURE 1: Schematic representation of the complement activation pathways.

TABLE 1: Complement receptors, their ligands and cellular distribution

LIGAND	RECEPTOR	CD NUMBER	CELL TYPE
C1q	C1q receptor		platelets
C3a	C3a receptor		neutrophils, monocytes/macrophages, eosinophils, basophils, endothelial cells?
C5a	C5a receptor		neutrophils, monocytes/macrophages
C3b	CR1	CD35	eosinophils, B lymphocytes, some T lymphocytes, monocytes/macrophages, glomerular phagocytes, erythrocytes, neutrophils, dendritic cells
C3dg	CR2	CD21	B and T lymphocytes
iC3b	CR3	CD11b/CD18	neutrophils, monocyte/macrophages, large granular lymphocytes

COMPLEMENT AND THE ENDOTHELIUM

Regulation

In addition to being a non-thrombogenic surface under normal healthy conditions, the endothelial lining of blood vessels also appears to be resistant to complement-mediated attack. This is due in part to the fact that endothelial cells themselves represent a non-activating surface for complement. In addition, cultured human endothelial cells have been shown to possess cell membrane proteins that inhibit C3 convertase and C5b-9 formation on their cell surface. One of these is decay-accelerating factor (DAF) which inhibits C3 convertase formation by interfering with factor B binding to C3b (*Fujita et al., 1987*) and thereby down regulates C3 activation. Another is membrane cofactor protein (MCP), which assists in the factor I-mediated degradation of C3b into iC3b (*Seya et al., 1986*). Since iC3b cannot participate in C3 convertase formation (in contrast to C3b), activation is downregulated. As mentioned above, HRF 20 prevents C5b-9 formation on the cell surface through inhibiting C8 and C9 binding to the C5b,6,7 complex. We have recently observed this regulatory protein in frozen sections of large and small human arterial blood vessels using a monoclonal antibody produced by Okada et. al., (1989). DAF has similarly been observed on the endothelium of human blood vessels by immunohistochemistry (*Kinoshita et al., 1985*). To date, MCP has only been described on cultured human umbilical vein endothelium (*McNearney et al., 1989*), yet there is no reason to expect that it is not a feature of the endothelium *in vivo*. At present, nothing is known of the regulatory expression of these molecules except that DAF expression can be altered by phorbol esters via activation of protein kinase C under *in vitro* conditions (*Bryant et al., 1990*).

Activation

In certain instances, endothelial cells have been observed to become complement reactive. For example, Vlaciu *et al.*, (1985) published a photomicrograph of a human atherosclerotic

TABLE 2: Stimulatory effects of the anaphylatoxins

C3a	C5a
Histamine release from mast cells	Histamine release from mast cells and basophils
Lysosomal degranulation	Lysosomal degranulation
Interleukin-1 production by monocytes	Interleukin-1 production by monocytes
	Respiratory burst
	Chemotaxis of neutrophils, monocytes and fibroblasts
	Contraction of smooth muscle
	Regulation of CRI and DAF expression on neutrophils and CRI, CR3 and Fc receptor expression on monocytes

lesion wherein a lumenal cell in the position of the endothelium was positively stained with anti-C5b-9 antibodies. Luminal cells (presumably endothelium) staining positively for the C5b-9 terminal complex have been observed in cholesterol-fed rabbits in the pre-fatty streak stage of lesion development (*Seifert et al., 1989*). Rabbit aortic endothelial cells rendered ischemic have been shown to activate autologous serum complement generating C5a(desArg) (*Seifert et al., 1988*). Vimentin intermediated filaments contained in endothelial cells activate complement *in vitro* (*Linder et al., 1979; Hansson et al., 1987*), as do mitochondria (*Pinckard et al., 1975; Kagiyama et al., 1989*). The data up to this point indicate that endothelial cell complement reactivity occurs only if the cell is irreversibly damaged.

The pathophysiological effects of complement activation on damaged endothelium remain unknown. Endothelial cell-mediated generation of C5a(desArg) and C3b/iC3b deposition would be expected to promote monocyte adhesion to the endothelium (*Doherty et al., 1987*) since C5a(desArg) upregulates adhesion proteins, e.g. the C3b/iC3b receptors on monocytes. It is interesting to note that monocytes preferentially adhere to the periphery of forming lesions (*Taylor and Lewis, 1986*), and that this area also corresponds to the region of increased endothelial cell turnover (*Walker et al., 1986*). One might speculate, therefore, that dying but undesquamated endothelial cells activate complement and thereby attract and promote monocyte adhesion in this area.

Receptors

In regard to complement receptors, human endothelium appears to be negative for CR1, CR2, CR3, and CR4. Endothelial cells have been shown to actively internalize C3a in a time and dose-dependent manner *in vitro*, yet, a specific saturable receptor could not be identified (*Denny and Johnson et al., 1979*). Human and bovine pulmonary arterial endothelium binds Clq (*Andrews et al., 1981*), a subunit of C1, and when treated with neutrophil-derived lysosomal enzymes, influenza or cytomegalovirus, bovine endothelium was shown to bind C3b-coated erythrocytes (*Ryan et al., 1981*). Cultured human endothelial cells treated with herpes simplex virus exhibit C3b-binding activity as well (*Cines et al., 1982; Smiley et al., 1985*). In this case, it was found to be due to the expression of a viral glycoprotein on the cell surface and not to the induction of cellular CR1 or CR3 (*Freidman et al., 1984; McNearny et al., 1987*).

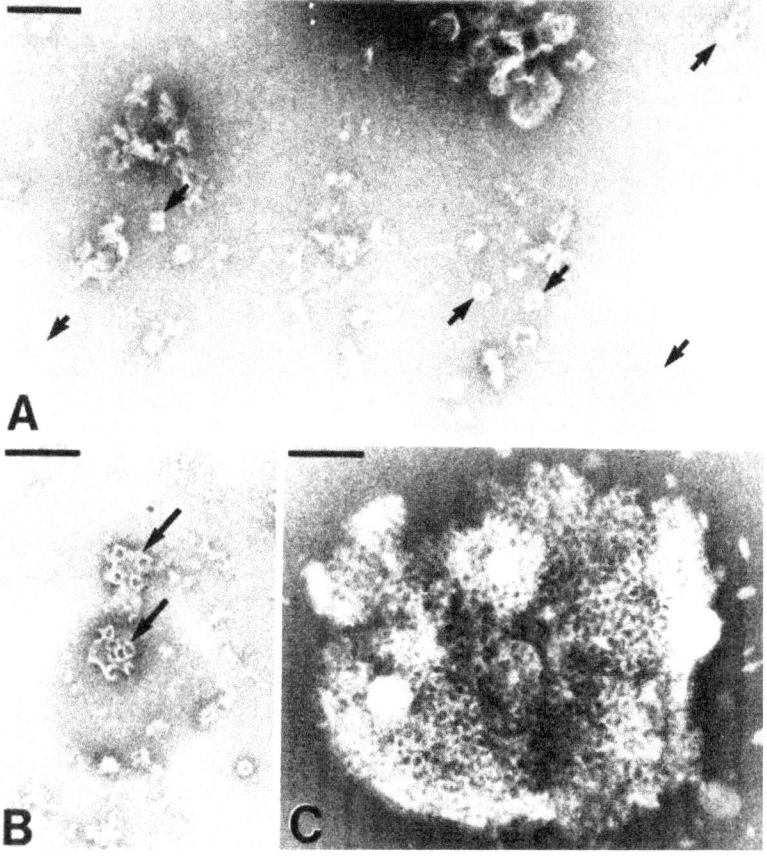

FIGURE 2: C5b-9 complexes isolated by sucrose density gradient centrifugation from saline-extracts of human aortic atherosclerotic lesion, as seen by transmission electron microscopic examination after negative staining. Cylindrical C5b-9 complexes (arrows) can be seen singly (A) in small clusters (B) and in massive amounts on spherical particles (C). Scale bar indicates 100 nm. (Reproduction from *J. Exp. Med.* vol. 172, 1990)

COMPLEMENT IN HUMAN ATHEROSCLEROSIS

C5b-9 Terminal Complement Complexes

It was not until 1985, when *Vlaciu et al.*, published a report on C5b-9 terminal complement complexes in human atherosclerotic lesions, that convincing data for complement activation in atherosclerosis was presented. Earlier studies had demonstrated the presence of most of the native complement components (*Hollander et al., 1979; Vlaciu et al., 1985*). However, as the antibodies could not distinguish native from activated proteins, it remained questionable whether *in situ* activation actually took place. In contrast, demonstration of the C5b-9 complex rather conclusively proves *in situ* activation in the arterial wall since: antibodies to this complex recognize neoantigenic epitopes found only on the assembled complex and not on the native proteins, C5b-9 formation implies the prior presence of C3 and C5 convertases and because of the large size of this complex (approximately one million daltons), it is unlikely that it arrives in the arterial wall by another mechanism other than *in situ* assembly.

Further evidence for the existence of terminal complement complexes in human atherosclerotic lesions came from experiments wherein they were isolated and semi-purified from saline extracts of lesions (*Seifert et al., 1990*). Extracts of lesions, dissected from aortic and carotid tissue samples, were centrifuged through a sucrose gradient and collected in ten equivolume fractions. Purified SC5b-9 was run simultaneously for comparison. SC5b-9 sediments at a slower rate than C5b-9(m), thereby allowing one to distinguish between these two forms; the antibodies currently available are unable to make this distinction. The majority of the gradient fractionated lesional C5b-9 sedimented as C5b-9(m). When the C5b-9 peak from the sucrose density gradient was examined by electron microscopy, it was found to contain the characteristic C5b-9(m) cylinders (Figure 2). The C5b-9 cylinders existed singly, as well as in massive numbers on spherical particles, perhaps cell or mitochondrial membranes. In addition, when analyzed by SDS PAGE followed by immunoblotting, it was found that plaque extract contained C9 dimers. C9 dimers are characteristic of C5b-9(m), but not C5b-9, complexes (*Ware and Kolb, 1981*). This data should not be taken to imply that lesional C5b-9 is exclusively C5b-9(m), but rather that a significant quantity of C5b-9(m) does exist in plaques and raises questions as to the pathophysiologic consequences of its formation. Although this question has not yet received investigative attention, it should be noted that cholesterol-fed rabbits deficient in C6 were reported to exhibit a reduced incidence and severity of atherosclerosis when compared to complement normal rabbits, despite comparable levels of serum cholesterol (*Geertinger and Sorensen, 1975*). This, of course, argues that C5b-9 formation promotes lesion development, yet it remains to be determined by what mechanisms.

Complement Activation

Examination of atherosclerotic lesion material by immunohistochemistry invariably reveals massive widespread staining for C5b-9 complexes, thus rendering accurate interpretations of activating structures difficult, if not impossible. In addition, it is currently believed that terminal C5b-9 complexes remain at their sites of formation for extended periods of time since they are relatively resistant to proteolytic digestion they would not be expected to exhibit a significant diffusion rate bcause of their large size and the affinity for S protein, which is found not only on fluid-phase SC5b-9 complexes but also on C5b-9(m), due to post binding of S protein (*Bhakdi et al., 1988*) to connective tissue elements. Thus, over the years that atherosclerotic lesion development occurs, accumulation and retention of terminal complexes would not necessarily be expected to coincide spatially with the potentially more ephemeral nature of the complement activator.

One way to circumvent the above mentioned difficulties in studying complement activation by C5b-9 immunohistochemistry is to examine terminal complex deposition shortly after it occurs. This was taken advantage of in a study aimed at investigating at what point in time the complement activation take place in atherosclerotic lesion development (*Seifert et al., 1989*). Rabbits were fed a 0.3% cholesterol-supplemented diet and the aortas examined for lipid and C5b-9 deposits at 2, 6 and 10 weeks following diet commencement. Aortic intimal C5b-9 deposits were observed as early as 2 weeks into the diet. These deposits were small and focal and always coincided with oil red 0 lipid stain. After 6 weeks, the deposits of both lipid and C5b-9 increased in extent to form confluent masses within the intima. Up to this time point, few if any monocytes or macrophages were observed on the endothelium or within the intima. In contrast, at the 10 week time point an overt macrophage-filled intimal fatty streak had developed. Lipid was now found to be located throughout the intima, largely in cells within the fatty streak itself and as finer droplets extracellularly about the internal elastic lamina. C5b-9 staining at this stage was found almost exclusively in the area of the internal elastic lamina, corresponding to the location of extracellular lipid.

This study thus revealed that in experimentally-induced fatty streak lesion formation, complement participates from the earliest stages. Therefore, complement is an integral feature of atherogenesis since it is also observed in advanced plaques. Secondly, the results were highly suggestive that lipids are involved in triggering complement activation in the arterial wall. Lastly, since C5b-9 deposition preceded overt monocyte infiltration into the intima the C5a desArg generated during complement activation may be responsible, at least in part, for attracting monocytes to lipid-rich intimal areas (*Marder et al., 1985*).

That lipid is the agent responsible for triggering lesional complement activation is an attractive hypothesis since cholesterol has been previously implicated in complement activation. For example, crystalline cholesterol activates human serum complement *in vitro* in a manner that appears to involve C3b-binding to the free hydroxyl group on unesterified cholesterol (*Vogt et al., 1985; Seifert and Kazatchkine, 1987*). Additionally, liposomes constructed with a high (greater than 50 mole percent) concentration of unesterified cholesterol also activate human serum complement *in vitro* (*Cunningham et al., 1975*). Supporting of a lipid-mediated mechanism for complement activation in atherosclerotic vessels is the fact that liposomal-like lipid particles rich in unesterified cholesterol have been

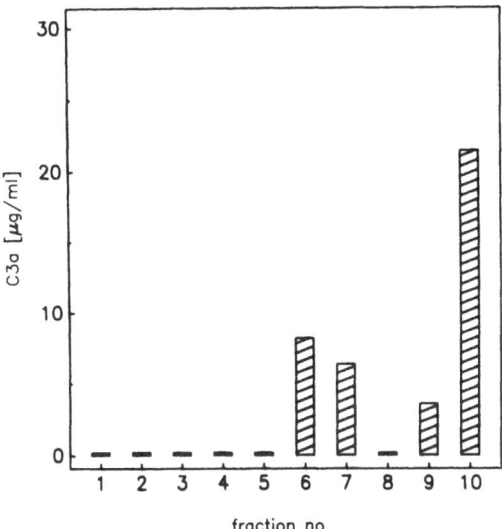

FIGURE 3: The complement-activating capacity of human atherosclerotic lesion extract fractionated by sucrose density gradient centrifugation. All 10 fractions collected were dialyzed against veronal-buffered saline and incubated with normal human serum 60 minutes at 37°C. The specific C3a desArg generated was measured by radioimmunoassay. (*Reproduction from J. Exp. Med. vol. 172, 1990*)

observed equally early and are consistent features of both human and animal atherosclerotic lesions (*Kruth, 1985; Simionescu et al., 1986; Mora et al., 1987; Chao et al., 1988*).

Experiments were recently performed to confirm whether or not a specific lipid species exists in atherosclerotic lesions with the capacity to activate human complement (*Seifert et al., 1990*). Lesions were dissected from aortic or carotid tissue samples, finely minced and extracted into buffered saline. When the crude extract was mixed with normal human serum (NHS) for 45 minutes at 37°C, a specific activation of complement occurred as measured by the generation of C5a desArg.

Having established that the extract contains a complement-activating substance, it was fractionated by sucrose density gradient centrifugation. Ten equivolume fractions were collected, and each reacted with NHS to determine which fractions contained complement activating activity. The majority of complement-activating activity was located in fraction 10, the uppermost fraction, and to a lesser degree in fractions 6,7 and 9 (Figure 3). Analysis of fraction ten revealed that it contained a high amount of cholesterol, as well as some protein, a portion of which was identified as apoprotein B (apo B), the protein component of LDL.

To further purify the major complement-activating component, fraction 10 from the sucrose density gradient was applied to a Sepharose 2B gel filtration column. Biochemical analysis of the eluted fractions revealed two lipid peaks, one in the void volume fractions and another towards the end volume. The vast majority of protein eluted coincident to the second, or end volume peak. When each fraction was reacted with NHS, complement-activating activity was located exclusively in the void volume fractions (Figure 4). Pooled concentrates of these fractions were then further analyzed and found to exhibit the following characteristics. The unesterified to total cholesterol ratio was 0.58 indicating a high concentration of unesterified cholesterol. Electron microscopic examination revealed spherical particles 100 to 500 nm in diameter with smooth surfaces upon freeze fracture, indicating an absence of intramembranous protein (Figure 5). The only detectable protein was albumin, which is known not to activate complement. When reacted with NHS, a dose-dependent activation of complement was observed, which took place primarily according to the alternative pathway, that is, by direct activation of C3.

In contrast to the void volume peak, the second lipid peak failed to activate serum complement, and exhibited many of the characteristics of LDL. It was similar in size and

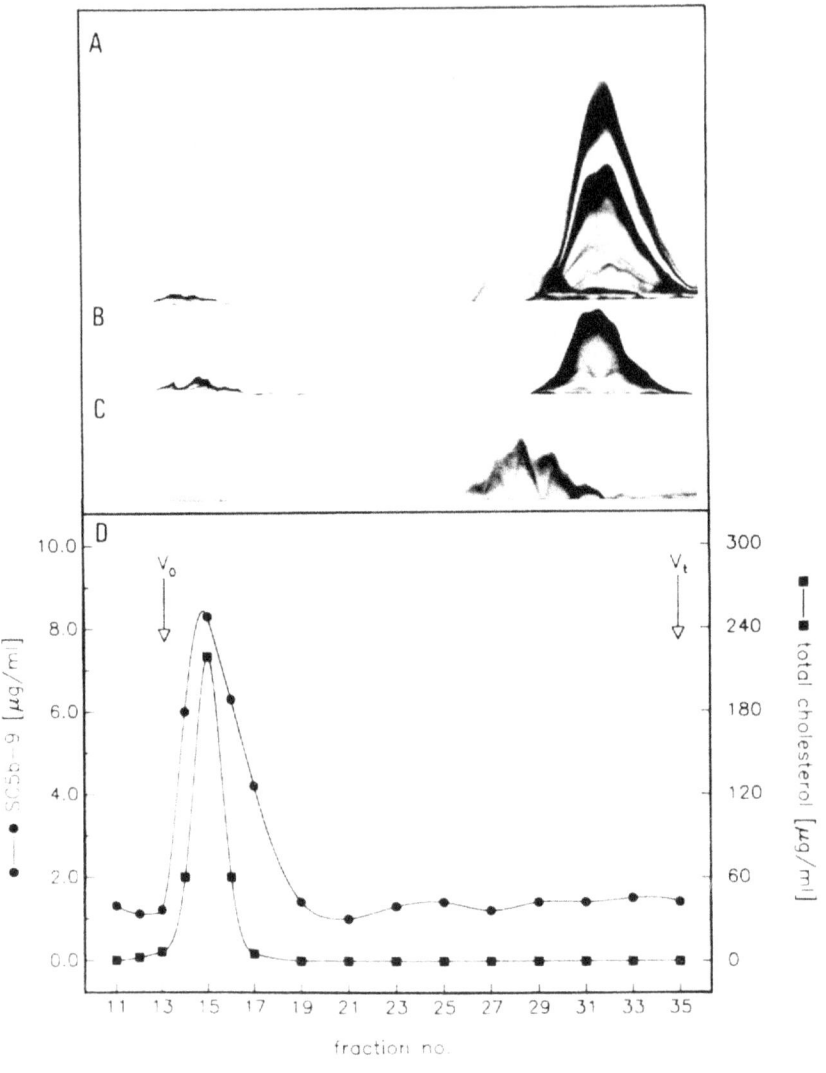

FIGURE 4: Elution profile of fraction 10 from the sucrose density gradient fractionation of human atherosclerotic lesion extract. A 20 x 100 cm Sepharose 2B column was used and eluted with veronal-buffered saline. Panel A: serum proteins; panel B: albumin and; panel C: apoprotein B detected by rocket immunoelectrophoresis using specific antibodies. Panel D demonstrates the cholesterol profile (filled circles) and the complement activity (filled squares) (*Reproduction from J. Exp. Med. vol. 172, 1990*)

cholesterol composition to LDL, eluted from the Sepharose 2B column in identical fractions to purified human plasma LDL and reacted with antisera against apo B in gel immuno-electrophoretic analysis.

This study thus demonstrated that a specific complement-activating lipid exists in human atherosclerotic lesions and appears to be the major, if not exclusive, soluble activating structure. Since this lipid species does not have any of the characteristic features of LDL, yet the vast majority of lesion lipid arises from the influx of plasma LDL, the complement-activating lipid must come about from a rather extensive metabolic modification of intimal LDL.

Complement Regulation

Many of the known complement regulatory proteins have been identified in human atherosclerotic lesions (*Seifert and Hansson, 1989*) (Table 3). Immunohistochemistry of carotid

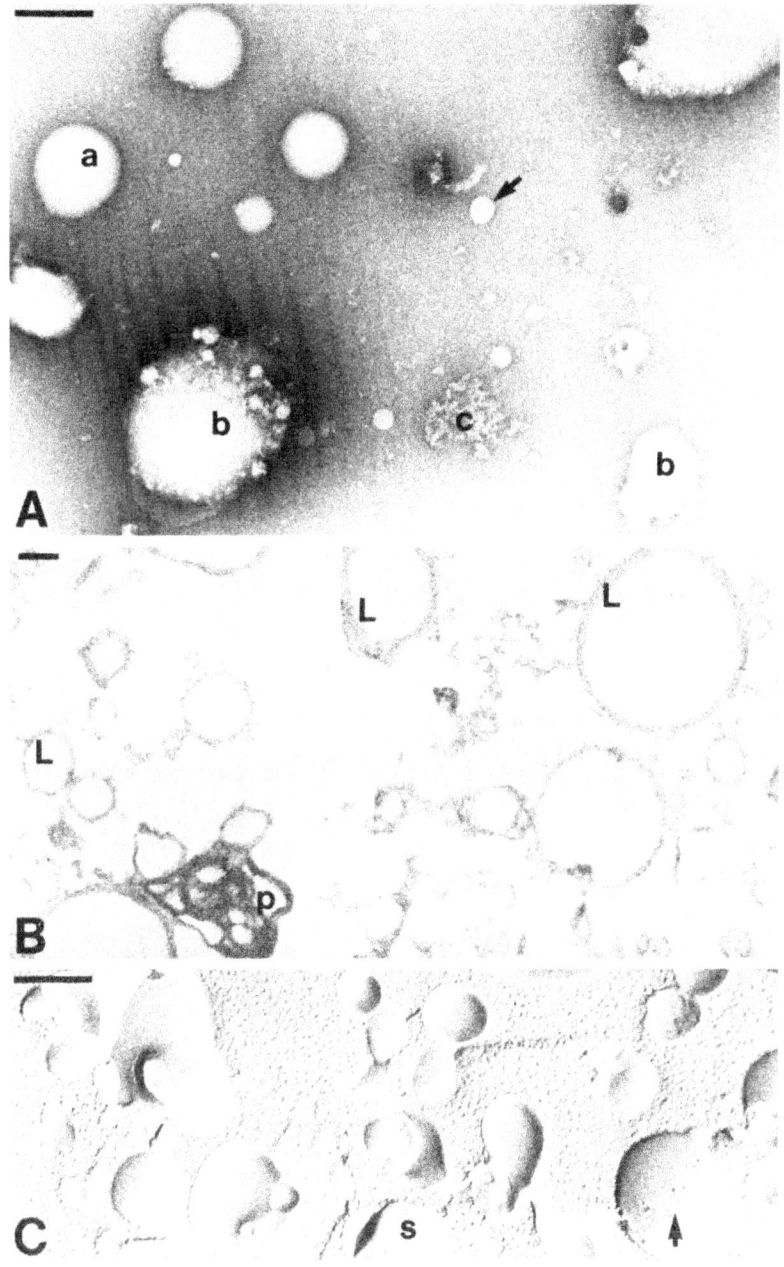

FIGURE 5: (A) Negatively stained, (B) thin-sectioned and (C) freeze-fracture replication preparations of the Sepharose 2B column void volume fractions (see Figure 4). Spherical lipid particles varying from 100 to 500 nm in diameter are evident. Scale bars indicate 200 nm. (*Reproduction from J. Exp. Med. vol. 172, 1990*)

TABLE 3: Complement regulatory proteins in human atherosclerotic lesions

Regulatory protein	Presence?	Comment
C1inh	yes	cellular and extracellular present in low amounts
C4bp	unknown	
properdin	trace	
factor H	yes	extracellular; possibly some cellular; prevalent
factor B	yes	extracellular; prevalent
factor D	unknown	
factor I	unknown	
DAF	yes	smooth muscle cells, macrophages, extracellular, prevalent
MCP	unknown	
CRI	yes	macrophages
HRF	unknown	
sp4o/40	unknown	
S protein	yes	extracellular

Other complement proteins

CR2	no	
CR3	yes	macrophages
CR4	unknown	

lesions revealed weak and sporadic C1q and C1 inhibitor staining, suggesting that participation of the classical complement pathway may be minor in comparison to that of the alternative pathway. Complement components C3 and factor B for example, were found in large amounts. Factor H, a plasma protein that provides cofactor activity for factor I-mediated degradation of C3b, was a common and relatively perfuse component of advanced lesions. S protein, a regulator of the functional activity of C5b-9, has also been shown to be pervasively present in human lesions (*Niculescu et al., 1987*). The complement receptor type CRI, which can inhibit C3 convertase formation, has been identified on macrophages in lesions (*Seifert and Hansson, 1989*). Decay-accelerating factor (DAF), which also inhibits C3 convertase formation, albeit by a different mechanism from CRI, was found to be present on lesional macrophages and smooth muscles cells but not on smooth muscle cells of normal arterial wall (*Seifert and Hansson, 1989a; Seifert and Hansson, 1989b*). DAF was also observed in the extracellular matrix. Normal uterine arteries used as controls were negative for all components tested except for small amounts of DAF and C5b-9 occasionally in intimal thickenings. Complement regulatory proteins in lesions probably come from several sources such as insudation of plasma proteins, local synthesis by macrophages and *de novo* gene induction, such as in the case of smooth muscle cell DAF.

Clinical Atherosclerosis and Complement

Several clinically important consequences of atherosclerotic vascular disease include myocardial infarction, stroke, and ischemia. The complement system is known to become activated during myocardial infarction. This has been shown by the immunohistochemical identification of complement components, including the C5b-9 complex, in human myocardial infarcts (*Schafer et al., 1986*); by deposition of complement components in experimentally--induced infarcts (*McManus et al., 1983; Crawford et al., 1988; Rossen et al., 1988; Kagiyama et al., 1989*) and the measurement of activation products in patients serum following coronary ischemia/infarction (*Langlois and Grawryl, 1988; Mollnes et al., 1988; Yasuda et al., 1990*). Activation is likely to take place on ischemic cells since ischemia increases cell membrane permeability thereby allowing access of complement components to the interior of the cell where they can be activated on intermediate-sized cytoskeletal filaments and mitochondria.

The importance of complement activation in myocardial infarction has been demonstrated in animal models wherein artificial depletion of complement prior to induction of ischemia results in a reduced size and severity of the infarct (*Marko et al., 1978*). This may be due to the proinflammatory stimulatory effects of complement activation products on neutrophils (*Rossen et al., 1981*) since complement depletion prior to induction of ischemia reduces the extent of neutrophil infiltration into the infarcted area (*Hill and Ward, 1971*). Since complement anaphylatoxins stimulate the respiratory burst in neutrophils, complement may play an indirect role in tissue damage via the generation of toxic oxygen radicals by neutrophils. A neutrophil-independent effect of complement activation in myocardial infarction has also been demonstrated (*Crawford et al., 1988*). The mechanism behind this effect is unknown. Generation of C3a and C5a anaphylatoxins has been shown to induce cardiac arrythmia, alter ventricular contractile force and diminish coronary blood flow (*Del Balzo et al., 1985; 1989*). These disturbances in flow and function appear to be mediated via leukotrienes D4 and E4, thromboxane A2 and prostaglandin H2 (*Ito et al., 1990*).

In summary, complement activation takes place throughout the spectrum of what is recognized as atherosclerotic vascular disease. It is activated on the early lipid deposits in fatty streak formation as well as during the clinically manifest advanced stages of atherosclerosis, i.e. ischemia and infarction. Lipid-mediated complement-activation supports the concept that atherosclerosis exhibits an inflammatory component. Complement activation products likely influence lesion development through their numerous effects on monocyte-/macrophage function, e.g. cell adhesion, chemotaxis, phagocytosis, lysosomal degranulation, toxic oxygen and cytokine production.

REFERENCES

Andrews, B.S., Shadforth, M., Cunningham, P. and Davis IV, J.S. Demonstration of a C1q receptor on the surface of human endothelial cells. *J. Immunol.* 127:1075-1080, 1981.

Bhakdi, S. Functions and relevance of the terminal complement sequence. In *Bailliere's Clinical Immunology and Allergy* 2:363-385, 1988.

Bhakdi, S., Kaflein, R., Halstensen, T.S., Hugo, F., Preissner, K.T. and Mollnes, T.E. Complement S-protein vitronectin is associated with cytolytic membrane-bound C5b-9 complexes. *Clin. Exp. Immunol.* 74:459-464, 1988.

Bryant, R.W., Granzow, C.A., Siegle, M.I., Egan, R.W. and Billah, M.M. Phorbol esters increase synthesis of decay-accelerating factor, a phosphatidylinositol-anchored surface protein, in human endothelial cells. *J. Immunol.* 144:593-598, 1990.

Chao, F.F., Amende, L.M., Blanchette-Mackie, E.J., Skarlatos, S.I., Gamble, W., Resau, J.H., Mergner, W.T., Kruth, H.S. Unesterified cholesterol-rich lipid particles in atherosclerotic lesions of human and rabbit aortas. *Am. J. Pathol.* 131:73-83, 1988.

Cines, D.B., Lyss, A.P., Bena, M., Corkey, R., Kefalides, N.A. and Friedman, H.M. Fc and C3 receptors induced by herpes simplex virus on cultured endothelial cells. *J. Clin. Invest.* 69:123-128, 1982.

Crawford, M.H., Grover, F.L.. Kolb, W.P., McMahan, C.A., O'Rourke, R.A., McManus, L.M. and Pinckard, R.N. Complement and neutrophil activation in the pathogenesis of ischemic myocardial injury. *Circulation* 78:1449-1458, 1988.

Cunningham, C.M., Kingrette, M., Richards, B.L., Alving, C.R., Lint, T.F. and Gewurz, H. Activation of human complement by liposomes: a model for membrane activation of the alternative pathway. *J. Immunol.* 122:254-259, 1975.

Del Balzo, U.H., Levi, R. and Polley, M.J. Cardiac dysfunction caused by purified human C3a anaphylatoxin. *Proc. Natl. Acad. Sci. U.S.A.* 82:886-891, 1985.

Del Balzo, U., Polley, M.J. and Levi, R. Cardiac anaphylaxis. Complement activation as an amplification system. *Circ. Res.* 65:847-857, 1989.

Denny, J.B. and Johnson, A.R. Uptake of ^{125}I-labeled C3a by cultured human endothelial cells. *Immunology* 36:169-177, 1979.

Doherty, D.E., Haslett, C., Tonnesen, M.G. and Hensen, P.M. Human monocyte adherence: a primary effect of chemotactic factors on the monocyte to stimulate adherence to human endothelium. *J. Immunol.* 138:1762-1769, 1987.

Freidman, H.M., Cohen, G.H., Eisenberg, R.J., Seidel, C.A. and Cines, D.B. Glycoprotein C of herpes simplex virus 1 acts as a receptor for the C3b complement component in infected cells. *Nature* 309:633-636, 1984.

Fujita, T., Inoue, T., Ogawa, K., Iida, K. and Tamura, N. The mechanism of action of decay-accelerating factor (DAF). DAF inhibits the assembly of C3 convertases by disassociating C2a and Bb. *J. Exp. Med.* 166:1221-1228, 1987.

Geertinger, P. and Sorensen, H. On the reduced atherogenic effect of cholesterol feeding in rabbits with congenital complement C6 deficiency. *Artery* 1:177-184, 1975.

Hansson, G.K., Lagerstedt, E., Bengtsson, A. and Heideman, M. IgG binding to cytoskeletal intermediate filaments activates the complement cascade. *Exp. Cell Res.* 170:338-350, 1987.

Hill, J.H. and Ward, P.A. The phlogistic role of C3 leukotactic fragments in myocardial infarcts of rats. *J. Exp. Res.* 133:885-896, 1971.

Hollander, W., Colombo, M.A., Kirkpatrick. B. and Paddock. J. Soluble proteins in the human atherosclerotic plaque. With spectral reference to immunoglobulins, C3-complement component, a1-antitrypsin and a2-macroglobulin. *Atherosclerosis* 34:391-405, 1979.

Imagawa, D.K., Osifchin, N.E., Ramm, L.E., Koga, P.G., Hammer, C.H., Shin, H.S. and Mayer, M.M. Release of arachidonic acid and formation of oxygenated derivatives after complement attack on macrophages: role of channel formation. *J. Immunol.* 136:4637-4643, 1986.

Ito, B.R., Roth, D.M. and Engler, R.L. Thromboxane A_2 and peptidoleukotrienes contribute to the myocardial ischemia and contractile dysfunction in response to intracoronary infusion of complement C5a in pigs. *Circ. Res.* 66:596-607, 1990.

Kagiyama, A., Savage, H.E., Michael, L.H., Hanson, G., Entman, M.L. and Rossen, R.D. Molecular basis of complement activation in ischemic myocardium: identification of specific molecules of mitochondrial origin that bind human C1q and fix complement. *Circ. Res.* 64:607-615, 1989.

Kazatchkine, M.D. and Nydegger, U.E. The human alternative complement pathway: biology and immunopathology of activation and regulation. *Prog. Allergy.* 30:193-234, 1982.

Kinoshita, T., Medof, M.E., Silber, R. and Nussenzweig, V. Distribution of decay-accelerating factor in the peripheral blood of normal individuals and patients with paroxysmal nocturnal hemoglobinuria. *J. Exp. Med.* 162:75-92, 1985.

Kruth, H.S. Subendothelial accumulation of unesterified cholesterol. An early event in atherosclerotic lesion development. *Atherosclerosis* 57:337-341, 1985.

Langlois, P.F. and Grawryl, M.S. Detection of the terminal complement complex in patient plasma following acute myocardial infarction. *Atherosclerosis* 70:95-105, 1988.

Linder, E., Lehto, V.-P. and Stenman, S. Activation of complement by cytoskeletal intermediate filaments. *Nature* 278:176-178, 1979.

Marder, S.R., Chenoweth, D.E., Goldstein, I.M. and Perez, H.D. Chemotactic responses of human peripheral blood monocytes to the complement-derived pepti 5a and C5a desArg. *J. Immunol.* 134:3325-3331, 1985.

Maroko, P.R., Carpentier, C.B., Chiariello, M., Fishbein, M.C., Radvany, P., Knustman, J.D. and Hale, S.L. Reduction by cobra venom factor of myocardial necrosis after coronary artery occlusion. *J. Clin. Invest.* 61:661-668, 1978.

McManus, L.M., Kolb, W.P., Crawford, M.H., O'Rourke, R.A., Grover, F.L. and Pinckard, R.N. Complement localization in ischemic baboon myocardium. *Lab. Invest.* 48:436-442, 1983.

McNearney, T., Ballard, L., Seya, T. and Atkinson, J.P. Membrane cofactor protein of complement is present on human fibroblast, epithelial, and endothelial cells. *J. Clin. Invest.* 84:538-545, 1989.

McNearney, T.A., Odell, C., Holers, V.M., Spear, P.G. and Atkinson, J.P. Herpes simplex virus glycoproteins gC-1 and gC-2 bind to the third component of complement and provide protection against complement-mediated neutralization of viral infectivity. *J. Exp. Med.* 166:1525-1537, 1987.

Mollnes, T.E., Tambs, K.E., Myreng, Y. and Engebretsen, L.F. Acute phase reactants and complement activation in patients with acute myocardial infarction. *Complement* 5:33-39, 1988.

Mora, R., Lupu, F. and Simionescu, N. Prelesional events in atherogenesis. Colocalization of apolipoprotein B, unesterified cholesterol and extracellular phospholipid liposomes in the aorta of hyperlipidemic rabbit. *Atherosclerosis* 67:143-154, 1987.

Morgan, B.P., Dankert, J.R. and Esser, A.F. Recovery of human neutrophils from complement attack: removal of the membrane attack complex by endocytosis and exocytosis. *J. Immunol.* 138:246-253, 1987.

Muller-Eberhard, H.J. Molecular organization and function of the complement system. *Ann. Rev. Biochem.* 57:321-347, 1988.

Niculescu, F., Rus, H.G. and Vlaicu, R. Immunohistochemical localization of S-protein, C3d and apolipoprotein B in human arterial tissues with atherosclerosis. *Atherosclerosis.* 65:1-11, 1987.

Ojcius, D.M., Jiang, S. and Young, J.D.-E. Restriction factors of homologous complement: a new candidate? *Immunol. Today.* 11:47-49, 1990.

Okada, N., Horada, R., Fujita, T. and Okada, H. Monoclonal antibodies capable of causing hemolysis of neuraminidase-treated human erythrocytes by homologous complement. *J. Immunol.* 143:2262-2268, 1989.

Pinckard, R.N., Olson, M.S., Giclas, P.C., Terry, R., Boyer, J.T. and O'Rourke, R.A. Consumption of classical complement components by heart subcellular membrane *in vitro* and in patients after acute myocardial infarction. *J. Clin. Invest.* 56:740-749, 1975.

Ramm, L.E., Whitlow, M.B., Koski, C.L., Shin, M.L. and Mayer, M.M. Elimination of complement channels from the plasma membranes of U937, a nucleated mammalian cell line: temperature dependence of the elimination rate. *J. Immunol.* 131:1411-1415, 1983.

Rossen, R.D., Swain, J.L., Michael, L.H., Weekley, S., Giaamini, E. and Entman, M.L. Selective accumulation of the first component of complement and leukocytes in ischemic canine heart muscle. A possible initiator of an extra myocardial mechanism of ischemic injury. *Circ. Res.* 57:557-565, 1981.

Rossen, R.D., Michael, L.H., Kagiyama, A., Savage, H.E., Hanson, G., Reisberg, M.A., Moake, J.N., Kim, S.H., Self, D., Weakley, S., Giannini, E. and Entman, M.L. Mechanism of complement activation after coronary artery occlusion: evidence that myocardial ischemia in dogs causes release of constituents of myocardial subcellular origin that complex with human Clq *in vivo. Circ. Res.* 62:572-584, 1988.

Ryan, U.S., Schultz, D.R. and Ryan, J.W. Fc and C3b receptors on pulmonary endothelial cells: induction by injury. *Science* 214:557-558, 1981.

Schafer, H., Mathey, D., Hugo, F. and Bhakdi, S. Deposition of the terminal C5b-9 complement complex in infarcted areas of human myocardium. *J. Immunol.* 137:1945-1949, 1986.

Seifert, P.S. and Kazatchkine, M.D. Generation of complement anaphylatoxins and C5b-9 by crystalline cholesterol oxidation derivatives depends on hydroxyl group number and position. *Mol. Immunol.* 24:1303-1308, 1987.

Seifert, P.S., Catalfamo, J.L. and Dodds, W.J. Complement C5a desArg generation in serum exposed to damaged aortic endothelium. *Exp. Mol. Pathol.* 48:216-225, 1988.

Seifert, P.S. and Hansson, G.K. Complement receptors and regulatory proteins in human atherosclerotic lesions. *Arteriosclerosis* 9:802-811, 1989a.

Seifert, P.S., Hugo, F., Hansson, G.K. and Bhakdi, S. Prelesional complement activation in experimental atherosclerosis. Terminal C5b-9 complement deposition coincides with cholesterol accumulation in the aortic intima of hypercholesterolemic rabbits. *Lab. Invest.* 60:747-754, 1989b.

Seifert, P.S., Hugo, F., Tranum-Jensen, J., Zahringer, U.. Muhly, M. and Bahkdi, S. Isolation and characterization of a complement-activating lipid extracted from human atherosclerotic lesions. *J. Exp. Med.*, 1990, 172:547-557, 1990.

Seya, T., Turner, J.R. and Atkinson, J.P. Purification and characterization of a membrane protein (gp45-70) that is cofactor for cleavage of C3b and C4b. *J. Exp. Med.* 163:837-855, 1986.

Simionescu, N., Vasile, E., Lupu, F., Popescu, G. and Simionescu, M. Prelesional events in atherogenesis. Accumulation of extracellular cholesterol-rich liposomes in the arterial intima and cardiac valves of the hyperlipidemic rabbit. *Am. J. Pathol.* 123:109-125, 1986.

Smiley, M.L., Hoxie, J.A. and Friedman, H.M. Herpes simplex virus type 1 infection of endothelial, epithelial and fibroblast cells induces a receptor for C3b. *J. Immunol.* 134:2672-2678, 1985.

Suttorp, N., Seeger. W., Zinsky, S. and Bhakdi, S. Complement complex C5b-8 induces PGI_2 formation in cultured endothelial cells. *J. Physiol.* C13-C21, 1987.

Taylor, R.G. and Lewis, J.C. Endothelial cell proliferation and monocyte adhesion to atherosclerotic lesions of white carneau pigeons. *Am. J. Pathol.* 125:152-162, 1986.

Yasuda, M., Takeuchi, K., Hiruma, M., Ida, H., Tahara, A., Itagana, H., Toda, I., Akioka, K., Teragaki, M., Oku, H., Kenayama, Y., Takeda, T., Kolb, W.P. and Tamerius, J.D. The complement system in ischemic heart disease. *Circulation* 81:156-163, 1990.

Vlaciu, R., Niculescu,- F., Rus, H.G. and Cristea, A. Immunohistochemical localization of the terminal C5b-9 complement complex in human aortic fibrous plaque. *Atherosclerosis* 57:163-177, 1985.

Vlaicu, R., Rus, H.G., Niculescu, F. and Cristea, A. Immunoglobulins and complement components in human aortic atherosclerotic intima. *Atherosclerosis* 55:35-50, 1985.

Vogt, W. Anaphylatoxins: possible roles in disease. *Complement* 3:177-188, 1986.

Vogt, W., von Zabern, I., Damerau, B., Hesse, D., Luhmann, B. and Nolte, R. Mechanisms of complement activation by crystalline cholesterol. *Mol. Immunol.* 22:101-106, 1985.

Walker, L.N., Reidy, M.A. and Bowyer, D.E. Morphology and cell kinetics of fatty streak lesion formation in the hypercholesterolemic rabbit. *Am. J. Pathol.* 125:450-458, 1986.

Ware, C.F. and Kolb, W.P. Assembly of the functional membrane attack complex of human complement: formation of disulfide-linked C9 dimers. *Proc. Natl. Acad. Sci. U.S.A.* 78:6426-6432, 1981.

THE SPECTRUM OF ATHEROSCLEROSIS IN THE HUMAN

Allan D. Callow

Department of Surgery
Washington University
4960 Audubon Avenue
St. Louis, Missouri 63110, U.S.A.

INTRODUCTION

Of the approximately 57 million people in the United States estimated to have cardiovascular disease, ten million are symptomatic and 75% are less than 65 years of age. In 1986 there were an estimated 768,000 deaths due to coronary artery disease; 50,000 due to peripheral vascular disease, and pulmonary embolism accounted for an additional 30,000. (*Hawiger, M.D., 1988*) Cardiovascular disease is the leading cause of death in Western society and the most important single cause of this mortality is atherosclerosis. During the last four decades, more money and effort have been expended for the development of technology to diagnose, localize, and treat by surgical or other interventional methods the end stage lesions of atherosclerosis than has been spent for its prevention. From 1983 to 1989 *Index Medicus* cites some 16,000 articles on the subject illustrating the controversy and confusion which surround both etiology and pathogenesis. The purpose of this essay is to describe the distribution of atherosclerosis in the human population, identify the vulnerability of certain arterial beds and the resistance of others to the disease, and to call attention to the limitations of current interventional efforts to control the consequences of advanced disease.

EPIDEMIOLOGY

Aortic fatty streaks are found in many children under three years, in almost all persons by the age of ten, (*Strong and McGill, Jr., 1969*) and, in the coronary arteries in approximately 90% of all individuals over 30 regardless of race, geography or gender. Fatty streak incidence is roughly similar in countries with high or low myocardial infarction mortality rates.

Advanced and late stage atherosclerosis develops from the fibrous plaque which is substantially different from the fatty streak. The plaque contains smooth muscle cells, white cells, necrotic cellular debris, lipid, cholesterol crystals, hemorrhage, thrombus and areas of calcification, the amount of each depending upon the stage of development and degeneration. Because fatty streaks are as common in females as males, occur throughout the entire aorta and coronary vessels, and advanced symptomatic lesions occur more frequently in the abdominal than the thoracic aorta, the fatty streak may not be the precursor of the plaque. Data to the contrary are suggested by a study of the chronologic sequence of the cellular events in the arterial wall which develop during and following exposure to a low level of hypercholesterolemia of 200-400 mg/dl in *Macaca nemestrina*. After approximately six months, fatty streaks developed. These consisted of intimal accumulations of lipid laden macrophages and small numbers of T-lymphocytes. The developing irregular surface showed occasional

TABLE 1: Relative Risk of Developing Myocardial Infarction or Angina Pectoris Among Males by Area of Birth. (*Goldbourt, U. et al., 1986*)

Relative risk adjusted by risk factor levels

Area of Birth	MI	AP
Eastern Europe	1.18	1.09
Central Europe	1.22	0.81
Balkan countries	0.85	1.65
Israel	1.08	0.90
North Africa	1.04	1.00
Mideast	0.72	0.59
Total	**1.00**	**1.00**

(1.00 = average risk for study sample.) Death certificate diagnoses demonstrate substantial differences in coronary heart disease mortality from country to country. (*Goldbourt, U. et al., 1978*)

disruptions of endothelial cells, and formation of platelet microthrombi. Except for slower progression, these changes are virtually identical to those which develop in high level hypercholesterolemic animals (600-1000 mg/dl). Advanced lesions - fibrous plaques - are found in all animals at two and three years. These low level plaques contained monocyte/macrophages, T-lymphocytes and smooth muscle cells. The data suggest that many fatty streaks are converted to fibrofatty lesions, "some of which ultimately become converted to fibrous plaques." (*Masuda and Ross, 1990*) They are similar to plaques observed in hypercholesterolemic humans.

The events leading to the conversion of the fatty streak to a fibrous plaque and, that in turning from a stable plaque with a smooth surface to one with ulceration and necrotic debris, are speculative at best whatever the precise events and the factors participating in and possibly directing them. The stable plaque becomes unstable as it loses its endothelial covering to develop an ulcerated surface and possibly, in a second series of events, formation of an intramural, intraplaque hemorrhage and hematoma. Two terminal events, each with the potential for disaster on the target organ, may occur separately or as partners in the end stage lesion: embolization to a distant target and terminal thrombosis at the site of the unstable plaque. The classical description of cells of the arterial wall, viz. the endothelial cell, the smooth muscle cell, and the fibroblast, assigns functions of such specificity to each as to distinguish one cell type from another. The endothelial cell provides a nonthrombogenic surface on a highly selective barrier, the smooth muscle cell provides tone permitting contraction and relaxation as circumstances require, and the fibroblast provides a restraining sheath and a matrix and support structure for vasa vasorum and nerves. Under conditions of cell culture, these cells can be induced to express functions characteristic of the other. Within and around the developing plaque, aberrations of cell specific functions also occur: the endothelial cell becomes prothrombogenic, and the smooth muscle cell is transformed from a contractile to a synthetic mode. In the embryogenesis of the smooth muscle cell and the endothelial cell specific signals determine differentiation. Do these different cells retrace the steps of their embryonic development and lose their specialty functions? If pulsatile stretch is one of the determinants of specialization, why such a striking difference in functional outcome since both smooth muscle and endothelial cells are subjected to the same forces. Why does atherosclerosis spare smaller vessels, exposed as they are to the same alleged risk factors as the larger elastic and muscular arteries? (*Nugent and O'Connor, 1983*)

TABLE 2: Coronary Heart Disease Mortality Ratios in Selected Countries in 1977. (*Goldbourt U. and Neufeld, H.N., 1986*)

	Male/Female Ratios		
	Age-Adjusted Rates/100,000		1977 Age
Country	Male	Female	35-74
Norway	537	156	3.44
Finland	878	266	3.30
U.S.A.	670	262	2.56
Scotland	809	324	2.50
Japan	103	50	2.06
Israel (Jews)	527	290	1.82

RACE

Race appears to be a risk factor both in anatomic distribution of disease, as for example the cerebral, coronary and renal circulations, (*Strong, 1972; Berenson et al., 1984; Davis, et al., 1979; Maynard et al., 1986*) and in severity of lesions. In the infragenicular vessels, blacks display substantially more severe disease over whites by a factor of 2. (Figure 1) (*Sidaway, et al., 1990*) The black and white groups are comparable in terms of age, prevalence of diabetes, smoking history and hypercholesterolemia. Hypertension was more prevalent among the blacks, but when only nonhypertensive patients of both groups were considered, the mean severity score was still significantly higher with blacks prevailing by a factor of 2.1 vs. 1.42. In whites cerebrovascular disease is concentrated in the extracranial cervical carotid arteries while in blacks, the intracranial vessels are more often and more severely involved. (*Maxwell et al., 1989; Caplan et al., 1986; Solberg et al., 1972*)

Males of Middle Eastern birth appear to be protected against coronary disease. The sum of their five year risks remains much lower than that of European born men after risk factor adjustment. The genetically more heterogenous Jews born in Yemen, living in isolation for many years and with very rare intermarriage with non-Jews, have a substantially lower incidence of myocardial infarction and angina pectoris.

FAMILY

The influence of paternal and maternal coronary artery disease on siblings determined by arteriography is real. In a study of 1105 siblings of 411 males living in New York the overall prevalence of coronary heart disease was 16%. This fell to 12% when neither parent had CHD, rose to 19% when the mother was afflicted, to 31% with the father only afflicted, and when both parents were afflicted, the prevalence rose to 55%. (*Hamby, 1981; Shea et al., 1984*)

GENDER

Male mortality of CHD is approximately five times higher than in females in the age span of 35 to 44 years but falls to 2.4 times in the age span of 65 to 74. (Figure 2) (*Levy and Feinleib, 1984*) The female rate lags behind the male mortality rate by approximately ten years.

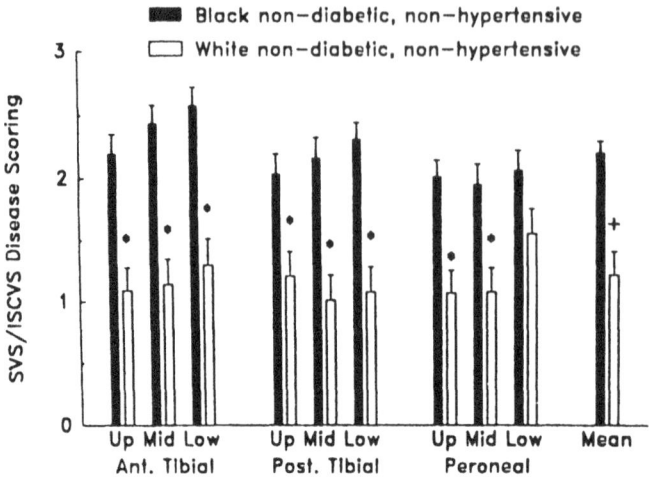

FIGURE 1: Comparison of the disease severity scoring between the black and white nondiabetic, nonhypertensive, patients. $^*p<0.05$, Mann-Whitney test. $^†p<0.05$, Student t test. (*With permission from Sidawy et al., 1990*)

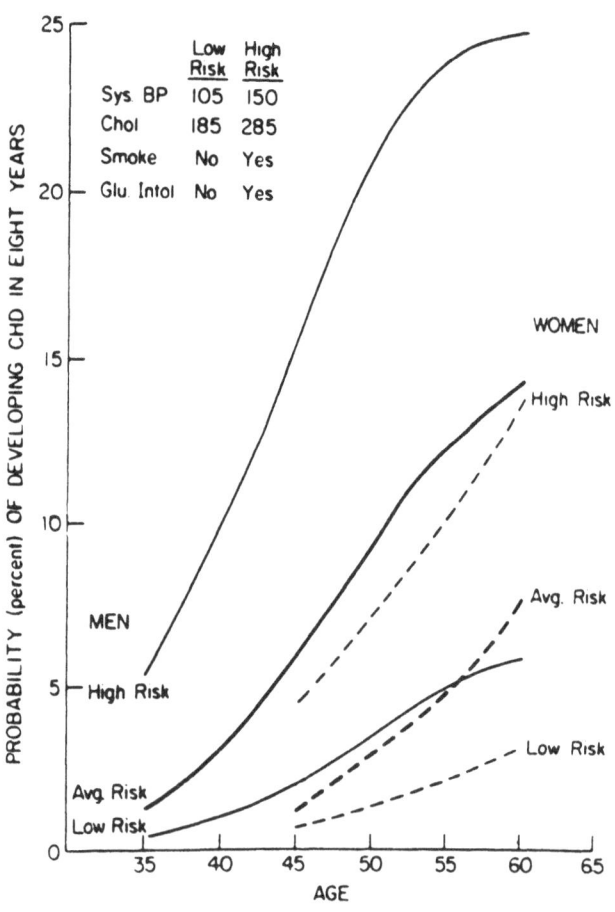

FIGURE 2: Probability of developing CAD in eight years according to age, sex, and risk category. (*Levy and Feinleib, 1984*)

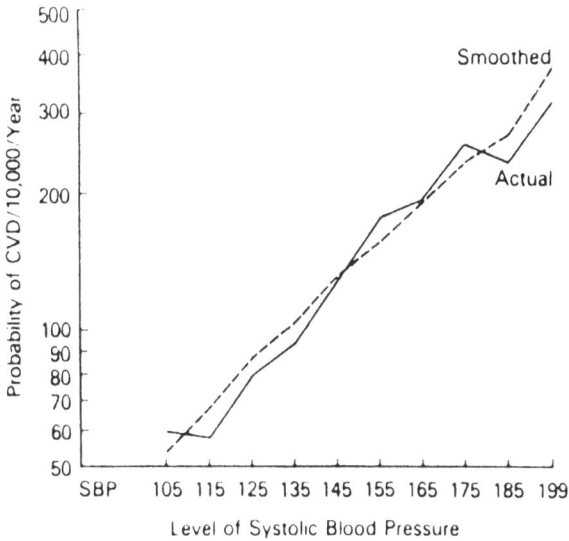

FIGURE 3: Probability of cardiovascular disease among men and women aged 45 to 64 years according to blood pressure level. Framingham Study. (*Kannel et al., 1976*)

In terms of myocardial infarction alone rather than mortality, the sex difference in the Framingham Study (*Kannel et al., 1976*) increases to approximately 20 years. This female/male sex difference, so striking in whites, is minimal in blacks.

HYPERTENSION

In men aged 45 to 62, with blood pressures exceeding 160/95, the incidence of CHD is five times that of the normotensive individual (140/90 or less). There is no sex differential favoring women in the relationship between blood pressure and coronary heart disease. (Figure 3) (*Kannel et al., 1976*)

GEOGRAPHY

Higher mortality rates for coronary heart disease have been consistently reported in Southeastern Atlantic coast states and the industrial states of the Midwest and Northeast compared to substantially lower rates in the Great Plains and the mountain states. No satisfactory explanation is known. (*Gordon et al., 1971*) (Figure 4) (*Levy and Feinleib, 1984*)

DIABETES MELLITUS

Men with glucose intolerance demonstrated a 50% greater chance of developing coronary artery disease than normal males. Women not only lose their favored status over men but experience more than double the incidence. Diabetes is also a significant independent contributor to the development of intermittent claudication (*Beach and Strandness Jr., 1980*) and their coexistence increases the risk of other cardiovascular events. (*Brand, F.N. et al., 1989*) Most studies fail to show that strict glucose level control prevents the development of peripheral vascular disease or improves the prognosis of coronary heart disease, stroke and cardiac failure (*Beach and Strandness Jr., 1980; Brand et al., 1989*).

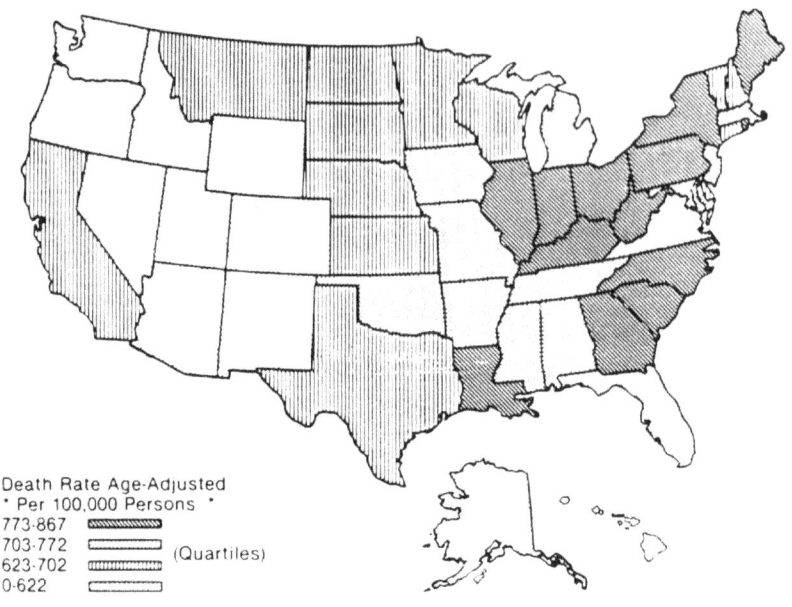

FIGURE 4: Mortality rates for cardiovascular diseases, white males ages 35 to 74, 1978. (*Levy and Feinleib, 1984*)

A large Veterans Administration Cooperative Study of the effect of aspirin and dipyridamole in male patients with non insulin dependent diabetes mellitus showed no favorable response as measured by progression of peripheral vascular disease. (Figure 5)

Alteration of large vessel endothelial cell function in diabetes mellitus are:
1. increased vWF formation,
2. decreased prostacyclin release,
3. decreased fibrinolytic activity,
4. decreased lipoprotein lipase activity, and
5. increased platelet activation and aggregability. (*Colwell et al., 1986*)

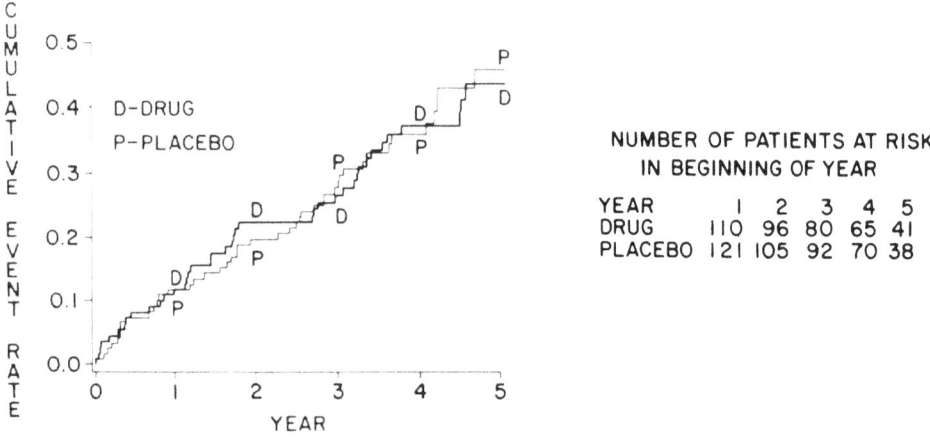

FIGURE 5: Failure of aspirin to influence vascular events in diabetic patients. (*Colwell et al., 1986*)

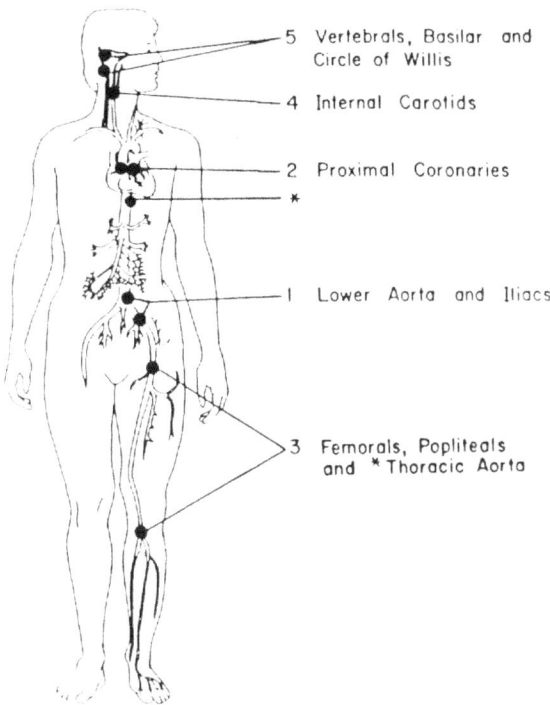

FIGURE 6: Major locations of arteriosclerotic lesion in rank order. (*Wissler, 1984*)

THE HUMAN LESION

Anatomic Distribution

The risk factors identified for coronary heart disease apply equally to cerebrovascular and peripheral arterial occlusive disease. Progress of the arteriosclerosis lesion to stenosis and expression of symptoms assumes definite anatomic patterns. Particularly vulnerable are:
1. the coronary circulation,
2. the infrarenal abdominal aorta
3. the major arteries of the lower extremity,
4. the brachiocephalic branches, and
5. visceral branches of the abdominal aorta. (Figure 6) (*Wissler, 1984*)

Mapping studies of the coronary circulation demonstrate that atherosclerosis commonly occurs at bifurcations such as the proximal branch point of the left anterior descending artery and the bifurcation of the left circumflex artery (*Solberg and Eggen, 1971*) and left coronary artery. Early lesions are usually highly localized as at the origin of the left anterior descending branch. (Figures 7,8,9) (*DeBakey, 1978*)

In the brachiocephalic system, the most frequent and severe disease occurs at the origin of the internal carotid artery. The external carotid, the common carotid, the cervical portion of the internal carotid, and the vertebral basilar vessels are less often involved. (Figure 10) (*Callow et al., 1968; Fields, 1974*)

The infrarenal abdominal aorta is especially prone to early and severe atherosclerosis with calcification, thrombus formation, cholesterol abscesses, and ulceration.

The upper extremity arteries are rarely diseased. Other arteries with relative resistance

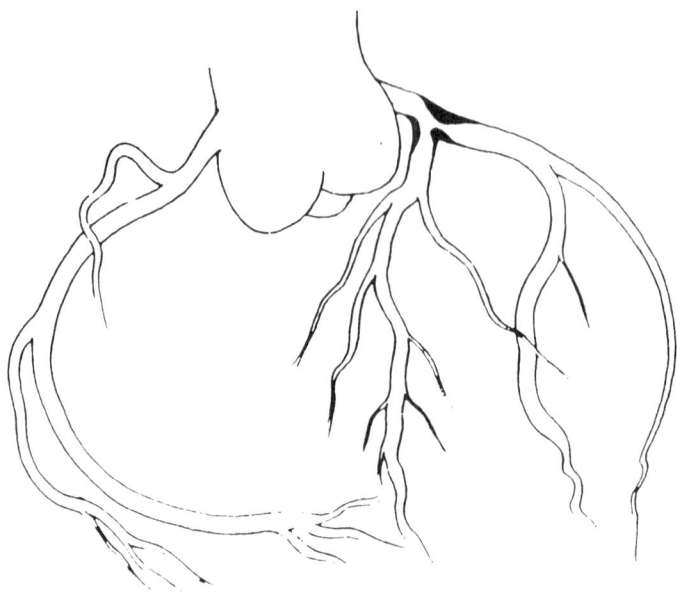

FIGURE 7: Localized stenotic lesion in main left coronary artery at origin of left anterior descending branch. (*DeBakey, 1978*)

to advanced atherosclerosis are certain visceral arteries such as the celiac axis and its branches, and the mesenteric, intercostal, and internal mammary arteries. (*DeBakey, 1978; DeBakey et al., 1985*)

THE PLAQUE

There is a substantial information gap concerning conversion of the fatty streak to the fibrous plaque and progression of the latter from a stable to an unstable lesion. The fibrous plaque may vary in color from a pale whitish hue to moderate yellow. Its cap is fibrous and rubbery; there is extracellular lipid deposition in its center, and the central core may be soft with semiliquid cholesterol or firm and resilient due to collagen and smooth muscle. Central core necrosis, intraplaque hemorrhage, intramural thrombus, and calcification characterize the end stage lesion. (Figure 11) (*Wissler, 1984*) Thrombus may be superimposed on the luminal surface. Stenosis slowly develops as the plaque slowly grows but intramural hemorrhage may cause an abrupt increase in size often accompanied by sudden, even critical diminution of the arterial lumen. The sequence of events leading to spontaneous plaque disruption are unknown, but presumably there is thinning of the fibrous cap overlying the necrotic core. (*Davies and Thomas, 1985*) This may be due to hemodynamic factors, release of proteolytic enzymes by macrophages (*Fuster et al., 1988*) or as a consequence of hemorrhage from the abnormally abundant vasa vasorum and the fragile vessels of neoangiogenesis. (*Barger et al., 1984; Adams and Shoen, 1989*)

Usually the plaque develops as a crescentic mass with its greatest bulk along the posterior or dorsal wall. The residual lumen is eccentric. (*Adams and Schoen, 1989; Brown et al., 1984*)

Although the stable fibrous plaque has an intact endothelial cover it should not be inferred that the endothelial cell is functionally intact. Other than loss of normal relaxation with paradoxical vasoconstriction to infused acetylcholine (*Ludmer, et al., 1986*) there is little information concerning the functions of the endothelial and smooth muscle cells in the atherosclerotic vessel. (*Ross, 1988; Munro et al., 1988; Reidy, 1985; Faggiotto et al., 1984; Walker et al., 1986*)

The smooth muscle cell, the macrophage, and the lymphocyte are all involved in the development of the advanced lesion. (*Wahl and Wahl, 1989; Adams and Schoen, 1989*) The lymphocyte constitutes approximately 20% of the cells in the fibrous cap. (*Jonasson et al., 1986*) Activated T-lymphocytes produce a number of lymphokines. Gamma Interferon (Figure 12) (*Jonasson et al., 1986*) induces the expression of class II histocompatibility anti-

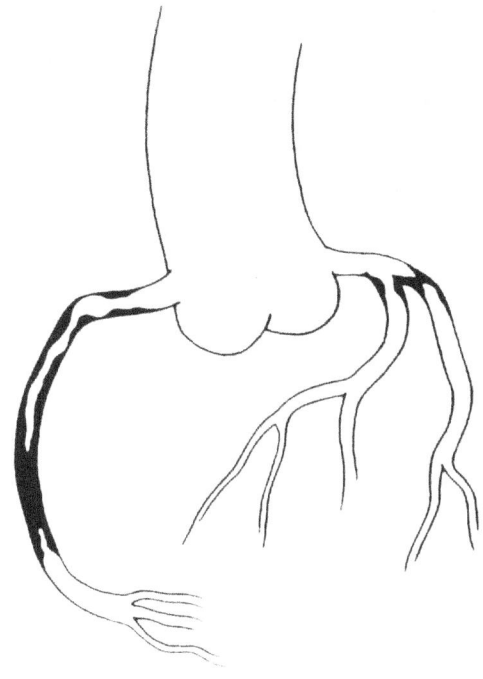

FIGURE 8: Moderate to severe stenotic and occlusive disease in right and left anterior descending arteries and circumflex coronary artery. (*DeBakey, 1978*)

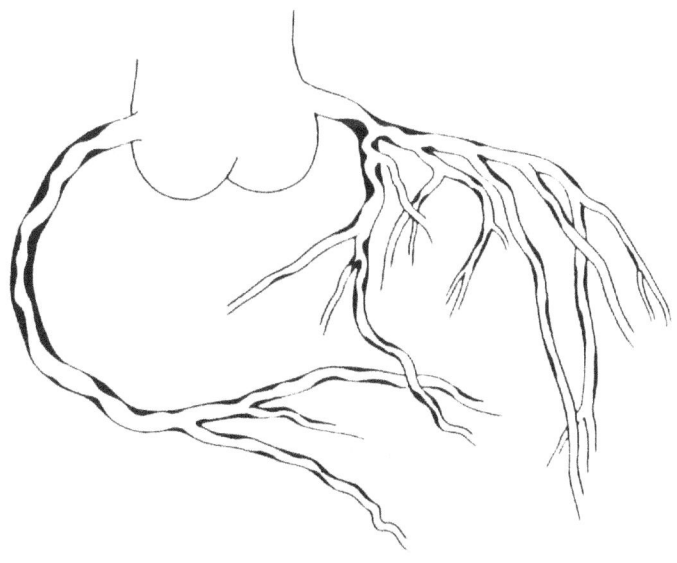

FIGURE 9: Diffuse atherosclerotic disease throughout entire coronary bed. (*DeBakey, 1978*)

gens on endothelial cells and smooth muscle cells suggesting that the atherosclerotic plaque may be the result of a delayed hypersensitivity reaction. (*Pober et al., 1983; Jonasson et al., 1985; Munro et al., 1987*)

CLINICAL MANIFESTATIONS

Although atherosclerosis is present throughout the arterial tree, its clinical manifestations depend upon the location of the plaque, its degree of instability, and whether the target organ is compromised by ischemia, embolization, or both. These differences are strikingly illustrated in the cervical carotid and the coronary arterial systems. Thrombosis superimposed upon ulcerated plaque resulting in occlusion of an internal carotid artery appears to be far less often the cause of transient ischemic attacks or cerebral infarction than is embolization.

Chronic and progressive ischemia of the cerebral hemisphere occurs but it is far less frequently a cause of neurologic deficit, either transient or permanent, than is sudden occlusion of an intracranial artery by an embolus originating in an ulcerated extracranial plaque. In this instance, the plaque in the internal carotid artery may undergo sudden change such as intraplaque hemorrhage with substantial increase in size and a corresponding reduction in vessel lumen.

In the coronary system, sudden occlusion due to thrombus formation, or severe spasm as in the left main stem coronary artery may result in massive ischemia, infarction and death. (*Schoen, 1980*) On the other hand, terminal thrombosis may be limited, spasm not overwhelming and the inadequate perfusion of the heart muscle is expressed as angina pec-

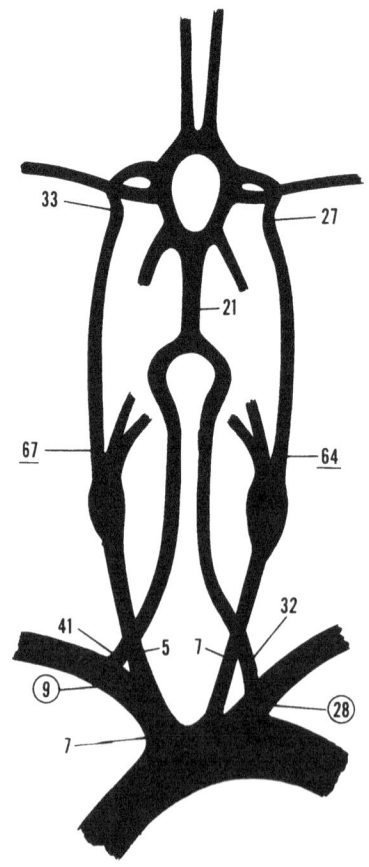

FIGURE 10: Location of segmental occlusive disease in 341 arteries in 100 patients. (*Callow et al., 1968*)

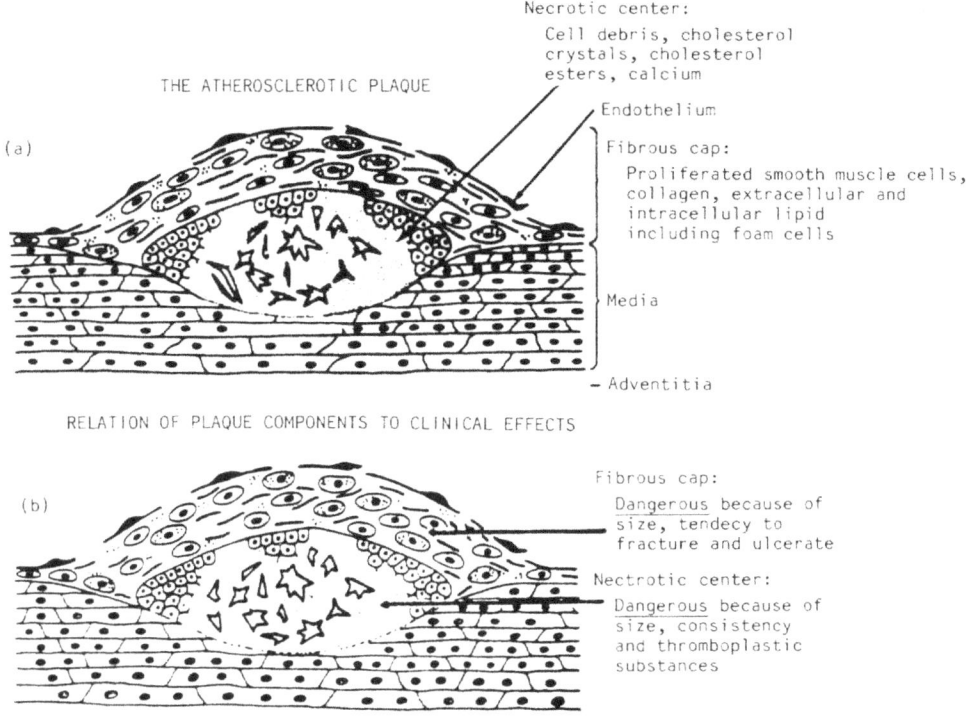

FIGURE 11: The atherosclerotic plaque. (*Wissler, 1984*)

toris. Ischemia of the skeletal muscles of the lower extremity, occurring with exercise and disappearing with rest, is the explanation for intermittent claudication.

PLAQUE NEOANGIOGENESIS

The rich vascularization seen in atheromatous lesions, as long ago as 1938 (*Winternitz et al., 1938*) may be the source of intraplaque and intramural hemorrhage. In the human atherosclerotic lesion blood vessels may penetrate the coronary artery media to enter the lumen. (*von Albertini, 1938; Morris, 1951; Paterson, 1952; Higginbotham et al., 1963; Osborn, 1963*) Evidence of bleeding from a plexus of new vessels within the coronary intima has been reported in 94% of 36 cases of sudden cardiac death. (*Reichenbach et al., 1977; Cliff et al., 1983*)

Barger and associates (*Barger et al., 1984*) utilizing cinematomographic recording of silicone polymer injections into cleared human hearts demonstrated neovascularization in and about plaques in the coronary artery. The microvessels were associated with the plaque and extended from the adventitia through the media and into the thickened intima. There was an abrupt change in the vascular pattern at the transition between injured and uninjured regions. (Figures 16 & 17) Where microvessels were not present arteries were free of atherosclerosis or showed only slight intimal thickening. (Figures 18, 19) (*Kamat et al., 1987*). This marked neovascularization is not seen in normal coronary arteries.

The implications of atherosclerosis and microvascular neoangiogenesis are obscure. (*Beeuwkes, III et al., 1990; Forman et al., 1985*) There may be preferential deliverance of vasoactive substances to the region of the plaque. A second and provocative observation is that flow from the lumen into the plaque is against the expected pressure gradient if the plaque contains blood vessels derived from an upstream source as demonstrated in injection specimens. (Figure 13) According to Bernoulli's law, a lower pressure in the area of narrowed lumen should be expected as compared to the adjacent capillaries. The gradient thus might be large. (*King, 1952*) With increase in flow, as during exercise, and a luminal area reduction of severe degree, the consequent relatively high pressure in new capillaries might lead to intraplaque hemorrhage and even rupture.

Plaque hemorrhage in the extracranial carotid circulation may have a significant relationship to onset of symptoms (*Lusby et al., 1982*). Intraplaque hemorrhage was present in most of the specimens from symptomatic patients and was absent in the plaques of asymptomatic patients. (*Fisher et al., 1987*) There now appears to be sufficient evidence to support a relationship between onset of symptoms, and plaque disintegration. (*Lusby et al., 1982; Fisher et al., 1987; Imparato, 1986; Persson, 1986; Imparato et al., 1983*)

Three special situations - the atherosclerosis of chronic renal failure, homocysteinemia, and organ transplantation - are worthy of note.

UREMIA

Because uremic levels of oxalic acid suppress replication and migration of human endothelial cells in culture, (*Levin et al., 1990*), it is suggested that atherosclerosis of chronic renal failure may be the consequence of a number of interrelated and complex factors which include the toxic effect of uremic levels of oxalic acid. (Figure 14) (*Levin et al., 1990*) Chronic renal failure and hemodialysis are associated with atherosclerotic lesions which occur at a younger age than in the general population.

HOMOCYSTEINEMIA

This rare inborn error of metabolism, the result of several enzymatic defects of which cystathionine synthetase is the primary one, results in an accumulation of homocysteine in the plasma and its increased urinary excretion. It is of interest for three reasons: the patchy desquamation of vascular endothelium with circulating endothelial cells seen in the baboon model of experimental homocysteinemia; (*Harker et al., 1976*) the increased platelet consumption and early onset of severe arteriosclerosis seen in homocysteinemic patients, and because high plasma lipid levels and other traditional risk factors combined, do not account for all cases of clinical atherosclerosis. (*Brattstrom et al., 1984; Boers et al., 1985*)

Homocysteinemia presents with a variety of signs and symptoms: ectopia lentis, mental retardation, skeletal deformities, and progressive and premature atherosclerotic cardiovascular disease. In a study of two groups of individuals: healthy volunteers, and peripheral vascular patients, basal plasma homocystine levels were higher in those patients with vascular disease. No correlation was found between the level of plasma homocystine and the usual risk factors

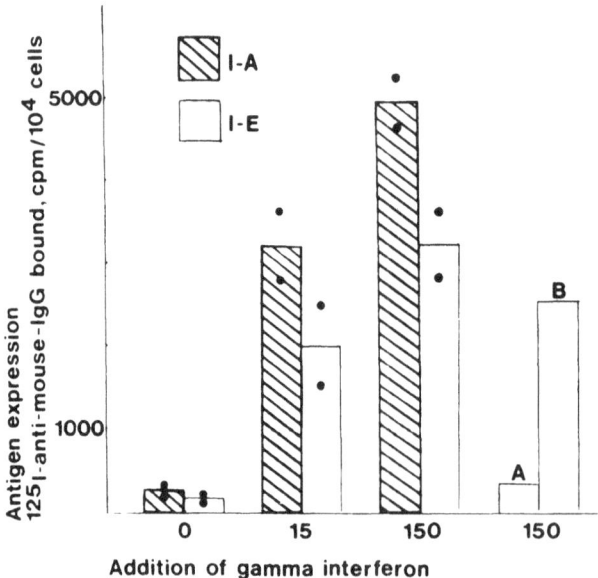

FIGURE 12: Gamma interferon induction of class II histocompatibility antigens. (*Jonasson et al., 1986*)

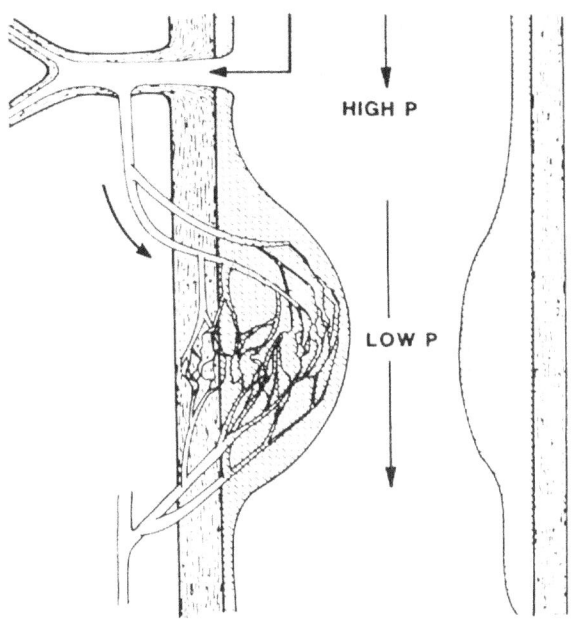

FIGURE 13: Schematic representation of high and low pressure regions in an area of stenosis. (*Beeuwkes et al, 1990*)

such as diabetes mellitus, hypertension, tobacco use, or plasma cholesterol levels. Elevated PHC may be an independent risk factor, an important observation for elevated plasma levels can be normalized with folic acid and pyridoxine. (*Harris et al., 1989*)

TRANSPLANTATION ATHEROSCLEROSIS

The accelerated atherosclerosis seen in cardiac transplantation and other organ transplants may be the most important long term complication. In organs undergoing chronic rejection, arterial intimal thickening with foam cells is common. (*Moore, 1981*)

Atherosclerosis occurring in the coronary arteries of the cardiac transplant recipient is a major cause of death. Cardiac retransplantation is the only therapy for, unlike the naturally occurring plaque, the accelerated plaque is concentric, and the disease is diffuse. Bypass grafting or balloon angioplasty cannot be performed. (*Gao et al., 1987; Billingham, 1987*)

IV THERAPY

There is no consistently effective medical therapy for the advanced complicated plaque. All current therapeutic efforts are based on the removal, reduction, or bypass of the obstructing plaque mass. This is true for percutaneous and laser balloon angioplasty, surgical endarterectomy, transluminal atherectomy, and bypass procedures with various synthetic and natural conduits. Autogenous grafts of the saphenous vein and the internal mammary artery and synthetic grafts of dacron and teflon are in world wide use. In large vessels such as the aorta and the iliofemoral tree, with their high flow rates and low resistance, all of these interventions are acceptably successful. However, in small caliber high resistance circuits such as the distal tibial vessels, patency rates are low and the incidence of failure is high. Aside from early technical failures which are usually recognizable and correctable, failure due to progression of atherosclerosis or restenosis of the bypassed, debulked or excised segment is a major problem. The restenosis is as often due to a cellular proliferative response as to atherosclerosis. If laser or thermal energy is utilized in conjunction with angioplasty (72- 74), the cellular proliferative response may be delayed and when it occurs may be less severe

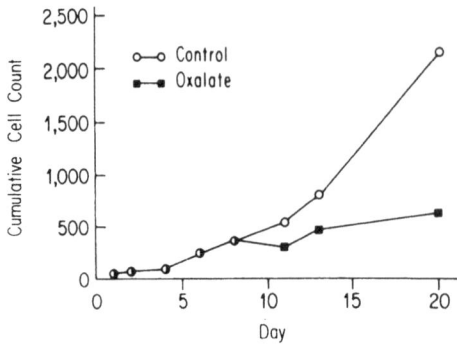

FIGURE 14: Influence of oxalic acid on endothelial cell replication. (*Levin et al, 1990*)

suggesting that the applied energy destroys smooth muscle cells and reduces the proliferative response. Radiofrequency balloon angioplasty and laser debulking of the mass are examples. (*Reidy et al., 1986; Callow, 1987; Callow, 1982; Callow and O'Donnell, 1986*) Contrariwise deep atherectomy specimens containing media as well as plaque are associated with a higher restenosis rate. (*Graor, 1990*) Both autocrine and paracrine factors have been implicated. (*Libby et al., 1987*) Inhibition of cellular proliferation is available in the experimental preparation but as yet is untested in the human. (*McGoff et al., 1989; Conte et al., 1989; Powell, et al., 1989*).

CAROTID ENDARTERECTOMY

Recurrent stenosis, both early and late, occurs in as many as 20% of operated patients. Possibly no more than 3%, however, are symptomatic and require re-operation. (*Nitzberg, et al., 1990*)

ANGIOPLASTY

Balloon angioplasty usually causes a nondistensible plaque to split in a variety of disruptive patterns which often extend into the intima, and occasionally generate new channels (*Schoen, 1989*). The uninvolved portion of the arterial wall opposite the eccentric plaque is usually stretched to its mechanical limit. Complications of angioplasty include elevation of an obstructing flap, intimal hemorrhage, endothelial avulsion, and intimal hyperplasia. Acute thrombosis occurs in probably no more than 5% of dilatations but it may be fatal. In the coronary circulation within about 6 months (*Callow, In Press; Kent, 1987; McBride et al., 1988*) restenosis of the angioplasty site occurs in about 30% of dilatations. Atherectomy consists of the intraluminal introduction of a device with a rotating burr or a sliding blade. Both chisel away portions of the plaque which extend into the lumen.

AUTOLOGOUS VEIN BYPASS GRAFT

The Aorto-Coronary Position

Autologous aorto-coronary saphenous vein bypass grafts fail at a rate of approximately 20% the first year and from 1-5% per year thereafter. Probably fewer than 50% are patent at the ten year period. (*Austin et al., 1985; Bourassa et al., 1984; Chesebro et al., 1984*) Myo-intimal hyperplasia and progression of atherosclerosis are the chief causes of restenosis. (*Schoen, 1989; Walts et al., 1987*) Acceleration of vein graft failure is associated with hyperlipidemia and cigarette smoking. (*Titus, 1988; Lie et al., 1977*) Internal mammary artery bypass grafts have a better rate of initial success and a lower incidence of later failure than the saphenous vein in the same position. Endothelial cell derived relaxation factor and prostacyclin production are reported as being produced in larger amounts by the internal mammary artery than the vein. (*Shelton et al., 1988; Luscher et al., 1988*)

THE SYNTHETIC GRAFT

Animal models reveal that the endothelial cell continues to show the metabolic activity of replication despite its having grown to confluence. The smooth muscle cell, even though covered by an endothelial cell layer, also may continue to proliferate. (*Clowes et al., 1985; Reidy et al., 1986; Callow, 1982; Owens and Reidy, 1985*) In large diameter grafts this response to chronic injury, if indeed this is the stimulus, appears to be controlled and limited by the high flow-low resistance circuit. In small diameter grafts, however, the hyperplastic response results in graft failure at a rate of approximately 60% by 36 months. In addition, progression of the proliferative response to an atheromatous lesion has been observed. (*Selman et al., 1980; DePalma, 1979*)

REGRESSION

Most animal model lesions are more like the fatty streak than the advanced fibrous plaque. Regression has been noted in these lesions when the atherogenic stimuli are discontinued. (*Selman et al., 1980; DePalma, 1979; Manilinow, 1982*) In the human, however, regression of intermediate and advanced lesions is yet to be convincingly documented. That examples of regression of early lesions are available is not disputed. (*Haimovici, 1977; Armstrong, 1976*) However, a medical regimen that induces regression of advanced and symptomatic lesions in repeatable fashion from patient to patient is not available.

The advanced unstable lesion is rarely, if ever, covered by an intact endothelial layer. (*Reidy, 1985*) It is doubtful that the endothelium can recover a denuded ulcerated atherosclerotic plaque and thus restore a smooth, nonthrombogenic surface to the ulcerated arterial wall. This is the major theoretical argument against regression of the advanced plaque in the human. Despite reduction of the bulk amount of plaque protruding into the arterial lumen and the arterial wall, meaningful regression must include restoration of a smooth surface which seals off the necrotic core from the circulating blood. The necrotic core of the ulcerated plaque seems particularly unsuited for cell migration and replication, be it the endothelial cell, smooth muscle cell or fibroblast. Clinical experience strongly indicates that advanced and end stage lesions in the human do not undergo meaningful regression with reduction or elimination of recognized risk factors. Even if a marked change in life style were embraced by Western society, there are still millions of adults with intermediate to advanced lesions which would probably progress to symptoms. Cardiac and peripheral vascular operations now number approximately one million yearly in the United States. For at least the next decade this number is unlikely to shrink appreciably. Efforts to control atherosclerosis should be directed toward prevention of progression of presently asymptomatic lesions. Preventive intervention must begin during the period of incubation a period of two, three, or four decades. The symptomatic lesion in one arterial bed is usually followed by symptoms in a second and even a third: intermittent claudication at age 50, angina pectoris at 60, and a neurologic deficit at 70. Compounding this problem is the recurrent stenosis which threatens all surgical and mechanical interventions. These therapeutic efforts are all directed toward end stage disease and have no influence whatsoever upon the atherosclerotic process throughout the rest of the large and medium sized arterial beds. Unless the progressive nature of the process can be halted, these lesions will inexorably become symptomatic. The forest fire smoulders almost undetected in the undergrowth of the forest floor - contained combustion. Its presence may not be suspected until the trees and leaves above burst forth into uncontrollable flame. The early years in which the disease is clinically undetectable are the years in which therapeutic intervention will be most effective. The end stage lesion is the point of no return.

REFERENCES

Adams, D.H., Schoen, F.J. Contemporary concepts in atherosclerosis pathology. In White, R.A. (ed). *Atherosclerosis and Arteriosclerosis* Boca Raton, CRC Press, pp49-86, 1989.

Armstrong, M.L. Regression of atherosclerosis, in *Atherosclerosis Reviews, Vol 1*, Paoletti, R. and Gotto, A.M. (eds). New York, Raven Press, 137, 1976.

Austin, C.E., Ratliff, N.B. and Hollman, J., et al. Intimal proliferation of smooth muscle cells as an explanation for recurrent coronary artery stenosis after percutaneous transluminal coronary angioplasty. *J. Am. Coll. Cardiol.* 6:369, 1985.

Barger, A.C., Beeukes, R., Lainey, L.L. and Silverman, K.J. Hypothesis: vasovasorum and neovascularization of human coronary arteries. *N. Engl. J. Med.* 31:175, 1984.

Barry, K.J., Kaplan, J., Connolly, R.J., Nardella, P., Lee, B.I., Becker, G.J., Waller, B.F., Callow, A.D. The effect of radiofrequency-generated thermal energy on the mechanical and histologic characteristics of the arterial wall *in vivo*: Implications for radiofrequency angioplasty. *Am. Heart J.* 117:332-341, 1989.

Beach, K.W., Strandness, D.E. Jr. Arteriosclerosis obliterans and associated risk factors in insulin-dependent and noninsulin-dependent diabetes. *Diabetes* 29:882-888, 1980

Beeuwkes, R. III, Barger, A.C., Silverman, K.J., Lainey, L.L. Cinemicrographic studies of the vasa vasorum of human coronary arteries. In: Glagov, S., Newman, W.P., Schaffer, S.A. (eds). *Pathobiology of the Human Atherosclerotic Plaque*. New York, Springer-Verlag, pp. 425-432, 1990.

Berenson, G.S., Webber, F.L.S., Srinivasan, S.R., Cresanta, J.L., Frank, G.C., Farris, R.P. Black-white contrasts as determinants of cardiovascular risk in childhood: precursors of coronary artery and primary hypertensive diseases. *Am. Heart J.* 108:672-683, 1984.

Billingham, M.E. Cardiac transplant atherosclerosis. *Transplant Proc.* 19(Suppl 5):19, 1987.

Boers, G.H.J., Smals, A.G.H., Trijbels, F.J.M. et al. Heterozygosity for homocystinuria in premature peripheral and cerebral occlusive arterial disease. *N. Engl. J. Med.* 313:709, 1985.

Bourassa, M.G., Enjalbert, M., Campeau, L., Lesperance, J. Progression of atherosclerosis in coronary arteries and bypass grafts: ten years later. *Am. J. Cardiol.* 53:102C, 1984.

Brand, F.N., Abbott, R.D., Kannel, W.B. Diabetes, intermittent claudication and risk of cardiovascular events. The Framingham Study *Diabetes* 38:504-509, 1989.

Brattstrom, L.E., Hardebo, J.E., Hultberg, B.L. Moderate homocystinemia: A possible risk factor for arteriosclerotic cerebrovascular disease. *Stroke* 15:1012, 1984.

Brown, B.G., Bolson, E.L., Dodge, H.T. Dynamic mechanisms in human coronary stenosis. *Circulation* 70:917, 1984.

Callow, A.D. Healing of the endarterectomy wound, in Vascular Diseases. *Current Research and Clinical Applications*.

Callow, A.D. Current status of vascular grafts. *Surg. Clin. North Am.* 62:501, 1982.

Callow, A.D., Moran, J.M., Kahn, P.C., Deterling, R.A. Jr. Patterns of atherosclerosis of extracranial cerebral arteries. In: Haimovici E (ed). Atherosclerosis: Recent Advances. *Ann. NY Acad. Sci.* 149:974-988, 1968.

Callow, A.D. and O'Donnell, T.F. Recurrent carotid stenosis: Frequency, clinical implications, and some suggestions concerning etiology. In: Bergan J.J., Yao JST (eds). *Reoperative arterial Surgery*. New York: Grune & Stratton 513-535, 1986.

Callow, A.D. Arterial injury and response. In: Proceedings of the 1st Congress of the International Society for Applied Cardiovascular Biology. Basel, Karger, in press.

Caplan, L.R., Gorelock, P.B and Hier, D.B. Race, sex and occlusive cerebrovascular disease: a review. *Stroke* 17:648-655, 1986.

Chesebro, J.H., Fuster, V., Elveback, L.R., et al. Effect of dipyridamole and aspirin on late vein-graft patency after coronary bypass operation. *N. Eng. J. Med.* 310:209, 1984.

Cliff, W.J. and Schoefl, G.I. Pathological vascularization of the coronary intima. In: *Development of the Vascular System*, Ciba, London, Pitman, 207-221, 1983.

Clowes, A.W., Gown, A.M., Hanson, S.R. and Reidy, M.A. Mechanisms of arterial graft failure. I. Role of cellular proliferation in early healing of PTFE prostheses. *Am. J. Pathol.* 118:43, 1985.

Colwell, J.A., Bingham, S.F., Abraira, C. et al. V.A. Cooperative Study on antiplatelet agents in diabetic patients after amputation for gangrene. II. Effects of aspirin and dipyridamole on atherosclerotic vascular disease rates. *Diabetes Care* 9:140-148, 1986.

Conte, J.V., Foegh, M.L., Calcagno, D., et al. Inhibition of angioplasty-induced myointimal hyperplasia with angiopeptin. In: Pannell, M. (ed.). *Surgical Forum, Vol. XL*. Chicago, American College of Surgeons, 302-303, 1989.

Davies, M.J. and Thomas, A.S. Plaque fissuring - the cause of acute myocardial infarction, sudden ischemic death, and crescendo angina. *Br. Heart J.K.* 53:363, 1985.

Davis, B.A., Crook, J.E., Vestal, R.E. and Oates, J.A. Prevalence of renovascular hypertension in patients with grade III or IV hypertensive retinopathy. *N. Engl. J. Med.* 301:1273--1276, 1979.

DeBakey, M.E. Patterns of atherosclerosis and rates of progression. In: Paoletti, R., Grotto A.M. Jr. (eds). *Atherosclerosis Reviews Vol 3*, New York, Raven Press, 1978.

DeBakey, M.E., Lawrie, G.M. and Glaeser, D.H. Patterns of atherosclerosis and their surgical significance. *Ann. Surg.* 201:115, 1985.

DePalma, R.G. Atherosclerosis in vascular grafts. *Atheroscler. Rev.* 6:147, 1979.

Faggiotto, A., Ross, R. and Harker, L. Studies of hypercholesterolemia in the non-human primate. I. Changes that lead to fatty streak formation. *Arteriosclerosis.* 4:323, 1984.

Fields, W.S. Clinical Symposia. Aortocranial Occlusive Vascular Disease. *Stroke* CIBA 26:1-31, 1974.

Fisher, M., Blumenfeld, A.M. and Smith, T.W. The importance of carotid artery plaque disruption and hemorrhage. *Arch. Neurol.* 44:1086, 1987.

Forman, M.B., Oates, J.A., Robertson D, Robertson R., Roberts L.J., Virmani R. Increased adventitial mast cells in a patient with coronary spasm. N. Engl. J. Med. 313:1138-1141, 1985.

Fuster, V., Badimon, L., Cohen, M., Ambrose, J.A., Badimon, J.J. and Cheseboro, J. Insights into the pathogenesis of acute ischemic syndromes. *Circulation* 77:1213, 1988.

Gao, S.Z., Schroeder, J.S., Alderman, E.L. et al. Clinical and laboratory correlates of accelerated coronary artery disease in the cardiac transplant patient. *Circulation* 76(Suppl 5):56, 1987.

Goldbourt, U., Neufeld, H.N., Medalie, J.H. and Oron, D. Disappearing differences between Yemenite-born and other Israeli: fact or fancy? *Harefuah (J. Isr. Med. Assoc.)* 94:1-5, 1978.

Goldbourt, U. and Neufeld, H.N. Genetic aspects of arteriosclerosis. *Arteriosclerosis* 6:357--377, 1986.

Gordon, T., Sorlie, P. and Kannel, W.B. *The Framingham Study. An Epidemiological Investigation of Cardiovascular Disease. Section 27.* Coronary Heart Disease, Atherothrombotic Brain Infarction, Intermittent Claudication - A Multivariate Analysis of Some Factors Related to their Incidence: Framingham Study, 16-year Follow-up. Washington DC. US Government Printing Office, 1971.

Graor, R. *Communication, International Society for Cardiovascular Surgery.* Los Angeles, California, June 6, 1990.

Haimovici, H. Atherogenesis. Recent biological concepts and clinical implications. *Am. J. Surg.* 134:174, 1977.

Hamby, R.I. Hereditary aspects of coronary artery disease. *Am. Heart J.* 101:639-649, 1981.

Harker, L.A., Ross, R., Slichter, S.J. and Scott C.R. The role of endothelial cell injury and platelet response in its genesis. *J. Clin. Invest.* 58:731, 1976.

Harris, E.J., Taylor, L.M., Malinow, M.R., Edwards, J.M. and Porter, J.M. The association between elevated plasma homocysteine and symptomatic peripheral arterial disease. *Surg. Forum* XL:307-309, 1989.

Hawiger, J. American Heart Association, Council on Thrombosis, *Newsletter*, November, 1988.

Higginbotham, A.C., Higginbotham, F.H. and Williams, T.W. Vascularization of blood vessel walls. In: Jones, R.J. (ed). *Evolution of the atherosclerotic plaque.* University of Chicago Press, Chicago, p 265-277, 1963.

Imparato, A.M. Presidential address: The carotid 30 bifurcation plaque - a model for the study of atherosclerosis. *J. Vasc. Surg.* 3:249, 1986.

Imparato, A.M., Riles, T.S., Mintzer, R. and Baumann, G. The importance of hemorrhage in the relationship between gross morphologic characteristics and cerebral symptoms in 376 carotid artery plaques. *Ann. Surg.* 197:195, 1983.

Jonasson, L., Holm, J., Skalli, O, Gabianni G, Hansson G.K. Expression of Class II transplantation antigen on vascular smooth muscle cells in human atherosclerosis. *J. Clin. Invest.* 176:125, 1985.

Jonasson, L., Holm, J., Skalli, O., Bondjero, G. and Hansson, G.K. Regional accumulations of T cells, macrophages, and smooth muscle cells in human atherosclerotic plaque. *Arteriosclerosis* 6;131, 1986.

Kamat, B.R., Galli, S.J., Barger, A.C., Lainey, L.L. and Silverman, K.J. Neovascularization and coronary atherosclerotic plaque: Cinematographic localization and quantitative histologic analysis. *Hum. Pathol.* 18:1036-1042, 1987.

Kannel, W.B. Role of blood pressure in cardiovascular disease: The Framingham Study. *Angiology* 26:1, 1975.

Kannel, W.B., McGee, D. and Gordon, T. A general cardiovascular risk profile: the Framingham study. *Am. J. Cardiol.* 38:46, 1976.

Kaplan, J. *Myointimal hyperplasia in cellular healing in the thermally injured arterial wall.* Thesis for the Degree of Master of Science in Surgery, Tufts University School of Medicine, Boston, 1988.

Kent, K.M. Reocclusions after percutaneous transluminal coronary angioplasty. *Transplant Proc.* 19(Suppl 5):44, 1987.

King, E.S.J. The hemodynamics of subintimal hemorrhage. *Australian Ann. Med.* 1:18-25, 1952.

Levin, R.I., Kantoff, P.W. and Jaffe, E.A. Uremic levels of oxalic acid suppress replication and migration of human endothelial cells. *Arteriosclerosis* 10:198-207, 1990.

Levy, R.I. and Feinleib, M. Risk factors for coronary artery disease and their management. In: *Braunwald E. Heart Disease. A Textbook of Cardiovascular Medicine.* Philadelphia, in W.B. Saunders, p1204-1234, 1984.

Libby, P., Birinyi, L.K. and Callow, A.D. Functions of endothelial cells related to seeding of vascular prostheses: The unanswered questions. In: Glover, J.L. and Herring, M. (eds). *Endothelial Seeding in Vascular Surgery,* New York, Grune & Stratton 17-35, 1987.

Lie, J.T., Lawrie, G.M. and Morris, G.C. Aortocoronary bypass saphenous veins graft atherosclerosis. Anatomic study of 99 vein grafts from normal and hyperlipoproteinemic patients up to 75 months postoperatively. *Am. J. Cardiol.* 40:906, 1977.

Ludmer, P.L., Selwyn, A.P., Shook, R.R., Wayne, G.L., Mudge, G.H. and Alexander, R.W. Paradoxical vasoconstriction induced by acetylcholine in atherosclerotic coronary arteries. *N. Engl. J. Med.* 315:1046, 1986.

Lusby, R.J., Ferrell, L.D., Ehrenfeld, W.K., Stoney, R.J. and Wiley, E.J. Carotid plaque hemorrhage, its role in production of cerebral ischemia. *Arch. Surg.* 117:1479-1488, 1982.

Luscher, T.F., Diedrich, D., Siebenmann, R. et al. Difference between endothelium--dependent relaxation in arterial and in venous coronary bypassed grafts. *N. Engl. J. Med.* 319:462-467, 1988.

Manilinow, M.R. Experimental Models of Atherosclerosis Regression. *Atherosclerosis,* 48:105-118, 1982.

Masuda, J.K. and Ross, R. Atherogenesis during low level hypercholesterolemia in the nonhuman primate. I. Fatty streak formation. II. Fatty streak conversion to fibrous plaques. *Arteriosclerosis* 10:164-177;178-187, 1990.

Maxwell, J.G., Rutherford, E.J., Covington, D. et al. Infrequency of blacks among patients having carotid endarterectomy. *Stroke* 20:22-26, 1989.

Maynard, C., Fisher, L.D., Passamani, E.R. and Pullman, T. Blacks in the Coronary Artery Surgery Study: risk factors and coronary artery disease. *Circulation* 74:64-71, 1986.

McBride, W., Lange, R.A. and Hillis, L.D. Restenosis after successful coronary angioplasty. Pathophysiology and prevention. *N. Engl. J. Med.* 318:1734, 1988.

McGoff, M.A., Allen, B.T., Sicard, G.A., Anderson, C.B. and Santoro, S.A. In vivo antithrombotic activity of arg-gly-asp-containing peptides in a canine model of small-diameter synthetic vascular graft thrombosis. *Surg. Forum* XL:321-323, 1989.

Moore, S. Injury mechanisms in atherogenesis, in Moore S (ed). Vascular Injury and *Atherosclerosis.* New York, Dekker, 1981.

Morris, J.N. Recent history of coronary disease. *Lancet* 1:1-7,69-73, 1951.

Munro, J.M. and Cotran, R.S. The pathogenesis of atherosclerosis: atherogenesis and inflammation. *Lab. Invest.* 58:249, 1988.

Munro, J.M., van der Walt, J.D., Munro, C.S., Chalmers, J.A.C. and Cox, E.L. An immunohistochemical analysis of human aortic fatty streaks. *Hum. Pathol.* 18:375, 1987.

Nitzberg, et al. Society of Vascular Surgery Meeting - June 6, 1990, Los Angeles, California.

Nugent, J. and O'Connor, M. (eds). Development of the Vascular System. *CIBA Foundation Symposium 100.* London, Pitman, 1-254, 1983.

Osborn, G.R. The incubation period of coronary thrombosis. London, Butterworths, 1963.

Owens, G.K. and Reidy, M.A. Hyperplastic growth response of vascular smooth muscle cells following induction of acute hypertension in rats by coarctation. *Circ. Res.* 57:695, 1985.

Paterson, J.C. Factors in the production of coronary artery disease. *Circulation* 6:732-739, 1952.

Persson, A.V. Intraplaque hemorrhage. *Surg. Clin. North Am.* 66:415, 1986.

Pober, J.S., Collins, T., Gimbrone, M.A. et al. Lymphocytes recognize human vascular endothelium and fibroblast Ia antigens induced by recombinant immune interferon. *Nature* 305:726, 1983.

Powell, J.S., Clozel, J-P, Muller, R.D.M. et al. Inhibitors of angiotensin-converting enzyme prevent myointimal proliferation after vascular injury. *Science* 245:186-188, 1989.

Reichenbach, D.D., Moss, N.S. and Meyer, R. Pathology of the heart in sudden cardiac death. *Am. J. Cardiol.* 39:865-872., 1977.

Reidy, M.A., Kossch, A.O., Kirkman, T.R. et al. Endothelial regeneration. VI. Chronic nondenuding injury in baboon vascular grafts. *Am. J. Pathol.* 123:432, 1986.

Reidy, M.A. A reassessment of endothelial injury and arterial lesion formation. *Lab. Invest.* 53:513, 1985.

Ross, R. The pathogenesis of atherosclerosis, in Braunwald E (ed). *Heart Disease*. A Textbook of Cardiovascular Medicine. Philadelphia, W.B. Saunders, 1135, 1988.

Schoen, F.J. Interventional and Surgical Cardiovascular Pathology. *Clinical Correlations and Basic Principles*. Philadelphia, W.B. Saunders, 1989.

Selman, S.H., Rhodes, R.S., Anderson, J.M. et al. Atheromatous changes in expanded polytetrafluoroethylene grafts. *Surgery* 87:630, 1980.

Shea, S., Ottman, R., Gabrieli, C., Stein, Z. and Nicholas, A. Family history as an independent risk factor for coronary artery disease. *J. Am. Col. Cardiol.* 4:793-801, 1984.

Shelton, M.E., Forman, M.B., Virmani, R., Bajaj, A., Stoney, W.S. and Atkinson, J.B. A comparison of morphologic and angiographic findings in long-term internal mammary artery and saphenous vein bypass grafts. *J. Am. Coll. Cardiol.* 11:297-307, 1988.

Sidawy, A.N., Schweitzer, E.J., Neville, R.F., Alexander, E.P., Temeck, B.K. and Curry, K.M. Race as a risk factor in the severity of infragenicular occlusive disease: Study of an urban hospital patient population. *J. Vasc. Surg.* 11:536-543, 1990.

Solberg, D.A. and Eggen, D.A. Localization and sequence of development of atherosclerotic lesions in the carotid and vertebral arteries. *Circulation* 43:711, 1971.

Solberg, L.A. and McGarry, P.A. Cerebral atherosclerosis in negroes and caucasians. *Atherosclerosis* 16:141-154, 1972.

Strandness, D.E., Didisheim, P., Clowes, A.W., Watson, J.T. (eds). Orlando, Grune & Stratton, 407, 1987.

Strong, J.P. Atherosclerosis in human populations. *Atherosclerosis* 16:193-201, 1972.

Strong, J.P. and McGill, H.C. Jr. The pediatric aspects of atherosclerosis. *J. Atheroscler. Res.* 9:251, 1969.

Titus, J.L. The heart after surgery for ischemic heart disease. *Am. J. Cardiovasc. Pathol.* 1:339, 1988.

von Albertini, A. Studien zur Aeteologie der Arteriosklerose. *Schweiz Z. Pathol. Bakteriol.* 1:3-22,163-187, 1938.

Wahl, S.M. and Wahl, L.M. Modulation of fibroblast growth and function by monokines and lymphokines. In: Pick, E. (ed). New York, Academic Press, 179, 1989.

Walker, L.N., Reidy, M.A. and Bowyer, D.E. Morphology and cell kinetics of fatty streak formation in the hypercholesterolemic rabbit. *Am. J. Pathol.* 125:450, 1986.

Walts, A.E., Fishbein, M.C. and Matloff, J.M. Thrombosed ruptured atheromatous plaques in saphenous vein coronary artery bypass grafts: ten years' experience. *Am. Heart J.* 114:-718, 1987.

Winternitz, M.C., Thomas, R.M. and Le Compte, P.M. The biology of arteriosclerosis. Springfield, Charles C. Thomas, 1938.

Wissler, R.W. Principles of the pathogenesis of atherosclerosis. In: Braunwald E (ed). *Heart Disease*. A Textbook of Cardiovascular Medicine, Philadelphia, W.B. Saunders, 1183-1204, 1984.

CEREBRAL ENDOTHELIAL INJURY IN STROKE, BRAIN TRAUMA AND HYPERTENSION

Hermes A. Kontos

Department of Internal Medicine
Medical College of Virginia
Richmond, Virginia 23298, U.S.A.

INTRODUCTION

Endothelial injury occurs in cerebral vessels in a variety of acute experimental interventions. These include ischemia followed by reperfusion, acute severe hypertension, traumatic brain injury, prolonged seizures, topical application of hydrogen peroxide or arachidonate or bradykinin, xanthine oxidase plus substrate, or hydrogen peroxide plus iron salts (for review, see *Kontos, 1989a*). Endothelial injury in these conditions is characterized by similar morphological, biochemical, and functional abnormalities.

There is substantial evidence that the endothelial injury in these conditions is due to oxygen radicals (*Kontos, 1989a*). Investigation of the mechanism of endothelial injury in these experimental situations may be fruitful, because it not only may be informative as to the basic mechanisms of endothelial cell damage, but also because it may provide clues concerning abnormalities in endothelial function in human disease such as thrombotic stroke, brain trauma, hypertensive encephalopathy and status epilepticus.

VASCULAR INJURY IN ISCHEMIA/REPERFUSION, ACUTE HYPERTENSION, OR TRAUMATIC BRAIN INJURY

The vascular injury which occurs in the abnormal experimental situations discussed above has been best studied in ischemia/reperfusion (*Kontos, 1989b*) in acute hypertension (*Kontos, 1985*) and in brain injury (*Kontos and Povlishock, 1986*). The characteristic features of this type of injury are listed in Table 1. Examination of the injured vessels in these conditions reveals discrete destructive lesions, which on scanning electron microscopy appear as blebs or as craters. By transmission electron microscopy, the bleb appears to be a localized destruction of the portion of the cell close to the surface which allows the ballooning out of the cell membrane without rupture. Rupture of the cell membrane at the site of the bleb creates the crater. The number of lesions is usually low, approximately 1-2 per 100 μm^2. In addition to these destructive lesions, there are also numerous microvilli which protrude into the lumen of the vessel as well as pinocytic vesicles. The destructive endothelial lesions, blebs and craters, appear to occur with equal frequency in the pial arterioles as well as in the intracerebral ones. On the other hand, the villi and pinocytic vesicles occur predominantly in the intracerebral vessels.

The injured vessels are dilated and display abnormal reactivity to both vasoconstrictor and vasodilator stimuli. For example, the vasoconstrictor response to hypocapnia is compromised (*Wei et al., 1980; Kontos et al., 1981*), the vasodilator response to arterial hypercapnia is also reduced (*Kontos et al., 1981*), and the ability of the vessels to respond by autoregulatory vasodilation to reductions in blood pressure is either reduced or eliminated completely (*Wei et al., 1980*). Endothelium-dependent responses are altered. For example, the normal vasodi-

TABLE 1: Features of Vascular Injury in Cerebral Ischemia/Reperfusion, in Hypertension and in Brain Trauma

1. Endothelial blebs and craters
2. Endothelial microvilli and pinocytic vesicles
3. Vasodilation
4. Abnormal reactivity
5. Compromised endothelium-dependent responses
6. Reduced oxygen consumption of the vessel wall
7. Enhanced platelet aggregability
8. Breakdown of the blood-brain barrier to macromolecules
9. Chemotaxis with accumulation of phagocytes

lator response to topical application of acetylcholine is usually converted to a vasoconstriction or is eliminated (*Wei et al., 1985; Kontos, 1989c*). Endothelium-dependent vasoconstrictor responses are also abnormal. For example, the normal vasoconstrictor response to topical application of serotonin, which is dependent on the production of an endothelium-derived contracting factor, is converted to vasodilation (*Kontos, 1989c*).

The injured vessels also display biochemical abnormalities. The oxygen consumption of the vessel wall of these vessels is reduced, reflecting damage to the mitochondria (*Wei et al., 1980; Kontos, et al., 1981*).

The blood-brain barrier, which ordinarily does not allow the exit of macromolecules from the intravascular compartment into the brain, is abnormal. The breakdown of the blood-brain barrier is reflected on extravasation of plasma albumin, IgG, as well as exogenous macromolecules such as horseradish peroxidase (*Povlishock et al., 1978; Povlishock et al., 1980*). In the intracerebral vessels, the extravasation of protein takes place into the vessel wall as well as in the perivascular space.

With rare exceptions, platelet aggregation does not take place spontaneously in these injured vessels (*Wei et al., 1980*). However, in the presence of endothelial injury, platelet aggregability is enhanced, as shown by the fact that if another proaggregatory stimulus is used platelet aggregation occurs with considerably greater ease in the injured vessels than in intact vessels in control animals (*Rosenblum et al., 1982*).

Accumulation of phagocytic cells has not been studied systematically in most of the conditions that cause this type of endothelial injury. Systematic studies have been done in the case of topical application of arachidonate and 15-HPETE. In response to these agents, there is leukocyte accumulation which begins three to four hours after application and reaches a peak 24 hours later (*Christman et al., 1984; Kontos, et al., 1985*).

MECHANISM OF VASCULAR INJURY IN ISCHEMIA-REPERFUSION, ACUTE HYPERTENSION, AND TRAUMATIC BRAIN INJURY

There is substantial evidence that the vascular injury in these conditions is due to the generation of oxygen radicals and, more specifically, to the formation and action of hydroxyl radicals. The evidence supporting this conclusion derives from several sources. First, the vascular injury and its functional consequences are minimized or completely inhibited by pretreatment with specific scavengers of products of univalent reduction of oxygen, i.e. superoxide, hydrogen peroxide, or hydroxyl radical (*Wei et al., 1981; Kontos et al., 1981; Wei et al., 1985; Kontos, 1985*). The most effective protection is provided by the combination of superoxide dismutase (SOD) and catalase or by deferoxamine, an agent which scavenges catalytic iron and, thereby, prevents the formation of hydroxyl radical from the Haber-Weiss reaction. It should be noted that the combination of SOD and catalase also effectively prevents the formation of hydroxyl radical by eliminating its precursors. Second, it has been shown that superoxide is generated in the brain and, more specifically, in the brain vessels in the course of reperfusion following ischemia (*Kontos 1989b*), following acute hypertension (*Wei et al., 1985*), and after traumatic brain injury induced by fluid percussion (*Kontos and Wei, 1986*). The generation of superoxide in these conditions has been quantified by measuring the SOD-inhibitable reduction of nitroblue tetrazolium. Similarly, histochemical studies using the oxidation of manganese from the divalent to the trivalent state by superoxide, and the subsequent oxidation of diaminobenzidine, which upon fixation with osmic acid produces an electron dense product identifiable to electron microscopy, have confirmed superoxide production in the vessel wall (unpublished observations). With this latter histochemical technique, no superoxide production in the brain parenchyma could be demonstrated, but this may be due to incomplete penetration of the reagents which are applied topically on the brain surface. Superoxide is generated intracellularly, very likely in endothelial cells, and possibly in vascular smooth muscle. Following its formation, superoxide exits into the extracellular space via the anion channel as shown by the fact that anion channel blockers inhibit its appearance in the extracellular space (*Kontos et al., 1985; Wei et al., 1985*). Because of the lower concentration of endogenous radical scavengers in the extracellular space, superoxide can dismutate to hydrogen peroxide and the two in the presence of catalytic iron may form hydroxyl radical. The fact that both the combination of SOD and catalase as well as deferoxamine have protective effects suggest strongly that hydroxyl radical is the mediator of endothelial injury in these conditions. The mediation of the endothelial injury by oxygen radicals is also supported by the fact that the endothelial injury seen in ischemia/reperfusion, hypertension, and brain trauma is mimicked by topical application of agents which generate oxygen radicals. This is the case with arachidonate and some of its products like PGG_2 (*Kontos et al., 1980; Wei et al., 1986*) and 15-HPETE (*Christman et al., 1980*), or bradykinin which activates phospholipases and accelerates endogenous arachidonate metabolism (*Kontos et al., 1984*), or exogenous hydrogen peroxide or hydrogen peroxide plus iron (*Wei et al., 1985*).

The morphological destructive lesions in the endothelium seen in the cerebral vessels are similar to those seen in cultured hepatocytes (*Jewell et al., 1982*) or cultured polymorphonuclear leukocytes (*Badwey et al., 1984*) exposed to oxidant injury. In the hepatocytes, the lesions have been ascribed the oxidation of thiols in the actin molecules of the cytoskeleton of the cells, which causes aggregation of these molecules with a resultant morphological disturbance in the cell membrane (*Orrenius et al., 1988*). The lesions in the cultured cells are reversible if the noxious stimulus is discontinued. Evidence from fluid-percussion brain injury suggests that the same is true in the cerebral vessels after this type of injury. The alteration in resting vascular tone, the consequent changes in caliber and the altered reactivity are due to the direct effects of oxygen radicals. It is well known that the products of univalent reduction of oxygen including superoxide, hydrogen peroxide, and the hydroxyl radical, relax vascular smooth muscle and dilate cerebral arterioles (*Wei et al., 1985*). The generation of superoxide in ischemia/reperfusion, hypertension, and following traumatic brain injury induced by fluid-percussion continues for at least one hour following the insult. It appears, therefore, that the abnormalities in vessel caliber and reactivity are the result of continued production of oxygen radicals.

Oxygen radicals can affect endothelium-dependent responses in a variety of ways. One or more of the following mechanisms may account for the abnormal endothelium-dependent responses in these conditions. Endothelial injury with inability to produce endothelium-derived relaxing or contracting factors, destruction of EDRF in the extracellular space, and damage to the vascular smooth muscle with either generalized depression of responsiveness or selective elimination of responses dependent on guanylate cyclase explain the abnormalities.

The reduction in the oxygen consumption of the vessel wall following exposure to oxygen radicals is undoubtedly due to mitochondrial damage and destruction of the oxygen metabolizing apparatus of the cell.

It is well known that oxygen radicals enhance platelet aggregability by mechanisms that are not fully understood (*Handin, Karabin and Boxer, 1977*). The enhanced platelet aggregability in fluid-percussion brain injury may be related to this effect of oxygen radicals.

The precise mechanism of the extravasation of plasma proteins and other types of macromolecules from injured vessels in ischemia/reperfusion, hypertension or brain trauma is not fully understood. Since the injured vessels whose blood-brain barrier is disrupted, display vesicles which contain protein, vesicular transport may be involved. An interesting aspect is the type of vessels which leak protein. By electron microscopy, the protein extravasation occurs in the intracerebral arterioles and veins (*Wei et al., 1986*) while by direct observation of the brain surface using fluorescent labels, it was concluded that the extravasation occurs first from small venules (*Mayhan and Heistad, 1986*). By electron microscopy, however, neither the pial veins nor the pial arterioles show evidence of extravasation of protein.

The precise mechanism of accumulation of phagocytic cells following vascular injury is not well understood. It is possible that the radicals may be involved by generating a chemotactic factor by interaction with a plasma component (*Petrone et al., 1980*).

The origin of superoxide anion in acute hypertension (*Kontos et al., 1981*), brain trauma (*Wei et al., 1981*), and ischemia/reperfusion (*Armstead et al., 1988*) is accelerated metabolism of arachidonate via cyclooxygenase. In all three conditions, pretreatment with indomethacin, a cyclooxygenase inhibitor, prevents the formation of the generation of superoxide. In hypertension and in trauma, pretreatment with indomethacin also prevents the development of vascular injury (*Kontos et al., 1981; Wei et al., 1982*). This has not been tested in ischemia/reperfusion. It is known that cyclooxygenase in the presence of appropriate cofactors produces superoxide (*Kukreja et al., 1986*).

It is possible that a polypeptide might be the mediator which leads to the generation of the radical in acute hypertension, brain injury, and ischemia/reperfusion. In brain injury, a specific bradykinin antagonist inhibits some of the vascular consequences (*Ellis et al., 1988*). Bradykinin applied topically on the brain surface generates superoxide (*Kontos et al., 1985*) and simulates many of the vascular effects seen in brain injury, in ischemia, and in acute hypertension. Finally, it is known that polypeptides from sensory fibers are released in response to acute hypertension and ischemia/reperfusion (*Sakas et al., 1989; Moskowitz et al., 1989*) and contribute to a major extent to the hyperemia seen in these conditions.

CONSEQUENCE OF VASCULAR INJURY IN ISCHEMIA/REPERFUSION, ACUTE HYPERTENSION, AND BRAIN TRAUMA

The vascular injury induced by oxygen radicals in these conditions may have both short-term as well as long-term consequences. The breakdown of the blood-brain barrier and the associated vasodilation favor the production of edema. Edema may compress tissue, increase intracranial pressure, and cause a secondary dysfunction. In the presence of endothelial injury, there is a tendency towards platelet aggregation. Any proaggregatory stimulus under these conditions is likely to induce platelet aggregation more easily than in normal vessels. The changes in endothelial cells may also have effects of its own. For example, the microvilli which develop in abundance in the intracerebral vessels in these conditions may restrict the lumen of the vessels and increase vascular resistance. The increase in resistance may be further enhanced if the endothelial injury activates leukocytes which might then adhere to the endothelium. The alteration in reactivity to endothelium-dependent vasoactive agents may contribute to changes in baseline vascular resistance. Under these conditions, the agents which ordinarily produce vasodilation via an indirect action on the endothelium may produce vasoconstriction because their direct vasoconstrictor effects on vascular smooth muscle are unmasked following injury to the endothelium and its inability to produce EDRF.

All these factors may contribute to the production of hypoperfusion which occurs with regularity following a period of ischemia followed by reperfusion. In some situations, the hypoperfusion is extreme so that it leads to the no reflow phenomenon. Most commonly however, blood flow returns to a value of 50% to 75% of baseline.

Less well known are the long-term consequences of the endothelial injury which occurs in these acute conditions. It is possible, for example, for the endothelial injury to contribute to accelerated atherosclerosis, particularly in larger vessels. The extravasation of protein and fibrin into the vessel wall may eventually lead to fibrinoid necrosis.

There are recognized abnormalities in reactivity in hypertension to endothelium-dependent vasodilators. It is well known that these responses are depressed and that the production of EDRF from large vessels from hypertensive animals studied *in vitro* is reduced (*Konishi and Su, 1983*). It is not known what connection there is between these late abnormalities and chronic hypertension to those seen in acute experiments.

There is little information about long-term alterations in reactivity to endothelium-dependent or other vasoactive agents following brain trauma or ischemia/reperfusion.

In connection with the long-term effects of vascular injury induced by hypertension, ischemia and trauma, it should be noted that it is likely that any effects on the blood vessels might be modified by preexisting vascular disease such as atherosclerosis. It is, of course, well known that in this condition, endothelium dependent responses are modified markedly. It is known that the vasodilator responses to acetylcholine and serotonin and ADP are depressed or converted to vasoconstriction (*Harrison et al., 1987*).

REFERENCES

Armstead, W.M., Mirro, R., Busija, D.W., and Leffler, C.W. Postischemic generation of superoxide anion by newborn pig brain. *Am. J. Physiol.* 255:H@01- H403, 1988.

Badwey, J.A., Curnutte, J.T., Robinson, J.M., Berdes, C.B., Karnovsky, M.J., and Karnovsky, M.L. Effects of free fatty acids on release of superoxide and on change of shape by human neutrophils. *J. Biol. Chem.* 259:7870-7877, 1984.

Christman, C.W., Wei, E.P., Kontos, H..A., and Povlishock, J.T. Effects of 15-hydroperoxy-eicosatetraenoic acid (15-HPETE) on cerebral arterioles of cats. *Am. J. Physiol.* 247:H631-H637, 1984.

Ellis, E.F., Holt, S.A., Wei, E.P., and Kontos, H.A. Kinins induce abnormal vascular reactivity. *Am. J. Physiol.* 255:H397-H400, 1988.

Handin, R.I., Karabin, R., and Boxer, G.J. Enhancement of platelet function by superoxide anion. *J. Clin. Invest.* 59:959-965, 1977.

Harrison, D.G., Freiman, P.C., Armstrong, M.L., Marcus, M.L., and Heistad, D.D. Alterations in vascular reactivity in atherosclerosis. *Circ. Res.* 61 (suppl. II):II-74- II-80, 1987.

Jewell, S.A., Bellomo, G., Thor, H., Orrenius, S. Bleb formation in hepatocytes during drug metabolism is caused by disturbances in thiol and calcium ion homeostasis. *Science* 217:1257-1258, 1982.

Konishi, M., and Su, C. Role of endothelium in dilator responses of spontaneously hypertensive rat arteries. *Hypertension* 5:881-886, 1983.

Kontos, H.A. Oxygen radicals in CNS damage. *Chem.-Biol. Interactions* 72:229- 255, 1989(a).

Kontos, H.A. Oxygen radicals in cerebral ischemia. In *Cerebrovascular Diseases*, ed. by M.D. Ginsberg and W.D. Dietrich, Raven Press, New York, pp. 365-371, 1989(b).

Kontos, H.A. Oxygen radicals in experimental brain injury. In *Intracranial Pressure VII*, ed. by J.T. Hoff, and A.L. Betz, Springer-Verlag, Berlin, Heidelberg, 787-798, 1989(c).

Kontos, H.A., and Povlishock, J.T. Oxygen radicals in brain injury. *Central Nerv. System Trauma* 3:257-263, 1986.

Kontos, H.A., Wei, E.P., Povlishock, J.T., Dietrich, W.D., Magiera, C.J., and Ellis, E.F. Cerebral arteriolar damage by arachidonic acid and prostaglandin G_2. *Science* 209:l242-1-244, 1980.

Kontos, H.A. Oxygen radicals in cerebral vascular injury. *Circ. Res.* 57:508-516, 1985.

Kontos, H.A., Wei, E.P., Dietrich, W.D., Navari, R.M., Povlishock, J.T., Ghatak, N.R., Ellis, E.F., and Patterson, J.L., Jr. Mechanism of cerebral arteriolar abnormalities after acute hypertension. *Am. J. Physiol.* 240:H511-H527, 1981.

Kontos, H.A., Wei, E.P., Ellis, E.F., Jenkins, L.W., Povlishock, J.T., Rowe, G.T., and Hess, M.L. Appearance of superoxide anion radical in cerebral extracellular space during increased prostaglandin synthesis in cats. *Circ. Res.* 57:142-ISI, 1985.

Kontos, H.A., and Wei, E.P. Superoxide production in experimental brain injury. *J. Neurosurg.* 64:803807, 1986.

Kontos, H.A., Wei, E.P., Povlishock, J.T., and Christman, C.W. Oxygen radicals mediate the cerebral arteriolar dilation from arachidonate and bradykinin in cats. *Circ. Res.* 55:295-303, 1984.

Kukreja, R.C., Kontos, H.A., Hess, M.L., and Ellis, E.F. PGH synthase and lipooxygenase generate superoxide in the presence of NADH and NADPH. *Circ. Res.* 59:612-619, 1986.

Mayhan, W.G., and Heistad, D.D. Role of veins and cerebral venous pressure in disruption of the blood-brain barrier. *Circ. Res.* 59:216-220, 1986.

Moskowitz, M.A., Sakas, D.E., Wei, E.P., Kano, M., Buzzi, M.G., Ogilvey, C., and Kontos, H.A. Postocclusive cerebral hyperemia is markedly attenuated by chronic trigeminal ganglionectomy. *Am. J. Physiol.* 257:H1736-H1739, 1989.

Orrenius, S., McConkey, D.J., Nicotera, P. Mechanisms of oxidant-induced cell damage. In *Oxy-Radicals in Molecular Biology and Pathology*, ed. by P.A. Cerutti, I. Fridovich, and J.M. McCord, Alan R. Liss, Inc., New York, pp. 327-339, 1988.

Petrone, W.F., English, D.K., Wong K., et al. Free radicals and inflammation: superoxide-dependent activation of a neutrophil chemotactic factor in plasma. *Proc. Natl. Acad. Sci. U.S.A.* 77:1159-1163, 1980.

Povlishock, J.T., Kontos, H.A., Rosenblum, W.I., Becker, D.P., Jenkins, L.W., and DeWitt, D.S. A scanning electron microscope analysis of the intraparenchymal brain vasculature following experimental hypertension. *Acta. Neuropathol. (Berl.)* 51:203-213, 1980.

Povlishock, J.T., Becker, D.P., Sullivan, H.G., and Miller, J.D. Vascular permeability alterations to horseradish peroxidase in experimental brain injury. *Brain Res.* 153:223-239, 1978.

Rosenblum, W.I., Wei, E.P., and Kontos, H.A. Platelet aggregation in cerebral arterioles after percussive brain trauma. *J. Texas Heart Inst.* 9:345-348, 1982.

Sakas, D.E., Moskowitz, M.A., Wei, E.P., Kontos, H.A., Kano, M., and Ogilvy, C.S. Trigeminovascular fibers increase blood flow in cortical gray matter by axon reflex-like mechanisms during acute severe hypertension or seizures. *Proc. Natl. Acad. Sci.* 86:1401--1405, 1989.

Wei, E.P., Christman, C.W., Kontos, H.A., and Povlishock, J.T. Effects of oxygen radicals on cerebral arterioles. *Am. J. Physiol.* 248:H157-H162, 1985.

Wei, E.P., Dietrich, W.D., Povlishock, J.T., Navari, R.M., and Kontos, H.A. Functional, morphological, and metabolic abnormalities of the cerebral microcirculation after concussive brain injury in cats. *Circ. Res.* 46:37-47, 1980.

Wei, E.P., Ellison, M.D., Kontos, H.A., and Povlishock, J.T. Oxygen radicals in arachidonate--induced increased blood brain barrier permeability to proteins. *Am. J. Physiol.* 251:H693-H699, 1986.

Wei, E.P., Kontos, H.A., Christman, C.W., DeWitt, D.S., and Povlishock, J.T. Superoxide generation and reversal of acetylcholine-induced cerebral arteriolar dilation after acute hypertension. *Circ. Res.* 57:781-787, 1985.

Wei, E.P., Kontos, H.A., Dietrich, W.D., Povlishock, J.T., and Ellis, E.F. Inhibition by free radical scavengers and by cyclooxygenase inhibitors of pial arteriolar abnormalities from concussive brain injury in cats. *Circ. Res.* 48:95-103, 1981.

II. GENERAL CONCEPTS IN ENDOTHELIAL CELL PATHOPHYSIOLOGY

FREE-RADICAL MEDIATED ACTIONS ON

ENDOTHELIAL CELLS OF THE INTACT LUNG

C. Norman Gillis, X. Chen and M. Merker

Departments of Anesthesiology and Pharmacology
Yale University School of Medicine
New Haven, Connecticut 06510, U.S.A.

INTRODUCTION

In addition to their crucial role in maintaining fluid balance, endothelial cells of the lung microvasculature manifest an array of metabolic and biosynthetic functions that can form the basis for regulating local blood flow, systemic vascular tone and inflammatory and immunological responses to injury. (*Furchgott, 1984; Gillis, 1986; Pober, 1990*). Among these processes (see Figure 1) are: a) the selective removal and/or degradation of biogenic amines (e.g. serotonin [5-HT]), nucleotides and kinins and b) the biosynthesis of molecules that dilate or constrict vascular smooth muscle and regulate blood cell interactions with the luminal endothelial membrane. It is clear that the delicate pulmonary endothelial vascular lining is severely compromised by a variety of drugs and by hyperoxia, endotoxemia and radiation (*Gillis, 1986; Brigham, 1987; Fanburg and Deneke, 1989*); in man, such injury can lead to the Adult Respiratory Disease Syndrome (ARDS). Each of these experimental injury models as well as ARDS involves the generation of free radicals of oxygen (*Fanburg and Deneke, 1989*). The purpose of this chapter is to describe effects of free radicals or radical-generating treatments on the above mentioned lung endothelial functions and to explore functional consequences of the former.

METABOLIC FUNCTIONS OF THE LUNG ENDOTHELIUM

The lung removes significant quantities of 5-HT and norepinephrine but interestingly, not epinephrine - presented to it either by continuous infusion or bolus (see *Gillis 1986; Pitt et al., 1989* for review). From early studies (*Gillis, 1973*) it was clear that these processes were selective and that uptake occurred into endothelial cells and was followed by intra-endothelial oxidative deamination by monoamine oxidase. Furthermore, uptake was saturable and was inhibited by cocaine and tricyclic antidepressants, suggesting a membrane transport mechanism perhaps co-existing with a binding protein similar to that of the platelet or serotonergic nerves (*Gershon et al., 1983*). However, the kinetics of 5-HT transport and elements of its drug-sensitivity profile (*Gillis, 1986; Pitt et al., 1989*) clearly distinguish the process in lung from that in either the platelet or serotonergic nerve ending.

Because the huge lung endothelial surface is in intimate contact with the entire cardiac output many times each minute and can effectively remove and degrade 5-HT, it appears (*Vane, 1969; Gillis, 1973*) that endothelium can serve as a metabolic barrier to protect the heart, systemic vasculature and perhaps also the pulmonary veins from potentially deleterious vascular effects of 5-HT and some other substrates. More recently, the same theme was echoed in observations that the lung (perhaps the endothelium) also takes up atrial natriuretic factor (*Turrin and Gillis, 1986*) and endothelin (*Nucci et al., 1988; Rimar and Gillis, 1989; Sirvi et al., 1990*). Some lung endothelial metabolic functions may be organ-specific, since the heart (presumably the coronary endothelium) does not take up injected serotonin (*Moffett et al., 1988*) or endothelin (*Rimar and Gillis, 1989*).

The physiological role of 5-HT uptake by lung endothelium has received much less attention than the impaired uptake which accompanies many forms of lung injury (*Gillis and Catravas, 1982; Gillis, 1986*). In the earliest demonstration, Block and Fischer (1977) reported that *in vitro* perfused lungs of rats exposed to hyperoxia had decreased 5-HT uptake capacity. We extended this observation to conscious, chronically catheterized animals exposed to 100% oxygen. In these experiments, single pass, multiple tracer techniques were used to determine lung 5-HT uptake (*Dobuler et al., 1982*). Significantly, we found that 5-HT removal was decreased in these animals at a time when lungs were morphologically indistinguishable from control. We therefore proposed (*Gillis and Catravas, 1982*) that measurement of 5-HT uptake might offer an early indication of impending lung injury - by reflecting the endothelial damage that was thought to initiate the sequence of events leading to severe permeability edema - and ARDS. Several clinical studies (*Gillis et al., 1986; Morel et al., 1985*) have supported animal experiments, in that 5-HT (as well as norepinephrine or propranolol) uptake was depressed in the presence of acute respiratory failure. Particularly relevant is the report of Morel et al (1985) that the magnitude of 5-HT removal was inversely related to the severity of respiratory failure in these patients, thus supporting the proposal that the technique might be of prognostic value in the clinical management of ARDS. The measurement suffers from the limitation that determinants of 5-HT uptake include both impaired endothelial function and lung perfusion which are almost invariable accompaniments of clinically-evident ARDS. Thus, the problem becomes one of distinguishing the relative impact of endothelial from hemodynamic factors in the measured 5-HT removal. It seems likely that the value of measuring 5-HT removal will lie in making repeated measures in a given patient, which might yield information about endothelial function during the course of the disease, rather than as a single prognostic measurement early in the clinical course.

Much experimental work is currently devoted to *in vivo* measurement of the kinetics of 5-HT transport into lung endothelium (*Dawson et al., 1987; Pitt et al., 1989*) as well as that for hydrolysis of substrates for angiotensin converting enzyme (ACE) which lines the luminal surface of the endothelium (*Moalli et al., 1987; Toivonen and Catravas, 1986*). The apparent Km of 5HT transport or hydrolysis of ACE substrate is taken to reflect intrinsic endothelial function, while the derived apparent V_{max} indirectly reflects the number of transport or catalytic sites. Certain forms of experimental lung injury have been shown to affect mainly the Km, while V_{max} seems more responsive to hemodynamic factors such as altered transit time or perfusion heterogeneity (*Pitt et al., 1989; Dawson et al., 1989*). Although applicable experimentally, such techniques in their present form are unlikely to provide practical information in the clinical setting. More promising are measurements made with radio-iodinated

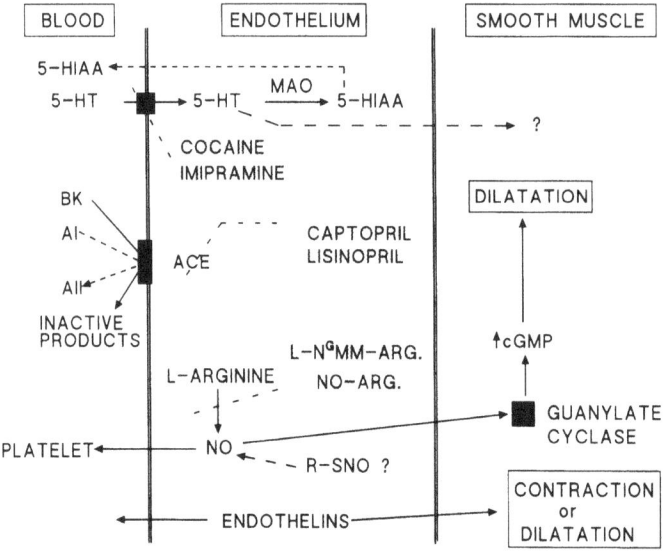

FIGURE 1: Diagram of processes occurring at the vascular wall.

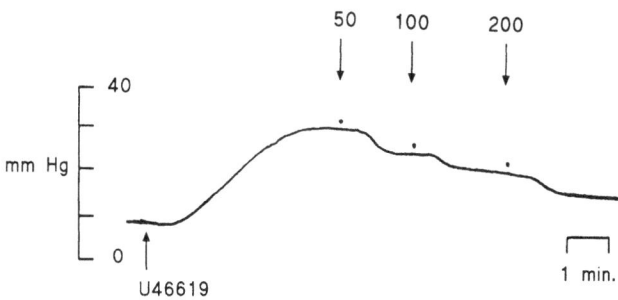

FIGURE 2: Typical vasodilator response to cumulative addition of ACh during recirculating perfusion (volume of medium = 50 ml) with a total of 30 nM U46619.

substances with the potential for application of external imaging techniques to document early evidence of lung endothelial injury (*Slosman et al., 1988*).

ENDOTHELIAL DERIVED RELAXING FACTOR (EDRF) IN THE LUNG

Intrinsic pulmonary vascular tone, circulating autocoids or neurotransmitters are crucially important in the vascular response to lung injury. We have therefore determined whether there was evidence that factors derived from the endothelium, such as EDRF participate in maintaining the normal, low pulmonary vascular tone or contribute to the vasoconstriction associated with experimental lung injury (*Brigham, 1987*).[1]

EDRF and Vascular Tone in the Intact Lung

Martin (1988) has assembled evidence in support of a basal release of EDRF, mainly on the basis of studies with preparations of large arteries or veins *in vitro*. Relatively little information is available, however, concerning the possible role of EDRF in maintaining low resistance in the lung microvasculature. Acetylcholine (ACh) is known to cause tone dependent vasodilatation in dog lung (*Kadowitz et al., 1981*) and to relax pre-contracted rabbit pulmonary artery *in vitro*. Unexpectedly, however, ACh causes intense pulmonary vasoconstriction in perfused rabbit lung (*Catravas et al., 1984*), even if vascular tone is raised. We confirmed this observation (*Cherry and Gillis, 1987*), but found that if cyclooxygenase activity was blocked by indomethacin, ACh did indeed produce dose-dependent vasodilatation when vascular smooth muscle tone was raised by the stable thromboxane A_2 analog U46619 (Figure 2). Furthermore, ACh vasodilatation was reversed by the presence of quinacrine or ferrous hemoglobin in the perfusion medium, both of which antagonize effects of EDRF in a variety of isolated vascular preparations (*Furchgott, 1984*). Interestingly, neither agent altered base-line pulmonary artery perfusion pressure which, since the lung was perfused at constant flow, was directly related to vascular resistance. More recently, the work of Moncada and his colleagues established that the radical NO, synthesized from L-arginine in cultured endothelial cells grown on cytodex beads, or released from the perfused lung was indeed an important, if not the only "EDRF" (for a detailed discussion of evidence in support of NO as EDRF see (*Moncada et al., 1988*). Moncada and his colleagues also showed that L- (but not D-) N^G mono-methylarginine (L-NMMA) competitively inhibited the conversion of L-arginine to NO. Large doses of L-NMMA injected intravenously increased the blood

---

[1] The general term "Endothelial derived relaxing factor" (EDRF) will be used throughout this chapter to describe the nitric oxide radical (*Moncada et al., 1988*) as well as S-nitrosothiol derivatives (*Ignarro, 1988*) including S-nitrosocysteine (*Myers et al., 1990*) from which the radical may be derived, or which may act *per se* to relax vascular smooth muscle.

pressure of anesthetized rabbits (*Reese et al., 1989*), suggesting inhibition of basal release of NO in the intact animal. Thus, if release of NO (or an S-nitrosothiol) occurred in response to ACh, it might be anticipated that L-NMMA should further raise tone in the vasoconstricted lung or antagonize established ACh vasodilation. Pretreatment with L-NMMA (0.4mM) eliminated ACh-induced vasodilatation (*Chen and Gillis, 1990a*) although the drug failed to cause further vasoconstriction if added at the peak of U46619-induced tone; if added before U46619, tone generated by the latter was significantly reduced (P < 0.05 - Figure 3). This is in striking contrast to rat pulmonary artery (*Crawley et al., 1990*) or aortic (*Reese et al., 1989a*) rings in which L-NMMA caused pronounced dose dependent reversal of ACh-induced relaxation in pre-contracted preparations. In these cases, L-NMMA also lacked effect on basal tone. A recent report (*Hecker et al., 1990*) describes N-demethylation of L-NMMA, by a deiminase in rat kidney, with formation of L-arginine which presumably then act as precursor for NO synthesis. Extrapolating this observation to the perfused lung, one can postulate that L-NMMA reduces (rather than enhances) U46619-induced vasoconstriction because the former actually results in production of the putative dilator radical, NO. As pointed out by Hecker et al (1990), demethylation cannot occur with nitroarginine, another competitive inhibitor of NO formation, which therefore had a more pronounced and longer lasting pressor effect in anesthetized rats. In preliminary experiments, we found that nitroarginine, in contrast to L-NMMA, potentiated the vasoconstrictor effect of U46619, thus supporting the explanation advanced above to account for the lack of similar action with L-NMMA. As with L-NMMA pretreatment of lungs with nitroarginine blocked the vasodilator action of ACh.

In normal human volunteers, Vallance et al (1989b) established by means of venous occlusion plethysmography, that intra-arterial infusion of L-NMMA decreased forearm blood flow (i.e. caused vasoconstriction) and inhibited ACh vasodilatation of previously constricted forearm vessels. In sharp contrast, however, the same group (*Vallance et al., 1989a*) reported

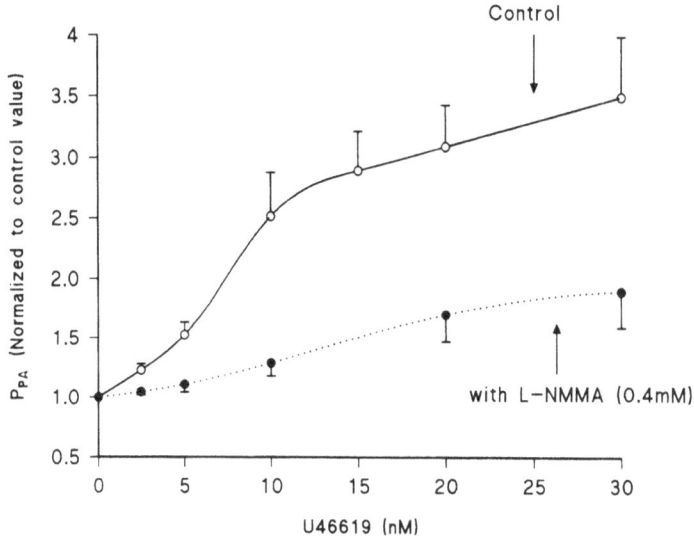

FIGURE 3: Effect of L-N^{G}MMA on U46619-induced vasoconstriction in 5 lungs perfused in a recirculating manner with Krebs medium at a rate of 20 ml/min. Perfusion pressure is normalized to the baseline perfusion pressure of each lung prior to beginning the addition of U46619. In all experiments, 30 uM indomethacin is added to the medium to inhibit cyclo-oxygenase (see text). U46619 was added cumulatively to 50 ml of medium to produce the indicated final concentrations. Each successive addition of U46619 was made after maximum vasoconstriction to the previous concentrations was reached. After assessing the control response to U46619, drugs were removed by single pass (i.e. non-recirculating) perfusion for 5-10 minutes. L-NMMA was added to a second 50 ml of Krebs and a second period of recirculating perfusion was started. Then cumulative addition of U46619 was again carried out.

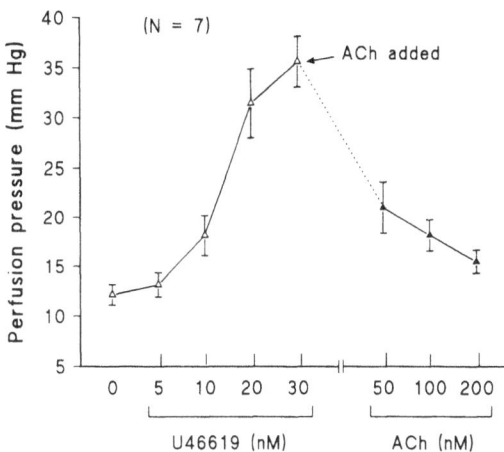

FIGURE 4: Effect of ACh added at the peak of U46619-induced vasoconstriction in 7 lungs perfused as described in Figure 2.

that while L-NMMA also antagonized ACh-vasodilatation of constricted veins, the drug failed to change blood flow in the dorsal veins of the hand in human volunteers. This interesting observation leads to the possibility that in vessels subject normally to low intravascular distending pressure (e.g. veins or the lung vasculature), the endothelium is "programmed" not to synthesize EDRF until tone is considerably elevated. Alternatively, rather than indicating functional heterogeneity, it may be that all endothelial cells can "express" release of EDRF, if conditions of shear stress, distending pressure and other environmental factors permit. In this regard, it is interesting that hypoxic vasoconstriction is reported by several groups to be modulated by EDRF release, and therefore to be augmented by inhibitors of EDRF including L-NMMA (*Archer et al., 1989, Brashers et al., 1988*). Furthermore, low basal tone in rat lung blood vessels has been attributed (*Hasunuma et al., 1990*) to release of a hyperpolarizing factor which activates membrane K^+ channels; it is unclear whether this observation applies to other species.

EDRF and Vascular Tone in the Injured Lung

A variety of highly reactive free radicals of oxygen are implicated in many forms of lung injury, including hyperoxia, endotoxin administration and radiation injury (*Fanburg and Deneke, 1989*). Several xenobiotics, including paraquat and bleomycin, both of which cause life threatening pulmonary toxicity, undergo redox cycling utilizing reducing equivalents of NADPH to produce damaging oxygen radicals. Finally, ischemia/reperfusion injury is known to cause free-radical mediated damage in coronary, (*Opie, 1989*) pulmonary (*Kennedy et al., 1989*) and other (*Brigham, 1987*) vascular beds. We, therefore, developed a model of direct, oxygen radical lung injury in order to extend our studies of EDRF in lung vascular reactivity. We used a modification of the technique described by Jackson et al (1986) to generate a variety of oxygen radicals within medium perfusing the isolated rabbit lung (*Chen and Gillis, 1990b*). Electrolysis in solution has been used in several laboratories (*Stewart et al 1988; Pi and Chen, 1989*) to study the mechanisms of reperfusion ischemia in the heart. Electrolytic stimulation (ES) allows delivery of several highly potent oxidizing species including superoxide and hydroxyl radicals, hydrogen peroxide and perhaps also hydrochlorous acid directly to the endothelial luminal membranes and avoids the necessity for prior ischemia or neutrophil activation in order to generate the radicals.

The protocol for these experiments is as follows. After a 20 min equilibration with single pass perfusion ([SP] - i.e. the lung effluent is discharged to waste) a control "challenge" or test of vascular reactivity was carried out involving a) development of vasoconstriction to the cumulative addition of U44619 to the recirculating reservoir (total volume of fluid = 50 ml) to produce a sustained perfusion pressure (P_{PA}) of 30-35 mm Hg and b) cumulative addition of ACh (50 - 200 knM final concentration) to produce vasodilatation (e.g. Figure 2). This took about 10 minutes. Figure 4 shows the results of 7 such challenges. Thereafter, perfusion is changed to SP for 5 to 10 minutes, thus removing all drugs from the vascular space. A 20

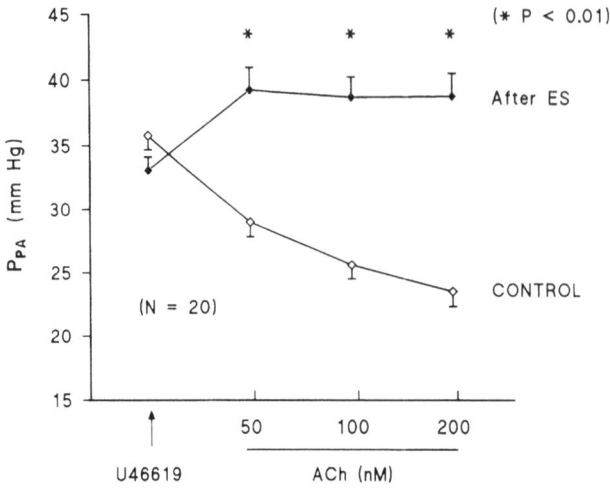

FIGURE 5: Effect of electrolysis (20mA for 2 minutes) on the vasodilator response to ACh during single-pass perfusion. ACh was added (before and after electrolysis in the same lung) to recirculating medium after vasoconstriction induced by cumulative addition of U46619 (see Figures 2 and 3 for further detail). Single pass perfusion for 10 minutes was used to clear drugs from the lung fascular space before electrolysis was carried out.

mA DC current is then passed for two minutes across electrodes cemented into the input tubing, such that the anode is 14.5 cm, and the cathode is 17.5 cm, proximal to the tip of the pulmonary artery catheter. It is important that this be done during SP perfusion. Some protocols required the presence of drugs (see below) during the application of ES. Next, recirculating perfusion is again started with fresh Krebs medium. The same drugs which potentially could modify the response to ES may also be added to the recirculation reservoir. A second "challenge" with U46619 and ACh is then carried out; some protocols included a third cycle of washout (SP perfusion), recirculation and a third challenge. In the absence of other treatment, two or three cycles of challenge and washout could be carried out without evidence of extensive edema.

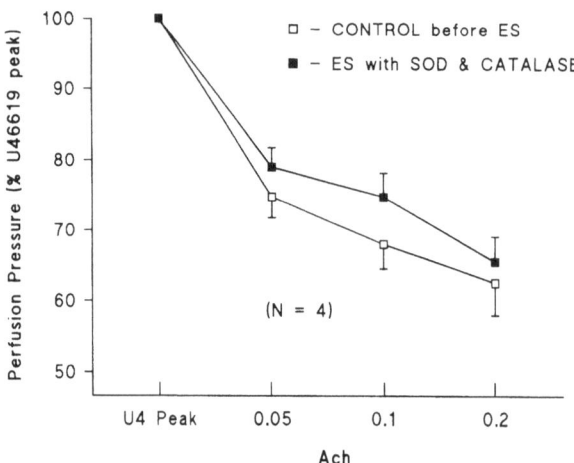

FIGURE 6: Effect of superoxide dismutase (SOD - 100 units/ml uM) and catalase (300 units/ml uM) on ACh response before and after electrolysis. See legend for Figure 5 for further details.

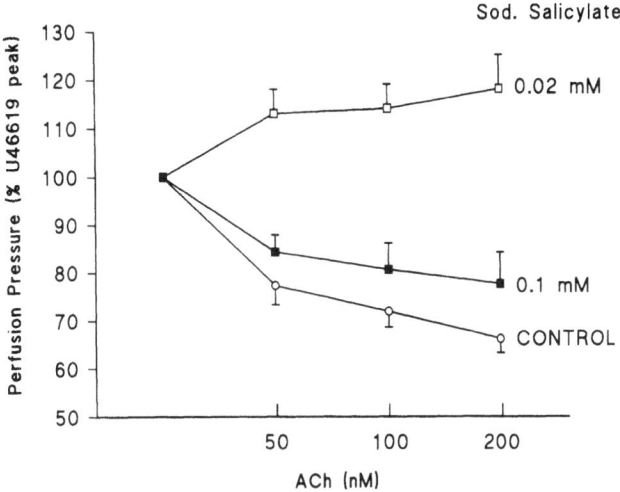

FIGURE 7: Effect of sodium salicylate on ACh response before and after ES (n=4). Notice that salicylate protection is concentration-dependent.

Electrolytic stimulation (ES) itself usually caused some vasoconstriction, increasing P_{PA} from 21 ± 2 to 31 ± 3 (n = 11); thereafter perfusion pressure returned to pre-ES levels. However, (Figure 5) the response to ACh was converted to vasoconstriction. This was due to the generation of free radicals, since a mixture of SOD and catalase present only during the ES completely preserved the ACh vasodilatation (Figure 6). Sodium salicylate (0.1 mM) also protected ACh-induced vasodilatation from the effects of ES (Figure 7). The salicylate effect presumably is based on trapping of hydroxyl radicals during the conversion of salicylate to dihydroxybenzoic acid (*Floyd et al., 1986*). The fact that sodium nitroprusside (SNP) retained its action after ES, while that of A23187 was eliminated (Figure 8) suggests that the vascular smooth muscle response to EDRF was unaltered, while the ability of endothelium to increase intracellular calcium (an integral step in EDRF release) is impaired. This conclusion is drawn because SNP generates NO directly within the vascular smooth muscle, while A23187 (a calcium ionophore) produces endothelium-dependent relaxation of vascular smooth muscle (*Furchgott, 1984*). If ES was carried out during recirculation rather that SP, there was much less modification of ACh vasodilatation and vasoconstriction was never observed (data not shown).

The effect of atropine was examined in lungs (n = 5) damaged by ES and in which ACh had caused vasoconstriction. Addition of atropine to the circulating medium decreased tone to approximately the peak U46619 response (Figure 9). Presumably, ES eliminated the normal response to ACh (see Figure 5) thus "exposing" classical (smooth muscle) muscarinic receptors by means of which ACh caused vasoconstriction. Atropine may displace ACh from these sites, allowing return to a level of tone reflecting the continuing presence of U46619 in the recirculating medium. If atropine is added before ACh, in the ES-damaged lung, then the level of tone generated by U46619 is maintained, again being consistent with prior blockade of M-receptors in the vascular smooth muscle.

Forms of lung injury which act via generation of free radicals (see Introduction above) impair endothelial uptake of serotonin. In earlier studies (*Cook et al., 1982*), we found that severe lung injury produced by xanthine/xanthine oxidase caused marked pulmonary edema and decreased 5-HT uptake. Accordingly, we sought evidence that the latter also might be altered by ES which caused much less edema. In these experiments, 3 separate determinations of ^{14}C-serotonin (^{14}C-5HT) uptake were made by a multiple indicator dilution method in routine use in this laboratory (*Merker and Gillis, 1988*). After a control ^{14}C-5HT uptake determination and ACh challenge (see above), ES was carried out as before and then recirculation with fresh medium was begun and the second ACh challenge was carried out. A third set of ^{14}C-5HT uptake measurements were made either 5 or 20 minutes after ES. In 12 of 14 experiments, ES converted ACh vasodilatation to vasoconstriction. ^{14}C-5HT uptake was significantly lowered ($P < 0.01$) 5 and 20 minutes after ES (Figure 10). In these experiments, the wet/dry weight ratio was 7.3 ± 0.7 (n = 12). In another series of experi-

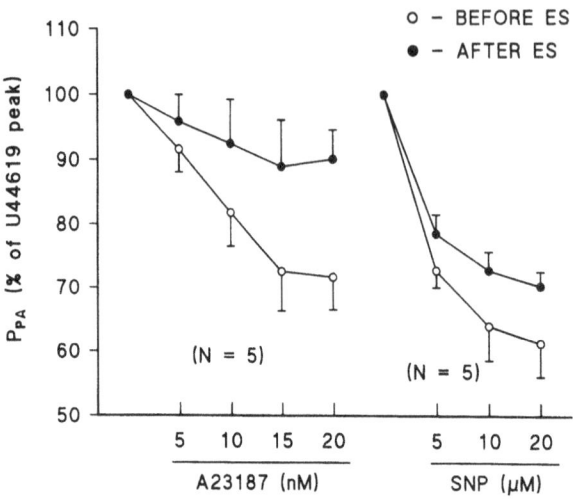

FIGURE 8: Effect of ES on responses to A23187 or sodium nitroprusside (SNP).

ments with lisinopril, which inhibits the effect of ES on ACh vasodilatation, 5-HT uptake was preserved (Figure 10) and the wet/dry ratio was 5.26 ± 0.2 (n = 5).

Effect of Lisinopril on Electrolytically-induced Injury

Several reports describe the protective action of not only free radical scavengers (*Opie, 1989*), but also ACE inhibitors against myocardial injury by reperfusion following ischemia (*Pi and Chen, 1989*). The latter authors consider that protection by captopril reflects release of prostacyclin since iloprost, a PGI_2 analog had similar effects, although they also note that some protection persisted after treatment with indomethacin. Since we routinely treated the perfused lungs with indomethacin in a high enough concentration (30uM) to completely block formation of prostacyclin, we also evaluated the possible protective action of captopril against the elimination of ACh vasodilatation after ES. Captopril (90uM) present during ES had some protective action (data not shown). This effect may reflect the sulfhydryl moeity of captopril which serves to keep glutathione in its reduced state, thus maintaining the radical

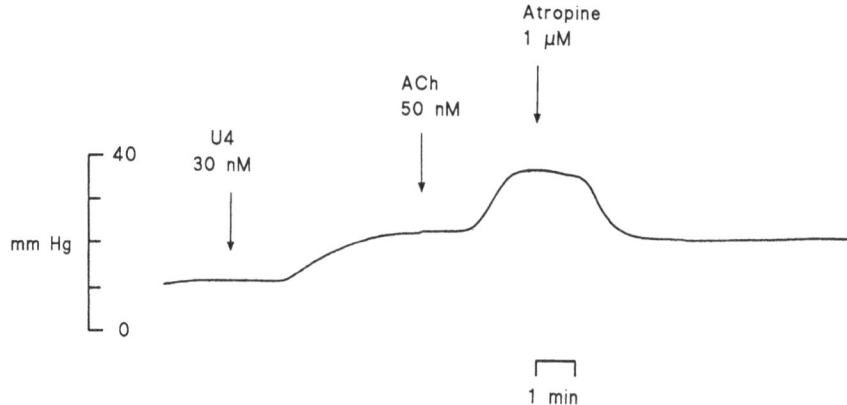

FIGURE 9: Typical vasoconstrictor response (*after* ES) to cumulative addition of 30 nM U46619 during recirculating perfusion with a total of 50 ml of Krebs medium. Further vasoconstriction occurs after ACh is added, but is reversed by atropine.

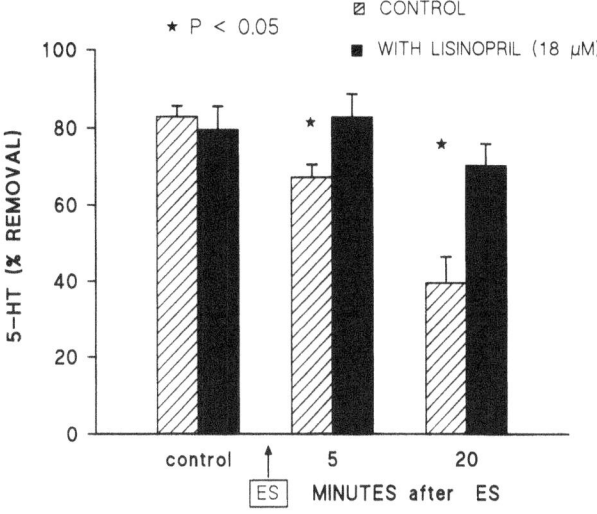

FIGURE 10: Single pass uptake of ^{14}C-serotonin (see text for details) before and 5 (n = 12) and 10 minutes (n = 6) after electrolysis. The solid bars reflect similar measurements in lungs treated for a total of 10 minutes prior to ES with 18 uM lisinopril (n = 5).

scavenging action of natural glutathione peroxides. However, we also found (Figure 11) that lisinopril (18 uM), a non-sulfhydryl containing ACE inhibitor, perfused during ES afforded complete protection of the ACh dilatation, whether challenged 1 minute or 20 minutes after ES. The mechanism of this protection is unknown but may involve an unanticipated anti-oxidant action or enzyme inhibition.

DISCUSSION AND CONCLUSIONS

Basal Release of EDRF

We have described a perfused lung model in which it has been possible to demonstrate ACh-vasodilatation provided the cyclooxygenase pathway of arachidonate metabolism is blocked. When EDRF synthesis is inhibited by nitroarginine (but not L-NMMA) vasoconstrictor responses to U46619 are enhanced, thus providing *in situ* equivalent of experiments with large vessel rings (*Martin, 1988*). The vascular resistance changes observed are likely to reflect tone in the microvascular bed since most of the resistance in lung is expressed in smaller vessels. Therefore, this model offers the opportunity for study of endothelial dependent vascular effects in functionally significant vessels of the lung. It will be interesting, for example, to use techniques of vascular occlusion (*Dawson et al., 1989*) to determine more definitively the sites at which EDRF mediated effects are manifest in the vascular tree.

EDRF in the Injured Lung - Effect of Electrolysis

Stewart et al. (1988) reported that direct electrolytic generation of free radicals in the coronary vascular bed reduced ACh vasodilatation induced during the passage of current. In our experiments, challenge with ACh was conducted at least 5 minutes after ES and/or washout of drugs used to inhibit the injury. This coupled with the fact that ES-generated radicals pass through the lung vasculature only once (i.e. ES is carried out only during SP) make it unlikely they could survive until ACh challenge and inactivate (by oxidation) released NO. We therefore suggest that release of EDRF is compromised by the injury, which may be due to radical-induced lipid peroxidation that could also account for the diminished 5-HT uptake we observed (Figure 10). In this context, it is interesting that Block (1987) proposed that hyperoxia diminished 5-HT uptake by endothelial cells in culture by means of a modification of membrane fluidity. Our results are consistent with such an effect. The fact

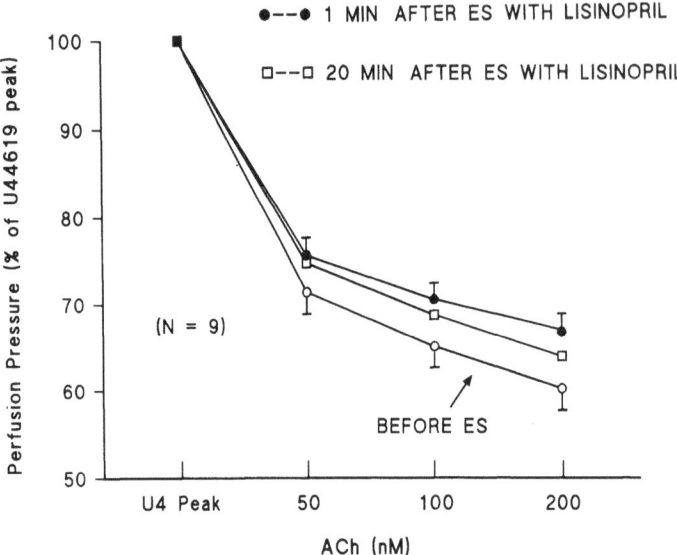

FIGURE 11: Protection of the ACh vasodilator response by 18 uM lisinopril added to medium for the 10 minutes before and during ES. Note that protection is complete both 1 and 20 minutes after injury.

that sodium nitroprusside retained its activity suggests that the smooth muscle dilatory mechanism involving guanylate cyclase stimulation is largely unaffected. Accordingly, the endothelial cell is the most likely site for the injury caused by electrolysis. Consistent with this proposal is elimination of the response to A23187 (Figure 8).

The fact that ES during recirculation is much less able to reverse the normal ACh dilatation (data not shown) suggests strongly that some antioxidant (or other protective substance) is released into the medium during this form of ES. Indeed, even when NO could be oxidized by free radicals generated by ES during recirculation, the dilator response to ACh is virtually intact. It should be noted that we were unable to reverse the effects of ES once established, suggesting permanent damage to the endothelium, as would be anticipated if peroxidation of lipid membrane structures is produced (*Jackson et al., 1986*). The cellular mechanism by which the ACh effect is eliminated is unknown. Experiments with atropine (e.g. Figure 9) suggest that endothelial cholinergic receptors are inactivated as vasoconstriction after ES reflects classical muscarinic smooth muscle vasoconstriction.

The mechanism of lisinopril protection of the ACh response (and 5-HT uptake) is unknown. It clearly does not depend on sulphydryl-related mechanisms but may reflect an unexpected antioxidant activity because the drug need be present only during ES to have a sustained protective action (Figure 11). Possibly its ACE inhibitory effect is necessary for protection although we know of no direct evidence to support the suggestion. Nevertheless it is intriguing that all ACE inhibitors tested, whether sulphydryl containing or not, exert protective action against experimental free radical-mediated injury. Studies to determine whether the pretreatment of intact animals with ACE inhibitors can protect against subsequent lung endothelial injury seem indicated.

To what extent do these data bear on the mechanism of pulmonary hypertension and ARDS? There is much evidence that pulmonary hypertension associated with endotoxin reflects, in its early stages, release of thromboxane. However, this effect can be blocked by catalase (*Brigham, 1987*), possibly indicating a peroxide-related mechanism, presumably via the cyclooxygenase pathway of arachidonate metabolism. Also, human lower torso ischemia has been described as causing a pulmonary ischemia reperfusion injury (*Klauser et al., 1989*) with pulmonary hypertension and non-cardiogenic pulmonary edema. Finally, Kennedy et al (1989) reported that ischemia in blood-perfused rabbit lungs, followed by reperfusion and expansion of the lungs caused a slowly developing pulmonary hypertension and extensive edema which were blocked by allopurinol or by a tungsten diet (Dr. Aubrey Taylor - personal communication), both of which implicate xanthine oxidase-linked free radical generation with these effects. However, in each of these studies and probably also in patients, the cyclooxygenase pathway was intact; in our model the latter was blocked and this may have unduly emphasized the role of other potential vasomodulator molecules. Nevertheless, we believe the technique of electrolytic injury in a perfused lung offers a new and useful model with which to study the effects of radical induced injury in the endothelium of the intact lung.

ACKNOWLEDGEMENTS

This work was supported by U.S. Public Health Service Grants HL-13315 and HL-40863 and a grant from the Eli Lilly Company. Several colleagues in the past have received support from U.S. Public Health Service Training Grant # HL-07410.

REFERENCES

Archer, S.L., Tolins, J.P., Raij, L. and Weir, E.K. Hypoxic pulmonary vasoconstriction is enhanced by inhibition of the synthesis of an endothelium-derived relaxing factor. *Biochem. Biophys. Res. Commun.* 164:1198-1205, 1989.

Block, E.R. Membrane Fluidity and Endothelial Function. In: *Pulmonary Endothelium in Health and Disease*, edited by Ryan, U.S. Marcel Dekker, New York, N.Y., 277-306, 1987.

Block, E.R. and Fisher, A.B. Depression of serotonin clearance by rat lungs during oxygen exposure. *J. Appl. Physiol.* 42:R33-R38, 1977.

Brashers, V.L., Peach, M.L. and Rose, C.E. Augmentation of hypoxic pulmonary vasoconstriction in the isolated perfused rat lung by in vitro antagonists of endothelium-dependent relaxation. *J. Clin. Invest.* 82:1495-1502, 1988.

Brigham, K.L. Mechanisms of endothelial injury. In: *Pulmonary Endothelium in Health and Disease*, edited by Ryan, U.S. Marcel Dekker, New York, N.Y., 207-236, 1987.

Catravas, J.D., Buccafusco, J.J. and Kashef, H.E. Effects of acetylcholine in the pulmonary circulation of rabbits. *J. Pharmacol. Exp. Ther.* 231:236-241, 1984.

Chen, X. and Gillis, C.N. Antagonistic effects of N(G)-monomethyl-L-arginine on acetylcholine induced vasodilatation and U46619 induced vasoconstriction in lungs. *Amer. Rev. Resp. Dis.* 141:A480, 1990a.

Chen, X. and Gillis, C.N. Effect of free radicals on pulmonary vascular response to acetylcholine. *J. Appl. Physiol.* (In press).

Cherry, P.D. and Gillis, C.N. Evidence for the role of endothelium-derived relaxing factor in acetylcholine induced vasodilatation in the intact lung. *J. Pharmacol. Exp. Ther.* 241: 516-520, 1987.

Cook, D.R., Howell, R.E. and Gillis, C.N. Xanthine oxidase-induced lung injury inhibits removal of 5-hydroxytryptamine from the pulmonary circulation. *Anesth. Analg.* 61: 666-670, 1982.

Crawley, D.E., Liu, S.F., Evans, T.W. and Barnes, P.J. Effects of L-arginine and L-NMMA on relaxation and contraction in rat pulmonary artery. *Amer. Rev. Resp. Dis.* 141:A480, 1990.

Dawson, C.A., Linehan, J.H. Rickaby, D.A. and Bronikowski, T.A. Kinetics of serotonin uptake in the intact lung. *Ann. Biomed. Eng.* 15:217-227, 1987.

Dawson, C.A., Roerig, D.L. and Linehan, J.H. Evaluation of endothelial injury in the human lung. *Clinics in Chest Medicine* 10:13-24, 1989.

Dobuler, K.J., Catravas, J.D. and Gillis, C.N. Early detection of oxygen-induced lung injury in conscious rabbits: Reduced activity of angiotensin converting enzyme and removal of 5-hydroxytryptamine in vivo. *Amer. Rev. Resp. Dis.* 126:534-539, 1982.

Fanburg, B.L. and Deneke, S.M. Hyperoxia: toxicity and adaptation. In: *Lung Cell Biology*, edited by Massaro, D. Marcel Dekker, Inc. New York, 1199-1226, 1989.

Floyd, R.A., Henderson, R., Watson, J.J. and Wong, P.K. Use of salicylate with high pressure liquid chromatography and electrochemical detection as a sensitive measure of hydroxyl free radicals in adriamycin-treated rats. *J. Free Radical Biol. Med.* 2:13-18, 1986.

Furchgott, R.F. Role of endothelium in the responses of vascular smooth muscle to drugs. *Ann. Rev. Pharmacol.* 24:175-797, 1984.

Gershon, M.D., Liu, K.P., Karpiak, S.E. and Tamir, H. Storage of serotonin in vivo as complex with serotonin-binding protein in central and peripheral serotonergic neurones. *J. Neurosci.* 3:1901-1911, 1983.

Gillis, C.N. Metabolism of vasoactive hormones by lung. *Anesthesiology* 39:626-632, 1973.

Gillis, C.N. Pharmacological aspects of metabolic processes in the pulmonary circulation. *Ann. Rev. Pharmacol.* 26:183-200, 1986.

Gillis, C.N. and Catravas, J.D. Altered removal of vasoactive substances in the injured lung: Detection of lung microvascular injury. *Ann. N.Y. Acad. Sci.* 384:458-474, 1982.

Gillis, C.N., Pitt, B.R., Wiedemann, H.P. and Hammond, G.L. Depressed Prostaglandin E1 and 5-hydroxytryptamine removal in patients with adult respiratory distress. *Amer. Rev. Resp. Dis.* 134:739-744, 1986.

Hasunuma, K., Yamaguchi, T., Rodman, D.M., O'Brien, R.F. and McMurtry, I.F. Effects of inhibitors of EDRF and EDHF on the vasoreactivity of perfused rat lungs. *J. Appl. Physiol.* In Press, 1990.

Hecker, M., Mitchell, J.A., Harris, H.J., Katsura, M., Thiemermann, C. and Vane, J.R. Endothelial cells metabolize NG-monomethylarginine to L-citrulline and subsequently to L-arginine. *Biochem. Biophys. Res. Commun.* 167:1037-1043, 1990.

Ignarro, L.J. Endothelium-derived nitric oxide: actions and properties. *FASEB J.* 3:31-36, 1989.

Jackson, C.V., Mickelson, J.K. Stringer, K., Rao, P.S. and Lucchesi, B.R. Electrolysis-induced myocardial dysfunction. *J. Pharmacol. Meth.* 15:305-320, 1986.

Kadowitz, P.J., Nandiwada, P.A. and Hyman, A.L. Parasympathetic neurohumoral control of pulmonary vascular bed. *Circ. Supp.* IV 180, 1981.

Kennedy, T.P., Rao, N.V., Hopkins, C. and Pennington, L. Role of reactive oxygen species in reperfusion injury of the rabbit lung. *J. Clin. Invest.* 83:1326-1335, 1989.

Klausner, J.M., Paterson, I.S., Mannick, J.A., Valeri, R., Hechtman, H. and Shepro, D.R. Reperfusion pulmonary edema. *J. Amer. Med. Ass.* 261: 1030-1035, 1989.

Lefer, A.M. Significance of lipid mediators in shock states. *Circ. Shock* 27:3-12, 1989.

Martin, W. Basal release of Endothelium-derived relaxing factor. In: *The Endothelium: Relaxing and Contracting Factors*, edited by VanHoutte, P. Humana Press, Inc., Clifton, N.J., 159-178, 1988.

Merker, M. and Gillis, C.N. Propranolol and serotonin removal in lung injury. *J. Appl. Physiol.* 65:2579-2584, 1988.

Moalli, R., Pitt, B.R. and Gillis, C.N. Effect of flow and surface area on angiotensin converting enzyme activity in isolated rabbit lungs. *J. Appl. Physiol.* 62:2042-2050, 1987.

Moffett, T.C., Chan, I.S. and Bassingthwaighte, J.B. Myocardial serotonin exchange: negligible uptake by capillary endothelium. *Amer. J. Physiol.* 254:H570-H577, 1988.

Moncada, S., Radomski, M.W. and Palmer, R.M.J. Endothelium-derived relaxing factor - Identification as nitric oxide and role in the control of vascular tone and platelet function. *Biochem. Pharmacol.* 37:2495-2501, 1988.

Morel, D.R., Dargent, F. Bachmann, M. Suter, P.M. and Junod, A.F. Pulmonary extraction of serotonin and propranolol in patients with adult respiratory distress syndrome. *Amer. Rev. Respir. Dis.* 132:479-484, 1985.

Myers, P.R., Minor, R.L., Guerra, R., Bates, J.N. and Harrison, D.G. Vasorelaxant properties of the endothelium-derived relaxing factor more resemble S-nitrosocyteine than nitric oxide. *Nature (Lond.)* 345:161-163, 1990.

Nucci, G.L., Thomas, R., D'Orleans-Juste, P., Antunes, E., Walder, C., Warner, T.D. and Vane, J.R. Pressor effects of circulating endothelin are limited by its removal in the pulmonary circulation and by release of prostacyclin and endothelium derived relaxing factor. *Proc. Nat. Acad. Sci.* 85:9797-9800, 1988.

Opie, L.H. Reperfusion injury and its pharmacologic modification. *Circulation* 80:1049-1062, 1989.

Pi, X. and Chen, X. Captopril and ramiliprat protect against free radical injury in the isolated working rat heart. *J. Mol. Cell. Cardiol.* 21:1261-1271, 1989.

Pitt, B.R., Lister, G. and Gillis, C.N. Metabolic Functions of the Lung and Cardiopulmonary Physiology. In: *Heart Lung Interactions in Health and Disease*, edited by Scharf, S.M. and S. Cassidy. Marcel Dekker, Inc., New York, N.Y., 391-406, 1989.

Pober, J.S. and Cotran, R.S. Cytokines and endothelial cell biology. *Physiol. Rev.* 70:427-451, 1990.

Rees, D.D., Palmer, R.M.J., Hodson, H.F. and Moncada, S. A specific inhibitor of nitric oxide formation from l-arginine attenuate endothelium-dependent relaxation. *Br. J. Pharmacol.* 96:101-107, 1989a.

Rees, D.D., Palmer, R.M.J. and Moncada, S. Role of endothelium-derived nitric oxide in the regulation of blood pressure. *Proc. Natl. Acad. Sci. U.S.A.* 86:3375-3378, 1989b.

Rimar, S. and Gillis, C.N. Differential uptake of endothelin by rabbit coronary and pulmonary microcirculations. *Circulation* 80 (II):213, 1989.

Sirvi, M.L., Metsrinne, K., Saijanmaa, O. and Fyhrquist, F. Tissue distribution and half-life of ^{125}I-endothelin in the rat. *Biochem. Biophys. Res. Commun.* 167:1191-1195, 1990.

Slosman, D.O., Morel, D.R. and Alderson, P.O. A new imaging approach to quantitative evaluation of pulmonary vascular endothelial metabolism. *J. Thorac. Imag.* 3:49-52, 1988.

Stewart, D.S., Pohl, U. and Bassenge, E. Free radicals inhibit endothelium-dependant dilation in the coronary resistance bed. *Amer. J. Physiol.* 255:H765-H769, 1988.

Toivonen, H.J. and Catravas, J.D. Effects of alveolar pressure on lung angiotensin-converting enzyme function *in vivo*. *J. Appl. Physiol.* 61:1041-1050, 1986.

Turrin, M. and Gillis, C.N. Removal of atrial natriuretic factor by perfused rabbit lungs *in situ*. *Biochem. Biophys. Acta*. 140:868-873, 1986.

Vallance, P., Collier, J. and Moncada, S. Nitric oxide synthesized from L-arginine mediates endothelium dependent dilatation in human veins *in vivo*. *Cardiovasc. Res.* 23:1053-1057, 1989a.

Vallance, P., Collier, J. and Moncada, S. Effects of endothelium-derived nitric oxide on peripheral arteriolar tone in man. *Lancet I*:997-1000, 1989b.

Vane, J.R. The release and fate of vaso-active hormones in the circulation. *Brit. J. Pharmacol.* 35:209-242, 1969.

REGULATION OF VASCULAR FUNCTION BY VASCULAR PERMEABILITY FACTOR

Daniel T. Connolly

Health Sciences
Monsanto Company
800 N. Lindbergh Boulevard
St. Louis, Missouri 63167, U.S.A.

INTRODUCTION

The blood vessels in and around tumors can be distinguished from normal vasculature in several ways. Tumor-associated vessels are derived from surrounding host blood vessels as a result of neovascularization (*Folkman, 1985*) and are typically more permeable than normal vessels (*Dvorak et al., 1988, Gerlowski and Jain, 1983*). These observations lead to the hypothesis that tumors can alter blood vessel function by producing regulator substances. A polypeptide regulator known as vascular permeability factor (VPF) or vascular endothelial growth factor (VEGF) has now been purified and found to display a surprisingly wide spectrum of vascular activities, some of which are pro-inflammatory in nature. In addition to synthesis by tumor cells, VPF has now been found to be a product of monocytes and lymphocytes. Taken together, these results suggest that VPF may be a new inflammatory cytokine.

VPF IN THE TUMOR ENVIRONMENT

It has been suggested that increased vascular permeability in the tumor environment would lead to extravasation of plasma proteins into the tissue, and would result in a series of cascade events leading to extravascular coagulation and fibrinolysis (*Dvorak, et al., 1981*). These events do not occur under normal circumstances except during inflammation and wound healing. In fact, the environment around solid tumors resembles that of a healing wound in many respects (*Dvorak, 1986*). Dvorak and coworkers first identified a permeability enhancing protein, VPF, derived from guinea pig line 10 tumor cells grown *in vitro* (*Senger, et al., 1983*). After partial chromatographic purification, small amounts of highly purified VPF were obtained using preparative SDS-PAGE. Antibody against VPF was used to demonstrate that the accumulation of ascites fluid by line 10 tumor cells in mice was dependent upon VPF. This was an extremely important observation because it demonstrated that tumors are able to specifically alter their environment by the production of VPF, and that blockage of VPF activity apparently restored the normal vascular function.

STRUCTURE OF VPF

VPF has been purified from large scale cultures of guinea pig line 10 tumor cells (*Connolly, et al., 1989*) and from human U937 cells (*Connolly, & Olander, et al., 1989*). The latter cell line is a histiocytic lymphoma of the monocyte lineage that displays many macrophage functions. In two dimensional electrophoresis, guinea pig VPF and human VPF appear as a heterogeneous group of proteins with molecular weights in the range 34 kDa to 46 kDa, with pIs that range from about 7.5 to 9.0. Upon reduction of disulfide bonds, VPF migrates as a group of bands with molecular weights ranging from about 18 kDa to 24 kDa implying that VPF is a disulfide linked dimer.

The sequence of a VPF cDNA from U937 cells codes for a 189 amino acid polypeptide containing 16 cysteines, a single N-glycosylation site, and an N-terminal 26 amino acid secretion signal sequence (*Keck, et al., 1989*). Leung et al. (*Leung, et al., 1989*) have identified two additional cDNA clones of VPF/VEGF that code for polypeptides containing 165 and 121 amino acids. These isoforms differ from VPF_{189} in that internal segments are deleted (Figure 1), and evidently arise from differential splicing of mRNA. Homology searches have revealed that the three forms of VPF share distant homology to the platelet derived-growth factor (PDGF) family of proteins. Even though the overall identity of the VPF and PDGF proteins are not strikingly high (about 18% for VPF_{189}), all eight of the PDGF cysteines are conserved in the VPF proteins, implying that the proteins share similar tertiary structures. Note that in VPF_{121}, seven cysteines are deleted, but that all of these fall outside of the PDGF homology region, and that the PDGF-like cysteines are conserved. The VPF/VEGF and PDGF proteins are disulfide-linked dimers, are acid and heat stable, and function as growth factors. However, the cellular and receptor specificities of VPF are clearly distinct from PDGF.

The heterogeneity that has been observed in purified preparations of VPF probably arises from multiple causes. The polypeptides coded for by the three different mRNA forms could potentially combine to form heterodimers as well as homodimers. Furthermore, the C-terminal half of VPF is extremely rich in lysine and arginine residues and is thus a good target for serine proteases. Finally, glycosylation could contribute to heterogeneity. It remains to be determined if different isoforms of VPF display differences in activity.

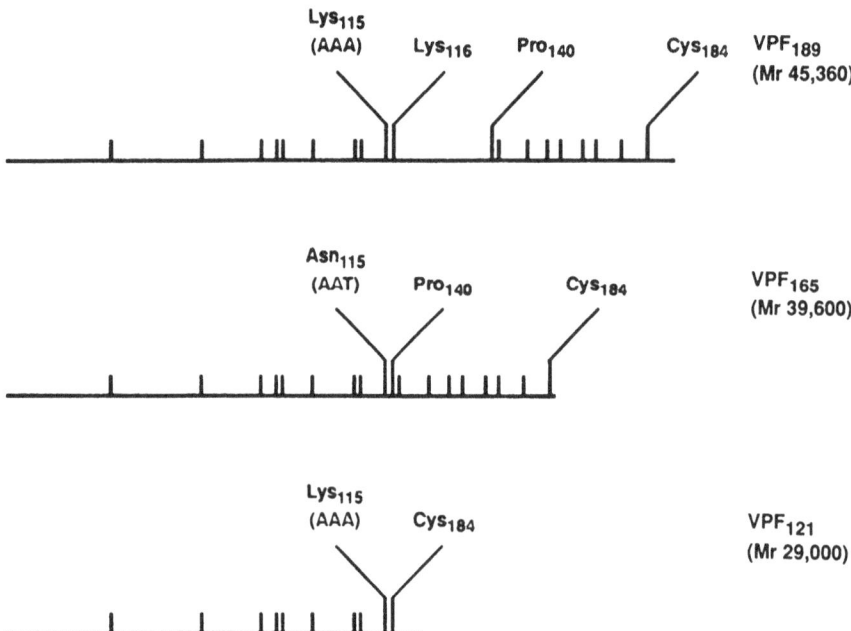

FIGURE 1: Structure of VPF/VEGF. Diagrammatic representations of the polypeptide chains corresponding to the three VPF/VEGF cDNAs identified by Keck et al. and Leung et al. are shown. The hatch marks represent cysteines. The first eight cysteines in the N-terminal regions are conserved in the PDGF family of proteins. The codons at amino acid 115 are shown in parentheses, and indicate that VPF_{165} and VPF_{121} could arise from differential splicing of mRNA within this codon. The molecular weights shown in parentheses are calculated based upon an average molecular weight of 120 for amino acids, and assuming that two identical chains associate to form a dimer.

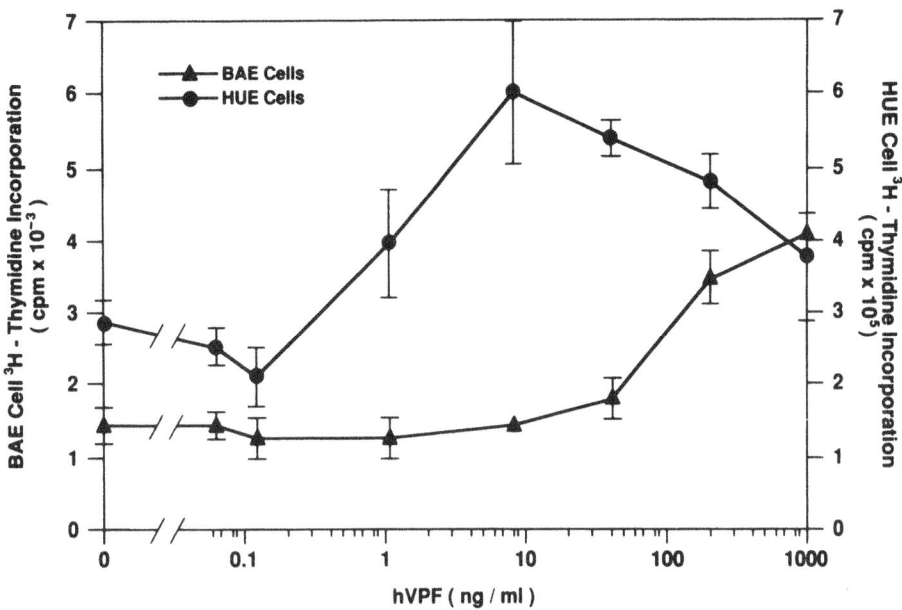

FIGURE 2: Effect of VPF on [^{3}H]Thymidine Incorporation by HUE and BAE Cells. Confluent cultures of either bovine aortic endothelial (BAE) cells or human umbilical vein endothelial (HUE) cells were incubated with different concentrations of hVPF prior to measurement of [^{3}H]thymidine incorporation into DNA. Each point represents the mean ± S.D. for triplicate cultures. (*Reprinted from Connolly and Olander with permission of The American Society for Biochemistry and Molecular Biology, Inc.*).

VPF PROMOTES ENDOTHELIAL CELL GROWTH AND ANGIOGENESIS

Soon after VPF was purified from guinea pig line 10 cell conditioned medium, we observed that VPF stimulated endothelial cell growth *in vitro* and angiogenesis *in vivo* (*Connolly, et al., 1989*). Figure 2 shows the [^{3}H]-thymidine responses of human umbilical vein (HUE) cells and bovine aortic endothelial (BAE) cells to different concentrations of purified human VPF. Half-maximal stimulation of [^{3}H]-thymidine incorporation occurred at about 50 pM VPF. VPF was thus an extremely potent mitogen for these cells. In contrast, other cell types including mouse 3T3 fibroblasts, human WI38 fibroblasts, and bovine smooth muscle cells did not respond under these conditions (*Connolly, et al., 1989*). Thus VPF has a unique target cell specificity that differentiates it from other known mitogens, including the heparin-binding fibroblast growth factors.

Several other groups have also purified an angiogenic endothelial cell mitogen referred to as vascular endothelial growth factor (VEGF) from pituitary follicular cells (*Ferrara and Henzel, 1989, Gospodarowicz, et al., 1989*), rat glioma-derived cells (*Conn, et al., 1990*), and a similar protein referred to as vasculotropin from AtT-20 pituitary cells (*Plouet, et al., 1989*). The biochemical properties, target cell specificities, amino acid sequences and cDNA sequences indicate that these molecules are very similar and probably identical to VPF (*Leung, et al., 1989, Tischer, et al., 1989, Conn, et al., 1990*).

It is well known that solid tumors recruit new blood vessels (*Folkman, 1985*). It is evident that VPF could participate in this process both directly and indirectly. First, the leakage of plasma proteins into the extravascular space could lead to events that promote angiogenesis. For example, extravascular fibrin deposition that would result from extravasation of plasma coagulation proteins into the tissue could stimulate angiogenesis (*Dvorak, et al., 1987*). Other extravasated plasma proteins, including growth factors, could also participate in this process or could attract other cell types such as neutrophils or macrophages that themselves could participate in promoting angiogenesis. Second, the mitogenic activity of VPF toward endothelial cells could provide a direct signal for blood vessel growth. It is not known if either of these mechanisms predominates, but it is clear that the multiple functions of VPF could provide multiple points for biological control.

Tumor necrosis factor (TNF) has been shown to inhibit the growth of endothelial cells, and to block their response to bFGF (*Frater-Schroder, et al., 1987*). We have therefore examined the effects of TNF on VPF stimulated mitogenesis (Pekala, et al., submitted for publication). Figure 3 shows that TNF blocks VPF stimulated mitogenesis, with half-maximal inhibition occurring at about 2 ng/ml TNF. It would appear from this result that TNF could be a general physiological antagonist toward endothelial cell mitogens. This result is somewhat perplexing, however, in light of the observation that TNF is itself angiogenic *in vivo*, and in fact acts synergistically in affecting other VPF responses (see below).

VPF AND TNF STIMULATE GLUCOSE TRANSPORT IN ENDOTHELIAL CELLS

In order to better understand the effects of VPF on endothelial cells, we have undertaken a study of various cellular processes. The first of these is glucose transport, a process central to metabolism. Chronic exposure of bovine aortic endothelial cells to VPF was found to stimulate glucose transport about 3-fold (*Pekala et al., 1990*). Interestingly, TNF was also found to stimulate glucose transport. In contrast to the antagonistic effect with regard to cell growth, the simultaneous addition of VPF and TNF led to an additive enhancement of glucose transport (Figure 4). Northern blot analysis indicated that the enhancement was due exclusively to *de novo* synthesis of the GLUT-1 isotype of glucose transporter. The aortic endothelial cells were completely unresponsive to insulin.

Why is the glucose transporter in aortic endothelial cells under the regulation of both VPF and TNF? It does not seem likely that the effect is related to mitogenic regulation since VPF stimulates cell growth and TNF inhibits cell growth. It is also unlikely that the effect is related to the control of systemic glucose levels since insulin had no effect on these cells and since mRNA for the insulin sensitive glucose transporter GLUT-4 could not be detected. Another possibility is that the induction of GLUT-1 transporter is related to an inflammatory type of signal elicited by TNF and VPF. It is well known that TNF is a general inflammatory mediator. Since sites of inflammation are often hypoxic, it might be expected that glucose utilization would increase at such sites, thus necessitating the need for increased glucose transport. Since the same effect is observed with VPF, this would be consistent with our current hypothesis that VPF acts as an inflammatory cytokine.

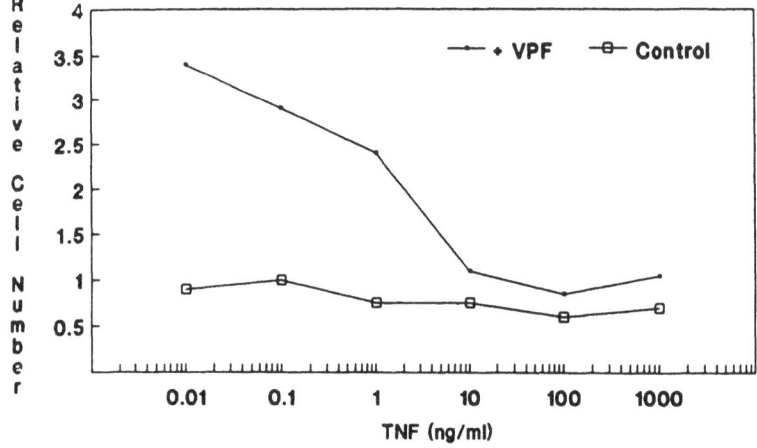

FIGURE 3: Inhibition of VPF-Stimulated BAE Cell Growth By TNF. BAE cells were plated in 96-well plates. After one day, fresh medium and TNF were added at various concentrations plus or minus VPF (100 ng/ml). Cell number was determined after five more days. (*Reprinted from Pekala et al., 1990, with the permission of The American Society for Biochemistry and Molecular Biology, Inc.*).

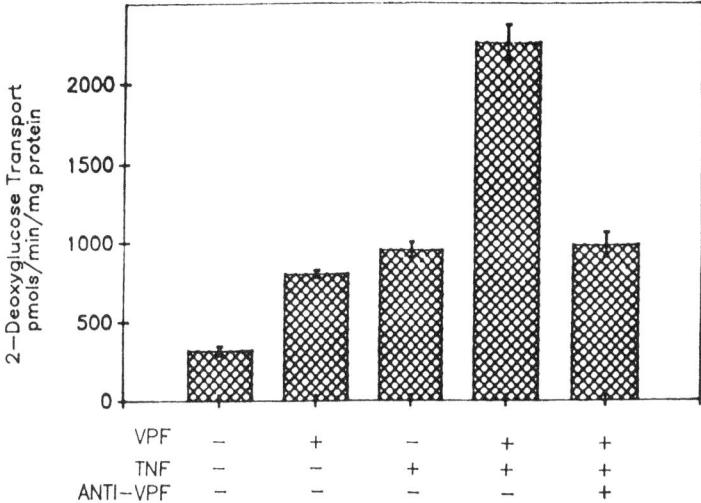

FIGURE 4: Effect of VPF and TNF on Hexose Transport in BAE Cells. At 4 days post-confluence, media were changed to DMEM supplemented with 0.5% (w/v) growth factor-free bovine serum albumin. Twenty-four hours later, the sample was added (VPF = 12.5 nM; TNF = 16.5 ng/ml) in fresh BSA containing DMEM. BAE cells were exposed to VPF, TNF, VPF and TNF together, or VPF and TNF together plus anti-VPF antiserum. After a 17 hour exposure, the cells were assayed for their ability to transport [^{3}H]2-deoxyglucose. (*Reprinted from Pekala et al., 1990, with the permission of The American Society for Biochemistry and Molecular Biology, Inc.*).

VPF AND TNF INDUCE TISSUE FACTOR EXPRESSION BY ENDOTHELIAL CELLS AND MONOCYTES

Certain tumors are hypersensitive to the effects of TNF (*Old, 1985*). Infusion of low doses of TNF into the circulation of mice that were infected with the Meth A sarcoma resulted in coagulation within the vessels of the tumor, but not in other vessels. Ultimately, thrombi and necrosis ensued (*Nawroth, et al., 1988*). In order to explain these results, Stern and coworkers hypothesized that the Meth A tumors produced a cofactor that acted synergistically with TNF to induce tissue factor expression on the endothelial surface (*Nawroth, et al., 1988*). Tissue factor is the key component in the initiation of clot formation, and its absence from the endothelial surface is one of the reasons that coagulation does not normally occur within blood vessels. It was supposed that the induction of tissue factor by TNF and the hypothetical tumor-derived cofactor could act in concert to induce intravascular coagulation.

The tumor-derived cofactor was purified and found to be the mouse homolog of VPF (*Clauss, et al., 1990*). Figure 5 shows that either the murine-derived Meth-A tumor VPF or the guinea pig-derived line 10 tumor VPF were able to stimulate tissue factor expression on human umbilical vein endothelial cells *in vitro*. The effect was dose-dependent, with 50% maximal stimulation occurring at 100 pM for each factor. Importantly, the addition of a very low dose of TNF (5 pM) enhanced the induction dramatically. Thus the presence of both factors together led to a higher expression of tissue factor than either factor alone.

Since tissue factor is also expressed on monocytes during inflammation, the effect of VPF and TNF on monocyte tissue factor expression was also examined. Results similar to those obtained with endothelial cells were obtained with monocytes. This is the first demonstration of an effect of VPF on a cell type other than endothelial cells.

VPF IS A CHEMOATTRACTANT FOR MONOCYTES

The effects of VPF on monocyte and neutrophil chemotaxis were examined using a standard trans-membrane type of assay in which VPF was placed in the bottom chamber and cells placed in the top chamber (*Clauss, et al., 1990*). Migration through the membrane was

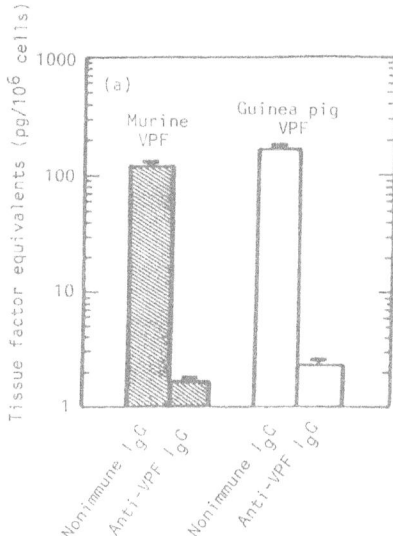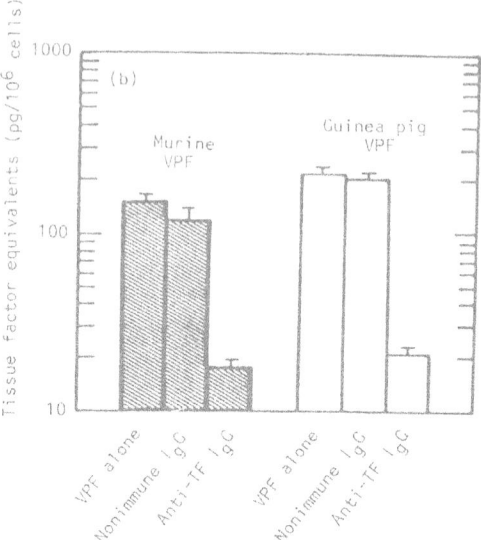

FIGURE 5: Effect of VPF and Antibody to VPF on Endothelial Cell Tissue Factor Activity. A. Induction of endothelial tissue factor by VPF. Murine VPF (5 ng/ml) or guinea pig VPF (5 ng/ml) was incubated with polyclonal IgG to VPF (anti-VPF; 10 µg) or nonimmune IgG (10 µg) adsorbed to protein G Sepharose (0.1 ml), and the resin was removed by centrifugation. Samples were then incubated with endothelial monolayers for 6 hr at 37°c, and tissue factor activity was assessed as described. The mean of triplicate determinations is shown. B. Identification of the endothelial procoagulant activity induced by VPF as tissue factor. Murine VPF (5 ng/ml) or guinea pig VPF (5 ng/ml) was incubated with endothelial monolayers for 6 hr at 37°C, and then after washing the cultures, anti-tissue factor IgG (anti-TF IgG; 5 µg/ml) or the same amount of nonimmune IgG was incubated with the cultures under serum-free conditions for 30 min at 37°C. Then, tissue factor activity was assessed as described. The mean of triplicate determinations is shown. (*Reprinted from Clauss et al., 1990, with permission of The Rockefeller University Press*).

measured. These experiments indicated that monocytes, but not neutrophils, were attracted to VPF.

Monocytes responded to doses as low as 30 pM. Furthermore, VPF was shown to attract monocytes through monolayers of endothelial cells *in vitro*.

VPF IS SYNTHESIZED BY MONOCYTES AND LYMPHOCYTES

What is the normal site of synthesis of VPF, and what is its function? We have examined guinea pig tissues and human cell sources for the presence of VPF mRNA using Northern blot analysis and PCR. At this point, VPF mRNA has been detected only in peripheral blood monocytes and lymphocytes (P. Keck and D. Connolly, unpublished observations). The message levels were too low to be detected by Northern hybridization and required PCR amplification. It is not known at this time if expression of VPF can be induced by inflammatory stimuli. The significance of this finding is unclear, but the expression of VPF by these inflammatory/immune cells is consistent with our current hypothesis that VPF is an inflammatory cytokine.

CONCLUSIONS

In addition to enhancing vascular permeability, VPF has also been shown to induce endothelial cell growth and angiogenesis, to stimulate glucose transport and expression of glucose transporter mRNA, and to convert the endothelial cell surface from a non-thrombogenic surface to a procoagulant surface through the induction of tissue factor, an important regulator of the coagulation cascade. The regulation of glucose transport and tissue factor expression can be co-regulated in a synergistic fashion by the immune modulator TNF.

The various biological activities of VPF suggest that it could normally be involved in inflammation. This hypothesis is supported by two recent observations. First, expression of VPF mRNA has been detected in monocytes and lymphocytes. Second, VPF has been found to be a potent chemoattractant for monocytes and is able to induce tissue factor activity on monocyte cell surfaces. Both of these effects are generally regarded as hallmarks of inflammatory mediators. It therefore appears that VPF is a multifunctional inflammatory cytokine that can regulate a variety of normal and pathological events in both vascular and humoral systems.

REFERENCES

Clauss, M., Gerlach, M., Gerlach, H., Brett, J. Wang, F., Familletti, P.C., Pan, Y.-C.E., Olander, J.V., Connolly, D.T., and Stern, D. Vascular permeability factor: a tumor-derived polypeptide which induces endothelial cell and monocyte procoagulant activity, and promotes monocyte migration. *J. Exp. Med.* 172:1535-1545, 1990.

Conn, G., Bayne, M.L., Soderman, D.D., Kwok, P.W., Sullivan, K.A., Palisi, T.M., Hope, D.A., and Thomas, K.A. Amino acid and cDNA sequences of a vascular endothelial cell mitogen that is homologous to platelet-derived growth factor. *Proc. Natl. Acad. Sci. U.S.A.* 87:2628-2632, 1990.

Conn, G., Soderman, D.D., Schaeffer, M., Wile, M., Hatcher, V.B., and Thomas, K.A. Purification of a glycoprotein vascular endothelial cell mitogen from a rat glioma-derived cell line. *Proc. Natl. Acad. Sci. U.S.A.* 87:1323-1327, 1990.

Connolly, D.T., Olander, J.V., Heuvelman, D., Nelson, R., Monsell, R., Siegel, N., Haymore, B.L., Leimgruber, R., and Feder, J. Human vascular permeability factor: isolation from U937 cells. *J. Biol. Chem.* 264:20017-20024, 1989.

Connolly, D.T., Heuvelman, D.M., Nelson, R., Olander, J.V., Eppley, B.L., Delfino, J.J., Siegel, N.R., Leimgruber, R.M., and Feder, J. Tumor vascular permeability factor stimulates endothelial cell growth and angiogenesis. *J. Clin. Invest.* 84:1470-1478, 1989.

Dvorak, A.M. Fibrin containing gels induce angiogenesis. *Lab. Invest.* 57:673-686, 1987.

Dvorak, H.F., Orenstein, N.S., and Dvorak, A.M. Tumor-secreted mediators and the tumor microenvironment: relationship to immunological surveillance. *Lymphokines* 2:203-233, 1981.

Dvorak, H.F. Tumors: wounds that do not heal. *N. Eng. J. Med.* 315:1650-1659, 1986.

Dvorak, H.F., Harvey, V.S., Estrella, P., Brown, L.F., McDonagh, J., and Dvorak, H.F., Nagy, J.A., Dvorak, J.T., and Dvorak, A.M. Identification and characterization of the blood vessels of solid tumors that are leaky to circulating macromolecules. *Am. J. Pathol.* 133:95-109, 1988.

Ferrara, N. and Henzel, W.J. Pituitary follicular cells secrete a novel heparin-binding growth factor specific for vascular endothelial cells. *Biochem. Biophys. Res. Comm.* 161:851-858, 1989.

Folkman, J. Tumor angiogenesis. *Adv. Cancer Res.* 43:175-203, 1985.

Frater-Schroder, M., Risau, W., Hallmann, R., Gautschi, P., and Bohlen, P. Tumor necrosis factor type α, a potent inhibitor of endothelial cell growth *in vitro*, is angiogenic *in vivo*. *Proc. Natl. Acad. Sci. U.S.A.* 84:5277-5281, 1987.

Gerlowski, L.E. and Jain, R.K. Microvascular permeability of normal and neoplastic tissues. *Microvasc. Res.* 31:288-305, 1986.

Gospodarowicz, D., Abraham, J.A., and Schilling, J. Isolation and characterization of a vascular endothelial cell mitogen produced by pituitary-derived folliculo stellate cells. *Proc. Natl. Acad. Sci. U.S.A.* 86:7311-7315, 1989.

Keck, P.J., Hauser, S.D., Krivi, G., Sanzo, K., Warren, T., Feder, J., and Connolly, D.T. Vascular permeability factor, an endothelial cell mitogen related to PDGF. *Science* 246:1309-1312, 1989.

Leung, D.W., Cachianes, G., Kuang, W.-J., Goeddel, D.V., and Ferrara, N. Vascular endothelial growth factor is a secreted angiogenic mitogen. *Science* 246:1306-1309, 1989.

Nawroth, P., Handley, D., Matsueda, G., DeWaal, R., Gerlach, H., Blohm, D., and Stern, D. Tumor necrosis factor/cachectin-induced intravascular fibrin formation in Meth A fibrosarcomas. *J. Exp. Med.* 168:637-647, 1988.

Old, L.J. Tumor necrosis factor (TNF). *Science* 230:630-632, 1985.

Pekala, P., Marlowe, M., Heuvelman, D., and Connolly, D. Regulation of hexose transport in aortic endothelial cells by vascular permeability factor and tumor necrosis factor-α, but not by insulin. *J. Biol. Chem.* 265:18051-18054, 1990.

Plouet, J., Schilling, J. and Gospodarowicz, D. Isolation and characterization of a newly identified endothelial cell mitogen produced by AtT-20 cells. *EMBO J.* 8:3801-3806, 1989.

Senger, D.R., Galli, S.J., Dvorak, A.M., Perruzzi, C.A., Harvey, V.S., and Dvorak, H.F. Tumor cells secrete a vascular permeability factor that promotes accumulation of ascites fluid. *Science* 219:983-985, 1983.

Tischer, E., Gospodarowicz, D., Mitchell, R., Silva, M., Schilling, J., Lau, K., Crisp, T., Fiddes, J.C., and Abraham, J.A. Vascular endothelial growth factor: a new member of the platelet-derived growth factor gene family. *Biochem. Biophys. Res. Comm.* 165:1198-1206, 1989.

III. CHEMICAL MEDIATORS OF ENDOTHELIAL CELL INJURY

ENDOTHELIAL CELLS AS TARGETS FOR AND PRODUCERS OF CYTOKINES

E. Dejana, G. Bazzoni, I. Martin-Padura, S. Walter
and Alberto Mantovani

Instituto di Ricerche Farmacologiche Mario Negri
Via Eritrea 62
20157 Milan, Italy

INTRODUCTION

Endothelial cells (EC) have long been defined as a "passive" lining of vessels endowed with negative properties; the most important one being that of representing a non-thrombogenic substrate for blood. As such, EC were viewed to participate in tissue reactions essentially as target for injurious agents. The possibility to isolate and culture endothelial cells from various districts gave the tools for studying their complex reactions to a variety of activating stimuli. EC have, in this way, emerged as active participants in many physiological and pathological processes. It is now evident that hemostasis, inflammatory reactions, and immunity involve close interactions between immunocompetent cells and vascular endothelium. In particular, the ontogeny and function of white blood cells require an intimate relationship with vascular EC. Cytokines are mediators of these complex bidirectional interactions between leukocytes and vascular elements (for review see *Mantovani and Dejana, 1989*).

ENDOTHELIAL CELLS AS TARGETS FOR THE ACTION OF CYTOKINES

Cytokines are mediators of the complex bidirectional interactions between leukocytes and vascular elements. Cytokines produced by activated lymphoid and mononuclear phagocytes elicit a complex spectrum of responses in vascular cells. The responses elicited seem to follow discrete, essentially non overlapping patterns (for review see *Mantovani and Dejana, 1989*).

IL-1 and TNF are the prototypes of cytokines which induce proinflammatory-prothrombotic changes in EC. The antithrombotic properties of EC are profoundly altered by exposure to these inflammatory cytokines. IL-1 and TNF induce thromboplastin, which promotes activation of the coagulation system, and decreases the cell surface anticoagulant activity mediated by the thrombomodulin-protein C pathway (*Nawroth et al., 1986*). They induce production of "platelet activating factor" (PAF) (*Bussolino et al., 1986*) and augment production of a plasminogen activator inhibitor, thus impairing the ability of the cells to dissolve a fibrin clot (*Gramse et al., 1986*).

The only activity that is apparently in contrast with the "prothrombotic" pattern elicited by IL-1 and TNF is the production of prostacyclin (PGI_2) by activated EC (*Rossi et al., 1985*). PGI_2 is involved in vasodilation at sites of inflammation and cell-mediated immunity and in the hypotension associated with systemic administration of these mediators. Furthermore, the ability of the cells to produce increased amounts of PGI_2 lasts for several hours and is probably important as a defense mechanism in inhibiting platelet aggregation and thrombus formation.

Table 1: Cytokines produced by endothelial cells

Cytokine	Stimulus	Function
IL-1	LPS IL-1 TNF IFN-g	Lymphocyte activation; local and systemic inflammation; acute phase responses; hematopoiesis
IL-6	"spontaneous" LPS IL-1 TNF	Lymphocyte activation; acute phase responses, hematopoiesis
CSFs	LPS IL-1 TNF modified LDL	Leukocyte recruitment and activation; hematopoiesis; EC proliferation and migration
Chemotactic factors (IL-8 and MCP)[+]	LPS IL-1 TNF	Leukocyte recruitment and activation
PDGF	IL-1	Smooth muscle cell proliferation (atherosclerosis)

[+]See Table 2

At variance with IL-1 and TNF, interferon gamma (gIFN) induces a series of EC responses that do not essentially overlap with those induced by these inflammatory mediators. In general, gIFN induces EC to act as accessory cells. It augments the expression of MHC class I antigens and induces class II antigens and the invariant chain (*Pober et al., 1983; Collins et al., 1984*). In contrast to IL-1 and TNF, gIFN does not change EC properties in regulating coagulation and fibrinolysis.

Granulocyte- and granulocyte-macrophage-colony stimulating factors (G-CSF and GM--CSF, respectively) and fibroblast growth factors (FGF) all induce patterns of responses in EC that differ from those induced by IL-1 and TNF or gIFN. It is in particular noteworthy that these cytokines that regulate growth and differentiation of hematopoietic cells (G and GM-CSFs) induce proliferation and migration of EC (*Bussolino et al., 1989a, 1989b*). These phenomena are frequently accompanied by induction of lytic enzymes (as urokinase or collagenase) that facilitate cell migration and extracellular matrix digestion.

ENDOTHELIAL CELLS AS PRODUCERS OF CYTOKINES

EC produce cytokines that regulate the hematopoietic system, the proliferation and differentiation of T and B cells and the recruitment of leukocytes at sites of inflammation. As mentioned above, EC produce copious amounts of classical CSFs (G, M and GM). Two of these molecules (G and GM-CSF) in turn act on endothelial cells inducing migration and proliferation. Thus, EC are an important constituent of the hematopoietic microenvironment.

TABLE 2: An emerging superfamily of cytokines involved in the regulation of leukocyte recruitment in tissues

Family	Molecule(s)	Function	Source
Cys-X-Cys	PF4	chemotaxis, immunomodulation	platelets
	Platelet basic protein and derived peptides	Chemotaxis NAP-2	platelets
	IL-8	PMN and lymphocyte chemotaxis	monocytes, lymphocytes, endothelial cells, fibroblasts
	gro	growth of melano-cytic cells; PMN chemotaxis	transformed cells, endothelial cells
	MIP-2 (mouse)	PMN chemokinesis	macrophages
	KC (mouse)	?	fibroblasts, macrophages
	IPIO	?	macrophages
Cys-Cys	MCP	monocyte chemotaxis	various tumors, endothelial cells, fibroblasts
	MIP-1 mouse)	PMN chemokinesis	macrophages
	LD78 RANTES TCA3	?	T cells

Unless specified, human cytokines/genes. The two families are distinguished on the basis of the first 2 conserved cysteins Cys-X-Cys versus Cys-Cys, see text). There is no agreement as to the nomenclature of several of these factors/genes.

Upon exposure to inflammatory stimuli (as endotoxin, IL-1 or TNF), EC produce IL-1 and interleukin-6 (IL-6) (*Libby et al., 1986; Sironi et al., 1989*). These are pleiotropic cytokines which affect activation and proliferation of different cell types, such as T and B lymphocytes and hemopoietic precursors. IL-1 also induces proliferation of arterial smooth muscle cells of the media, and in this way, possibly contributes to the vascular thickening during the atherosclerotic process (*Libby et al., 1988; Raines et al., 1989*). IL-1 and IL-6 also induce the release of acute phase proteins and, in this way, producing these peptides, EC take part in the regulation of specific immunity and acute phase responses (*Billiau, 1988*). Finally, EC have the capacity to produce CSFs and other cytokines (MCP and interleukin-8) which can activate and induce chemotactic migration of leukocytes, as discussed in detail in the following section.

LEUKOCYTE RECRUITMENT

EC express a series of adhesive structures on their membrane that are a substrata for monocyte, lymphocyte and polymorphonuclear cell adhesion. In resting conditions, EC express two integral membrane proteins denominated "intercellular adhesion molecules" (ICAM-1 and ICAM-2) (*Dustin and Springer, 1988*). These proteins belong to the large superfamily of immunoglobulins and are homologous in their amino acid sequences. They are expressed even without activation of EC and are probably responsible for the basal adhesion of leukocytes to the endothelial cell surface. The receptor of ICAM-1 and ICAM-2 is a member of the leukocyte integrin family; it has been denominated LFA-1 or CD11a/18 or $\alpha_1\beta_2$ and is present on monocytes, polymorphonuclear leukocytes, lymphoid cells and related leukemic lines (*Marlin and Springer, 1987; Staunton et al., 1989*).

When EC are activated by the inflammatory mediators interleukin-1 IL-1 and (TNF) 2 (and also to a minor extent by interferon g), the expression of ICAM-1 is augmented (*Dustin and Springer, 1988*). In addition to increasing ICAM-1, IL-1 and TNF, induce *de novo* expression of two novel adhesion structures denominated "endothelial leukocyte adhesion molecule 1" (ELAM-1) (*Bevilacqua et al., 1989*) and "vascular cell adhesion molecule" (VCAM) (*Osborn et al., 1989*).

The cDNA encoding ELAM-1 has recently been cloned (*Bevilacqua et al., 1989*) and shows it to be a member of an emerging family of cell adhesion molecules denominated "selectins". Other members defined so far are the MEL 14 homing receptor in lymphocytes (*Gallatin et al., 1983*) and the platelet and endothelial cell GMP140 (*McEver and Martin, 1984*). These molecules have a relatively high homology and share the common characteristic of containing a lectin domain, an epithelial growth factor domain and a complement binding domain. The lectin domain is probably the most important in determining the adhesive properties of these antigens through the binding to specific carbohydrate regions in the ligand cell or matrix proteins. VCAM, at variance, belongs to the immunoglobulin superfamily (*Osborn et al., 1989*) and is, therefore, unrelated to ELAM-1. Its receptor has been identified as the integrin molecule VLA-4 (or a_4b_1) (*Elices et al., 1990*). As a consequence of expression of leukocyte adhesion structures, EC exposed to inflammatory stimuli provide a better substratum for the adhesion of polymorphonuclear cells, monocytes and lymphocytes. This structural change on EC membrane lasts several hours. Thus, at sites of inflammation and cell-mediated immune reactions, EC participate in white blood cell recruitment by changing their adhesive properties.

In addition to expressing molecules that act as substrata for adhesion, EC participate in the regulation of leukocyte extravasation by producing cytokines which are chemotactic for leukocytes. These include CSFs (see above) and members of a novel, emerging superfamily of chemotactic polypeptides (see Table 2; reviewed in *Baggiolini et al., 1989; Mantovani, 1990; Wolpe and Cerami, 1989*). We and others have shown that upon stimulation with inflammatory signals e.g. IL-1), EC produce IL-8 and monocyte chemotactic protein (MCP) (*Strieter et al., 1989; Schroder and Christopher, 1989; Sica et al., 1990 a,b*). The superfamily to which IL-8 and MCP belong is characterized by a structural hallmark represented by 4 conserved cysteins, the first two in tandem, which are probably important for the three dimensional structure of these molecules. The same 4 Cys motif is shared by cytokines belonging to the platelet factor 4 (PF4) family, except that the first 2 Cys tandem is interrupted by an intervening amino acid. The PF4 family includes IL-8 and gro, which are chemotactic for neutrophils. IL-8 (also referred to as neutrophil activating protein, NAP-1) is a cytokine active on neutrophils, basophils, and lymphocytes (*Baggiolini et al., 1989*). IL-8 also induces migration of melanoma (*Wang et al., 1990*) and this, together with augmented binding to EC, may contribute to augmented metastasis after IL-1 administration in mice (*Giavazzi et al., 1990*). MCP is a cytokine selectively active on monocytes (*Yoshimura et al., 1989; Matsushima et al., 1989; Bottazzi et al., 1990; Van Damme et al., 1989*). It was originally identified in supernatants of tumor cells and is probably one important determinant of macrophage infiltration in tumors (*Bottazzi et al., 1983; Mantovani, 1990*). By producing MCP and IL-8, (*Strieter et al., 1989; Schroder and Christophers, 1989; Sica et al., 1990 a,b*), EC participate in chemotaxis of different leukocyte populations at sites of inflammation and cell-mediated immunity. Production of MCP by vessel wall elements (EC and smooth muscle cells) may in particular be relevant in the pathogenesis of atherosclerosis, in which monocyte extravasation is an early event, and vasculitis, where monocyte accumulate in and around vessel walls.

CONCLUDING REMARKS

EC constitute the barrier between circulating elements and tissues. Many observations indicate that EC cannot be considered as a passive surface but they can actively participate in immunity, inflammation and hemostasis. EC represent a source and a target of cytokines. Activation of EC by inflammatory stimuli or other modulatory peptides dramatically changes EC function and surface properties. Cytokine-induced EC functional reprogramming follows discrete patterns with limited overlapping. Some cytokines, such as IL-1 and TNF, induce a program related to inflammation and immunity; gIFN activates accessory cell function, FGF and G- and GM-CSF induce EC migration and proliferation.

REFERENCES

Baggiolini, M., Walz, A. and Kunkel, S.L. Neutrophil-activating protein-1/interleukin 8, a novel cytokine that activates neutrophils. *J. Clin. Invest.* 84:1045-1049, 1989.

Bevilacqua, M.P., Stengelin, S., Gimbrone, M.A., Jr. and Seed, B. Endothelial-leukocyte adhesion molecule 1: An inducible receptor for neutrophils related to complement regulatory proteins and lectins. *Science* 243:1160-1165, 1989.

Billiau, A. IL-6: Structure, production and actions. In: Monokines and other Non-lymphocytic Cytokines. Powanda, M.C., Oppenheim, J.J., Kluger, M.J., Dinarello, C.A., eds., New York, Alan R. Liss, *Progress in Leukocyte Biology* vol. 8:3-13, 1988.

Bottazzi, B., Colotta, F., Sica, A., Nobili, N. and Mantovani, A. A chemoattractant expressed in human sarcoma cells tumor-derived chemotactic factor, TDCF) is identical to monocyte chemoattractant protein-1/monocyte chemotactic and activating factor MCP-1/MCAF). *Int. J. Cancer* In press, 1990.

Bottazzi, B., Polentarutti, N., Acero, R., Balsari, A., Boraschi, D., Ghezzi, P., Salmona, M. and Mantovani, A. Regulation of the macrophage content of neoplasms by chemoattractants. *Science* 220:210-212, 1983.

Bussolino, F., Breviario, F., Tetta, C., Aglietta, M., Mantovani, A. and Dejana, E. Interleukin 1 stimulates platelet-activating factor production in cultured human endothelial cells. *J. Clin. Invest.* 77:2027-2033, 1986.

Bussolino, F., Wang, J.M., Defilippi, P., Turrini, F., Sanavio, F., Edgell, C.J., Aglietta, M., Arese, P. and Mantovani, A. Granulocyte- and granulocyte-macrophage-colony stimulating factors induce human endothelial cells to migrate and proliferate. *Nature* 337:471-473, 1989a.

Bussolino, F., Wang, J.M., Turrini, F., Alessi, D., Ghigo, D., Costamagna, C., Pescarmona, G., Mantovani, A. and Bosia, A. Stimulation of the Na^+/H^+ exchanger in human endothelial cells activated by granulocyte- and granulocyte-macrophage colony stimulating factor. Evidence for a role in proliferation and migration. *J. Biol. Chem.* 264:18284-18287, 1989b.

Collins, T., Korman, A.J., Wake, C.T., Boss, J.M., Kappes, D.J., Fiers, W., Ault, K.A., Gimbrone, M.A. Jr., Strominger, J.L. and Pober, J.S. Immune interferon activates multiple class II major histocompatibility complex genes and the associated invariant chain gene in human endothelial cells and dermal fibroblasts. *Proc. Natl. Acad. Sci. U.S.A.* 81:4917-4921, 1984.

Dustin, J.L. and Springer, T.A. Lymphocyte function-associated antigen-1 (LFA-1) interaction with intercellular adhesion molecule-1 (ICAM-1) is one of at least three mechanisms for lymphocyte adhesion to cultured endothelial cells. *J. Cell Biol.* 107:321-331, 1988.

Elices, M.L., Osborn, L., Takada, Y., Crouse, C., Luhowsky, S., Hemler, M.E. and Lobb, R.R. VCAM-1 on activated endothelium interacts with the leukocyte integrin VLA-4 at a site distinct from the VLA 4/fibronectin binding site. *Cell* 60:577-584, 1990.

Gallatin, W.M., Weissman, I.L. and Butcher, E.C. A cell-surface molecule involved in organ specific homing of lymphocytes. *Nature* 304:30-34, 1983.

Giavazzi, R., Garofalo, A., Bani, M.R., Chezzi, P., Boraschi, D., Mantovani, A. and Dejana, E. Interleukin-1-induced augmentation of experimental metastases from a human melanoma in nude mice. *Cancer Res.* 50:4771-4775, 1990.

Gramse, M., Breviario, F., Pintucci, G., Millet, I. and Dejana, E. Enhancement by interleukin-1 IL-1) of plasminogen activator inhibitor (PA-I) activity in cultured human endothelial cells. *Biochem. Biophys. Res. Commun.* 139:720, 727, 1986.

Libby, P., Ordovas, J.M., Birinyi, L.K., Augler, K.R. and Dinarello, C.A. Inducible interleukin-1 gene expression in human vascular smooth muscle cells. *J. Clin. Invest.* 78:143201438, 1986.

Libby, P., Warner, S.J.C. and Friedman, G.B. Interleukin 1: A mitogen for human vascular smooth muscle cells that induces the release of growth-inhibitory prostanoids. *J. Clin. Invest.* 81:487-498, 1988.

Mantovani, A. Tumor-associated macrophages. *Current Opinions in Immunology* In Press, 1990.

Mantovani, A. and Dejana, E. Cytokines as communication signals between leukocytes and endothelial cells. *Immunol. Today* 10:370-375, 1989.

Marlin, S.D. and Springer, T.A. Purified intercellular adhesion molecule-1 (ICAM-1) is a ligand for lymphocyte function-associated antigen 1 (LFA-1). *Cell* 51:813-819, 1987.

Matsushima, K., Larsen, C.G., DuBois, G.C. and Oppenheim, J.J. Purification and characterization of a novel monocyte chemotactic and activating factor produced by a human myelomonocytic cell line. *J. Exp. Med.* 169:1485-1490, 1989.

McEver, R.P. and Martin, M.N. A monoclonal antibody to a membrane glycoprotein binds only to activated platelets. *J. Biol. Chem.* 259:9799-9804, 1984.

Nawroth, P.P., Handley, D.A., Esmon, C.T. and Stern, D.M. Interleukin-1 induces endothelial cells procoagulant while suppressing cell-surface anticoagulant activity. *Proc. Natl. Acad. Sci. U.S.A.* 83:4360-3464, 1986.

Osborn, L., Hession, C., Tizard, R., Vassallo, C., Luhowskyi, S., Chi-Rosso, G. and Lobb, R. Direct expression cloning of vascular cell adhesion molecule 1, a cytokine-induced endothelial protein that binds to lymphocytes. *Cell* 59:1203-1211, 1989.

Pober, J.S., Gimbrone, M.A. Jr., Cotran, R.S., Reiss, C.S., Vurakoff, S.J., Fiers, W. and Ault, K.A. Ia expression by vascular endothelium is inducible by activated T cells and by human g-interferon. *J. Exp. Med.* 157:1330-1353, 1983.

Raines, E.W., dower, S.K. and Ross, R. Interleukin-1 mitogenic activity for fibroblasts and smooth muscle cells is due to PDGF-AA. *Science* 243:393-396, 1989.

Rossi, V., Breviario, F., Ghezzi, P., Dejana, E. and Mantovani, A. Prostacyclin synthesis induced in vascular cells by interleukin-1. *Science* 229:174-176, 1985.

Schroder, J.M. and Christophers, E. Secretion of novel and homologous neutrophil-activating peptides by LPS-stimulated human endothelial cells. *J. Immunol.* 142:244-251, 1989.

Schroder, J.M. and Christophers, E. Secretion of novel and homologous neutrophil-activating peptides by LP stimulated human endothelial cells. *J. Immunol.* 142:244-251, 1989.

Sica, A., Matsushima, K., Van Damme, J., Wang, J.M., Polentarutti, N., Dejana, E., Colotta, F. and Mantovani, A. IL-1 transcriptionally activates the monocyte-derived neutrophil chemotactic factor/IL-8 gene in endothelial cells. *Immunology* 69:548-553, 1990a.

Sica, A., Wang, J.M., Colotta, F., Dejana, E., Mantovani, A., Oppenheim, J.J., Larsen, C.G., Zachariae, C.O.C. and Matsushima, K. Monocyte chemotactic and activating factor gene expression induced in endothelial cells by IL-1 and TNF. *J. Immunol.* 144:3034-3038, 1990b.

Sironi, M., Breviario, F., Proserpio, P., Biondi, A., Vecchi, A., Van Damme, J., Dejana, E. and Mantovani, A. IL-1 stimulates IL-6 production in endothelial cells. *J. Immunol.* 142:549-553, 1989.

Staunton, D.E., Dustin, M.L. and Springer, T.A. Functional cloning of ICAM-2, a cell adhesion ligand for LFA-1 homologous to ICAM-1. *Nature* 339:61-64, 1989.

Strieter, R.M., Kunkel, S.L., Showell, H.J., Remick, D.G., Phan, S.H., Ward, P.A. and Marks, R.M. Endothelial cell gene expression of a neutrophil chemotactic factor by TNF-α, LPS, and IL-β. *Science* 243:1467-1469, 1989.

Van Damme, J., Decock, B., Lenaerts, J.P., Conings, R., Bertini, R., Mantovani, A. and Billiau, A. Identification by sequence analysis of chemotactic factors for monocytes produced by normal and transformed cells stimulated with virus, double-stranted RNA or cytokine. *Eur. J. Immunol.* 19:2367-2373, 1989.

Wang, J.M., Taraboletti, G., Matsushima, K., Van Damme, J. and Mantovani, A. Induction of haptotactic migration of melanoma cells by neutrophil activating protein/interleukin-8. *Biochem. Biophys. Res. Commun.* In Press, 1990.

Wolpe, S.D. and Cerami, A. Macrophage inflammatory protein 1 and 2: members of a novel superfamily of cytokines. *FASEB J.* 3:2565-2573, 1989.

Yoshimura, T., Robinson, E.A. Tanaka, S., Appella, E., Kuratsu, J.I. and Leonard, E.J. Purification and amino acid analysis of two human glioma-derived monocyte chemoattractants. *J. Exp. Med.* 169:1449-1459, 1989.

ROLES OF VASCULAR CELLS IN INFLAMMATION AND IMMUNOPATHOLOGY

Peter Libby

Vascular Medicine and Atherosclerosis Unit
Cardiovascular Division Department of Medicine
Harvard Medical School and Brigham and Women's Hospital
75 Francis Street
Boston, Massachusetts 02115, U.S.A.

INTRODUCTION

Blood vessels play important roles in most inflammatory and immune responses. Alterations in vascular function prove highly adaptive in host defense mechanisms, but can contribute to disease when triggered inappropriately. The last decade has witnessed enormous increases in the understanding of diverse aspects of the functions of vascular endothelial and smooth muscle cells, including growth control, maintenance of blood compatibility, involvement in mediating leukocyte adhesion and traffic, and interactions with the immune system. This review will illustrate how certain specific facets of vascular cell function may contribute to fundamental immunopathologic mechanisms important in human diseases. I will emphasize the interplay of mediators and leukocytes with vascular endothelial and smooth muscle cells in some selected examples of immunopathological processes. The goal of this volume is to help bridge the gap between basic science of vascular cell biology and the disease states that confront the clinician. Therefore, this brief overview does not aim to be comprehensive, but rather attempts to integrate some of the diverse points treated in detail in other chapters into the context of general immunopathologic mechanisms and to illustrate a few applications to examples of human disease states.

A time scale of immunopathologic reactions will provide the organizational thread for this discussion (Table 1). This framework will help to explain how various specific alterations in vascular function (e.g. thrombosis, leukocyte interactions with endothelium, and growth control) come into play during pathologic reactions. I will consider examples of an acute, subacute, intermediate, and finally chronic immunopathologic processes. Each example will highlight the practical significance of specific functions of vascular wall cells that have come to light in the last few years.

ROLES FOR BLOOD VESSELS IN
ACUTE IMMUNOPATHOLOGIC RESPONSES

Immediate hypersensitivity occupies the acute end of this time scale as it develops over a period of seconds to a few minutes. Major mediators of immediate hypersensitivity include low molecular weight autacoids, such as histamine, and lipid mediators, such as lipoxygenase products. The release of these autacoids depends upon reexposure to specific antigen and involves activation by IgE antibody of mast cells and basophils. Common clinical examples

Table 1: Time Scale of Immunopathologic Mechanisms

Acute:	immediate hypersensitivity (seconds to minutes)
Subacute:	Arthus reaction (2-4 hours)
Intermediate:	delayed-type hypersensitivity (1-2 days)
Chronic:	fibrosis, granuloma formation (weeks, months, years)

include the dramatic and life-threatening anaphylactic reaction. On the other extreme, acute allergic rhinitis, although not dangerous, can be extremely annoying. What are the roles for vascular wall cells in immediate hypersensitivity? Capillary leak with extravasation of fluid probably results from altered endothelial cell barrier function. Histamine can certainly provoke contraction of endothelial cells and produce other structural alterations of the vascular wall that likely contribute to the microvascular leak produced by mediators of immediate hypersensitivity (Majno et al., 1969; Ryan and Majno, 1977).

In addition to capillary leak due to endothelial derangements, altered smooth muscle tone also contributes to the clinical findings in anaphylaxis and explains other cardinal features of this syndrome. Histamine can cause not only increased capillary permeability, but vasodilatation and bronchoconstriction. The slow reacting substances of anaphylaxis, defined pharmacologically many years ago, were found in the last 15 years or so to be activities of the lipoxygenase pathway of arachidonate conversion (Samuelsson, 1983). The morbidity of anaphylaxis involves not only capillary leak, but the vascular collapse of vasodilatation provoked by these mediators and interference with gas exchange and ventilation due to bronchoconstriction. The mechanism of extrinsic asthma, another clinically important manifestation of immediate hypersensitivity, involves similar processes. Thus, altered smooth muscle and endothelial functions can explain key elements in the pathogenesis of immediate hypersensitivity.

ROLES FOR BLOOD VESSELS IN SUBACUTE IMMUNOPATHOLOGIC RESPONSES

The Arthus reaction provides an example of a subacute immunopathologic phenomenon which occurs over a period of two to four hours. The Arthus reaction is a local phenomenon elicited by antigen-antibody complexes that involves vascular injury, thrombosis, hemorrhage, and acute inflammation with granulocyte infiltration. The polymorphonuclear leukocytes appear to adhere to local microvascular endothelium and diapedese through the endothelial cells into the perivascular tissue. Subsequent endothelial injury or activation probably results from locally produced mediators derived from the recruited neutrophils, including such substances as platelet activating factor, toxic oxygen radicals, and arachidonic acid metabolites.

We have learned a great deal in the last few years about the structure of specific leukocyte adhesion molecules on the endothelial cell surface (Bevilacqua et al., 1989; Larsen et al., 1989; McEver et al., 1989). Two important generalizations emerge from the analysis of recent findings regarding specific leukocyte adhesion mechanisms. One is that specific cognate ligand receptor interactions can be selective for various leukocyte subclasses, such as granulocytes, lymphocytes, and mononuclear cells. Another point, which is important in the consideration of the Arthus reaction, is that these various leukocyte adhesion mechanisms may display distinct kinetics of expression after exposure of endothelial cells or leukocytes to an inductive stimulus during an inflammatory response. Maximum expression of the prototypic endothelial leukocyte adhesion molecule ELAM 1, a polymorphonuclear leukocyte selective mechanism, occurs about four hours after exposure to the inducer (Bevilacqua et al., 1985; Bevilacqua et al., 1987; Bevilacqua, Stengelin et al., 1989). After local injection of immune complexes, stimulation of ELAM 1 expression might fit in very nicely with the timing of the leukocyte accumulation that is characteristic of this process. The chemotactic mechanisms that cause penetration into the tissues of the adherent leukocytes very likely involve anaphylatoxin arising from activation of the complement cascade by antigen-antibody complexes.

ROLES FOR BLOOD VESSELS IN DELAYED IMMUNOPATHOLOGIC RESPONSES

Delayed-type hypersensitivity reactions provide an example of an immunopathologic mechanism intermediate in the time scale, requiring one to two days for lesion formation. Delayed-type hypersensitivity is an inflammatory reaction characterized by infiltration of mononuclear phagocytes and of T cells that are reacting to a specific antigen. A familiar example is the positive cutaneous response to extracts of tubercle bacillus in individuals who had previous experience with these antigens. Most readers will have had a tuberculin test at some time. Those who have had positive reaction know that this erythema (redness and swelling) and induration (hardening) occurs over a two day period and then slowly subsides. Mononuclear phagocytes and T lymphocytes activated during such a process release protein mediators of inflammation known as cytokines (*Dinarello, 1989, Le and Vilcek, 1987; Pober and Cotran, 1990*). Some of the cytokines characteristically released by an activated mononuclear phagocyte include interleukin 1 (IL1), interleukin 6 (IL6), and tumor necrosis factor (TNF). This subclass of cytokines are called monokines as they derive from mononuclear phagocytes. The T cells reacting to a specific antigen will release interleukin 2 (IL2), interferon γ (immune interferon) and lymphotoxin (a first cousin, both functionally and structurally, of tumor necrosis factor, and indeed, sometimes known as tumor necrosis factor β). This subclass of cytokines derived from T cells are called lymphokines.

What roles might vascular wall cells play in delayed-type hypersensitivity reactions? These cells may serve as important targets for the action of the cytokines elaborated by the immune-activated leukocytes in these lesions. Cytokines produced by the antigen-stimulated T lymphocytes and activated mononuclear phagocytes can promote thrombosis and leukocyte recruitment by venous endothelium. The morphology of delayed-type hypersensitivity reaction reveals not only mononuclear phagocyte and T cell infiltration but intravascular thrombosis and in some cases necrosis (*Dvorak et al., 1986*). Although polymorphonuclear leukocytes do not form a prominent part of the leukocytic infiltrate in delayed-type hypersensitivity, ELAM-1 does appear to be expressed in these lesions (*Cotran et al., 1986*), indicating that *in vivo* this endothelial-selective leukocyte adhesion molecule may aid mediate mononuclear cell as well as polymorph recruitment.

Another characteristic of chronic delayed-type hypersensitivity reaction, such as might occur in tuberculous granuloma or other kinds of granulomatous diseases, is an increase in smooth muscle cell or fibroblast proliferation and increased deposition of extracellular matrix. The cytokines may also contribute to the modulation of the mesenchymal cell growth and matrix gene expression. One of the major cytokines secreted by activated mononuclear phagocytes is IL1. The beta isoform of IL1 shares regions of significant sequence similarity at the protein level with acidic FGF (also known as heparin binding growth factor 1) (*Thomas et al., 1985*). We found those results quite interesting because at that very time we were testing the hypothesis that IL1, a major monokine which might be released in areas of vascular inflammation, might stimulate smooth muscle proliferation.

Inappropriate multiplication of smooth muscle cells is a constant feature of many arterial hyperplastic diseases of high interest to clinicians such as atherosclerosis, restenosis following arterial intervention such as balloon angioplasty, and anastomotic hyperplasia (*Libby et al., 1989; Ross, 1986; Ross and Glomset, 1973*). We found that prolonged incubation of cultured human smooth muscle cells with recombinant IL1 α or β caused a profound and concentration-dependent increase in proliferation (*Libby et al., 1988a*). The stimulatory effect of IL1 on smooth muscle cell proliferation may result in whole or in part by IL1-induced expression of one form of another protein known as platelet-derived growth factor (PDGF) (*Raines et al., 1989*). Smooth muscle cells themselves can produce the A isoform of PDGF, and this endogenous mechanisms of growth stimulation is an example of an "autocrine" pathway (*Libby et al., 1988b: Sejersen et al., 1986; Sjolund et al., 1988*).

Such cytokine-or growth factor-induced expression of endogenous growth factors is likely to be an increasingly recognized mechanism for cytokine effects on cell proliferation. Cytokines secreted at sites of delayed-type hypersensitivity reactions may not only stimulate smooth muscle cell proliferation, but inhibit this process as well. For example, interferon γ released by activated T cells can counteract the effect of IL1 on smooth muscle cell proliferation. High concentrations ($\leq$ 1000 U/ml) of interferon γ actually almost completely inhibit this effect of IL1 (*Warner et al., 1989a*). This property of interferon γ as a growth inhibitor is not limited to IL1 as a stimulus. Interferon γ also inhibits PDGF- and serum-induced proliferation of human smooth muscle cells.

Recent experiments by Helen Palmer in our laboratory, demonstrate that prolonged exposure of human smooth muscle cells in culture to tumor necrosis factor can cause smooth muscle cell growth. She has further found that interferon β, a nonimmune interferon, but one that can actually be induced in smooth muscle cells by tumor necrosis factor and other classic interferon-inducing stimuli, can also inhibiit. Thus an imbalance between stimulators and inhibitors of smooth muscle cell growth may result in deranged control of mesenchymal cell growth. On one hand, during delayed-type hypersensitivity reaction and many other kinds of cnronic inflammatory responses, mitogens such as PDGF, IL1, tumor necrosis factor, and transforming growth factor β can increase cell proliferation. On the other hand, interferon γ, or prostaglandins produced in response to some of the same stimuli can inhibit cell proliferation. Transforming growth factor β can either stimulate or inhibit smooth muscle cell growth because of variable induction of PDGF gene expression and/or modulation of PDGF receptor subtypes (*Majack et al., 1990*).

Delayed-type hypersensitivity reactions provoked by intracutaneous injection of an antigen usually regresses after several days without resultant permanent change in tissue architecture. If antigenic stimulation persists, as during a chronic infection, (e.g. tuberculosis) the same kind of mechanisms can yield fibrosis and granuloma formation over a period of weeks, months, or even years. Cytokines that are released during the immune and inflammatory responses can act on blood vessel wall cells to increase smooth muscle cell growth and interstitial collagen production (*Libby et al., 1989a*).

EXAMPLES OF SPECIFIC DISEASES THAT INVOLVE VASCULAR IMMUNOPATHOLOGIC PHENOMENA

Immune complex diseases

A few examples illustrate how immunopathologic mechanisms might involve blood vessel wall cells. Immune complex disease actually involves mechanisms that are extensions of those already discussed in the context of the Arthus reaction. Antigen-antibody complexes can form in response to injected soluble antigen (e,g. serum sickness), infectious agents (viral or bacterial), or in autoimmune states (e.g, lupus erythematosis). Antigen-antibody complexes deposited on the surface of microvessels can fix complement, cause endothelial injury, and activation. Immune-complex-induced glomerulonephritis involves local complement fixation by antigen antibody complexes, regional white blood cell recruitment due to C5a formation, and direct endothelial cell damage due to formation of the membrane attack complex or C5bC9. Altered endothelial permeability and vasomotion due to effects on smooth muscle cell tone of low molecular weight mediators released from white blood cells activated by anaphylatoxin (e.g. C3a, C5a). Over a more chronic time scale, altered growth of smooth muscle cells (or the related mesangial cells in the renal microvasculature, or pericytes in other vascular beds) may contribute to fibrosis.

Vasculitis

Another important category of vascular reactions that involve immune responses are the vasculitides. Vascular endothelial and smooth muscle cells have been classically viewed as targets of the immune system in vasculitides. However, these cells may also participate actively in the local stimulation of T and B lymphocytes by a variety of mechanisms including induction of surface expression of IL1, secretion of large amounts of the T cell stimulator IL6, and by histocompatibility gene expression. For example, endothelial or smooth muscle cells stimulated with either lipopolysaccharide (Gram negative endotoxin) or IL1 itself (a strong stimulus for IL1 gene expression in all cell types studied) express biologically active IL1 on their surfaces and secrete IL6 (*Jirik et al., 1989; Kurt-Jones et al., 1987; Loppnow and Libby, 1989a; Loppnow and Libby, 1989b; Sironi et al., 1989*), and (Loppnow and Libby, unpublished data). The surface expression of IL1 may be important in local activation of T lymphocytes as IL1 in combination with an antigenic or other kind of activating stimulus can lead to T cell activation of IL2 receptor expression, IL2 production, and secretion of interferon γ.

Interleukin 6 encompasses a variety of activities including interferon $β_2$, B cell stimulatory factor, hybridoma growth factor, hepatic stimulatory factor, and T cell activating factor (*Kishimoto and Hirano, 1988; Le and Vilcek, 1989*). This list of functions gives an idea of the wide spectrum of activities that this multipotent mediator expresses. The same stimuli which induce surface expression of IL1 cause the release of IL6 activity from both endothelial cells and particularly from smooth muscle cells (*Loppnow and Libby, 1989a; Loppnow and Libby, 1989b; Loppnow and Libby, 1989c*). Stimulated smooth muscle cells release enormous amounts of IL6 (picograms per cell). Quantitative immunoprecipitation experiments show that IL6 accounts for almost 4% of the secreted protein of an IL1-activated human smooth muscle cell (*Loppnow and Libby, 1989c*). Much of the thymocyte costimulatory activity released from endotoxin- or IL1 - stimulated smooth muscle or endothelial cells is actually due to IL6 rather than IL1 (Loppnow and Libby, unpublished data). Thus, vessel wall cells can play roles in vasculitis, not only by surface expression of biologically active IL1, but by secretion of large amounts of the B and T cell stimulator IL6. The surface associated IL1 probably acts at short distances by cell-cell contact, while the soluble IL6 molecule may act at greater distances.

Another important function of vascular cells in vasculitides is their ability to express histocompatibility genes. Class II histocompatibility antigens (denoted in humans HLA-DR, DP or DQ), are highly polymorphic protein families expressed on the surfaces of antigen-presenting cells. The population contains many variants of these proteins and each individual will express two specific variants (alleles) at each genetic locus. Each allele is inherited in a classical Mendelian fashion from one parent. Recognition of foreign antigens or tissues by the immune system absolutely requires expression of class II HLA molecules (*Benacerraf, 1985; Nossal, 1987*). Until recently, only activated bone marrow-derived leukocytes were thought capable of expressing class II HLA. Work from a variety of a laboratories, showed that human vascular endothelial and smooth muscle cells, although they express low levels of class II HLA under basal conditions, can synthesize these molecules when stimulated with interferon gamma. A wide variety of other cytokines tested (e.g. IL-1, IL2, IL4, IL6, interferons α and β, and GM-CSF) fail to induce class II HLA on vascular wall cells (*Libby et al., 1989c; Pober et al., 1986; Wagner et al., 1985; Warner et al., 1989b*). Gamma interferon regulates this increased surface expression due to control at the level of RNA (*Collins et al., 1984; Warner, Friedman et al., 1989b*).

How might the capacity of vascular wall cells to express class II HLA contribute to the pathogenesis of vasculitis? Cells capable of expressing class II antigens can act as antigen presenting cells to enable recognition of foreign antigens (e.g. viral proteins) by helper T lymphocytes. This is a key initiating event in generating the cellular immune response. In the case of a transplanted organ, foreign class II HLA, on arterial endothelial cells or smooth muscle cells, provide structures for recognition by the host immune system. We have hypothesized that a localized chronic cellular immune response in the blood vessel wall due to foreign HLA can account for the arteriosclerotic disease which occurs in the arteries of transplanted organs. This complication now presents the major limitation to the long term survival of heart transplant recipients (*Libby, Salomon et al., 1989c*). This example illustrates how expression of class II antigens may play a key role in various acute and chronic vasculitides, either by their ability to present foreign antigens, viral or self, or in the transplant situation.

Septic Shock

Septic shock is another pathologic state in which alterations in vascular cell functions contribute to the cardinal findings. Septic shock is a dreaded complication of infection with certain microorganisms, characteristically Gram negative bacteria. In this condition, circulatory disturbances result in maldistribution of blood flow and low blood pressure. Abnormalities of the blood clotting system lead to disseminated intravascular coagulation. Multiple organ systems fail as a consequence. The lungs develop adult respiratory distress syndrome, acute renal cortical necrosis due to the intravascular coagulation, and tubular necrosis due to hypotension cause kidney damage. A dramatic kind of hemorrhagic skin necrosis can occur, known as purpura fulminans, because of its rapidity of progression. A similar hemorrhagic thrombotic process occurring in the adrenal glands can cause the Waterhouse-Friderichsen syndrome characterized by acute adrenal insufficiency.

Cell wall lipopolysaccharides of Gram negative bacteria exhibit the endotoxin activity that can reproduce many of these effects associated with septic shock. Gram negative endotoxin has been used for decades in the laboratory to provoke the local and generalized Shwartzman reactions which share many pathologic features of clinical septic shock. Endotoxins are excellent stimuli for inducing the release of cytokines such as TNF or IL1 from leukocytes (*Loppnow et al., 1989*). These cytokine mediators promptly provoke the characteristic effects of Gram negative endotoxin that lead to septic shock by effects on vascular wall cells (*Libby *et al., 1989b*). Endotoxin can alter endothelial-blood compatibility by increasing tissue factor procoagulant (at least *in vitro*), decreasing fibrinolysis, increasing plasminogen activator inhibitor, decreasing thrombomodulin, and decreasing tissue plasminogen activator (*Bevilacqua et al., 1984; Emeis and Kooistra, 1986; Nawroth et al., 1986; Nawroth and Stern, 1986; Stern et al., 1985*). These coordinated changes increase the coagulant and thrombotic potential of the endothelial cell surface, and reduce its fibrinolytic capacity. Such cytokine-induced alterations in endothelial cell-blood compatibility underlie the development of disseminated intravascular coagulation in Gram negative sepsis.

Increased adhesivity of leukocytes can contribute to the leukostasis within microvessels in the lung, kidney, skin and other organs, and yield further vascular injury due to mediators derived from the inflammatory cells (e.g. toxic oxygen radicals, leukotrienes, platelet activating factor). The endotoxin- or cytokine-inducible endothelial-leukocyte adhesion mechanisms contribute to this leukocyte recruitment.

The vascular perturbations associated with Gram negative sepsis affect smooth muscle as well as endothelial cells. Altered smooth muscle cell vasomotor responses may contribute to the hemodynamic changes characteristic of this state. Interleukin 1 directly causes a decreased blood pressure and increased cardiac output *in vivo* (*Okusawa et al., 1988*). In acute experiments, IL1 neither contracts nor relaxes isolated arterial strips. However, rabbit arterial strips preincubated with IL1 show reduced responsiveness to the vasoconstrictor norepinephrine, a hormone whose levels are usually elevated in shock states (*Beasley et al., 1989*). Others have shown that administration of recombinant TNF reproduces many of the hemodynamic effects as well as the procoagulant effects of Gram negative endotoxin (*Tracey et al., 1986; Tracey et al., 1988*), and that administration of anti-TNF antibodies antagonize these pathologic effects (*Mathison et al., 1988*).

CONCLUSION

This discussion has presented a vascular biologist's view of certain immunopathologic mechanisms and their application to a few examples of disease processes. I have emphasized the interplay of mediators and leukocytes with vascular endothelial and smooth muscle cells. One of the main generalizations in this regard is that vascular wall cells can both produce and respond to locally acting growth factors and cytokines in ways that can contribute not only to the acute kind of responses but to the chronic fibrotic complications of prolonged or recurrent inflammation.

Pathologists of the last century already recognized the central role of blood vessels in immune and inflammatory responses. However, until recently, the underlying mechanisms remained obscure. The last few years have witnessed enormous progress in the unraveling of the relevant functions of vessel wall cells and the mediators that modulate the expression of these critical inducible or regulatable properties of endothelial and smooth muscle cells. Two important generalizations emerge from this recent progress. First, far from being passive bystanders in immunopathologic and other inflammatory processes, vessel wall cells actively participate in these responses. Second, pathologic states characteristically involve an alteration in dynamic balances in opposing aspects of vessel cell function, for example in the regulation of blood compatibility (e.g. coagulation vs. fibrinolysis) or of growth control (e.g. stimulation vs. inhibition). As basic vascular biologists continue to fill in gaps in our knowledge regarding these active functions of vascular endothelial and smooth muscle cells, more opportunities for rational intervention in clinically important disease states will emerge.

REFERENCES

Beasley, D., Cohen, R.A. and Levinsky, N.G. Interleukin 1 inhibits contraction of vascular smooth muscle. *J. Clin. Invest.* 83:331-335, 1989.

Benacerraf, B. Significance and biological function of class II MHC molecules. *Am. J. Pathol.* 120:333-343, 1985.

Bevilacqua, M.P., Pober, J.S., Majeau, G,R., Cotran, R.S. and Gimbrone, M.A., Jr. Interleukin-1 (IL-1) induces biosynthesis and cell surface expression of procoagulant activity in human vascular endothelial cells. *J. Exp. Med.* 160:618-623, 1984.

Bevilacqua, M.P., Pober, J.S., Majeau, G.R., Cotran, R.S. and Gimbrone, M.A., Jr. Interleukin-1 acts on cultured human vascular endothelium to increase the adhesion of polymorphonuclear leukocytes, monocytes and related leukocyte cell lines. *J. Clin. Invest.* 76:2003-2011, 1985.

Bevilacqua, M.P., Pober, J.S., Mendrick, D.L., Cotran, R.S. and Gimbrone, M.A. Jr. Identification of an inducible endothelial-leukocyte adhesion molecule. *Proc. Natl. Acad. Sci. U.S.A.* 84:9238-9242, 1987.

Bevilacqua, M.P., Schleef, R., Gimbrone, M.A., Jr. and Loskutoff, D.J. Regulation of the fibrinolytic system of cultured human vascular endothelium by IL-1. *J. Clin. Invest.* 78:587-591, 1986.

Bevilacqua, M.P., Stengelin, S., Gimbrone, M.A., Jr. and Seed, B. Endothelial Leukocyte Adhesion Molecule 1: An inducible receptor for neutrophils related to complement regulatory proteins and lectins. *Science* 243:1160-1165, 1989.

Collins, T., Korman, A.J., Wake, C.T., Boss, J.M., Kappes, D.J., Fiers, W., Ault, K.A., Gimbrone, Jr., M.A., Strominger, J.L. and Pober, J.S. Immune interferon activates multiple class II major histocompatibility complex genes and the associated invariant chain gene in human endothelial cells and dermal fibroblasts. *Proc. Natl. Acad. Sci. U.S.A.* 81:4917-4921, 1984.

Cotran, R.S., Gimbrone, M.A., Jr., Bevilacqua, M.P., Mendrick, D.L. and Pober, J.S. Induction and detection of a human endothelial activation antigen *in vivo*. *J. Exp. Med.* 164:661-666, 1986.

Dinarello, C.A. Interleukin 1 and its biologically related cytokines. *Adv. Immunol.* 44:153-205, 1989.

Dvorak, H.F., Galli, S.J. and Dvorak, A.M. Cellular and vascular manifestations of cell-mediated immunity. *Hum. Pathol.* 17:122-137, 1986.

Emeis, J.J, and Kooistra, T. Interleukin 1 and lipopolysaccharide induce an inhibitor of tissue-type plasminogen activator *in vivo* and in cultured endothelial cells. *J. Exp. Med.* 163:1260-1266, 1986.

Jirik, F.R., Podor, T.J., Hirano, T., Kishimoto, T., Loskutoff, D.J., Carson, D.A. and Lotz, M. Bacterial lipopolysaccharides and inflammatory mediators augment IL6 secretion by human endothelial cells. *J. Immunol.* 142:144-147, 1989.

Kishimoto, T. and Hirano, T. Molecular regulation of B lymphocyte response. *Ann. Rev. Immunol.* 6:485-512, 1988.

Kurt-Jones, E,A,, Fiers, W. and Pober, J.S. Membrane interleukin 1 induction on human endothelial cells and dermal fibroblasts. *J. Immunol.* 139:2317-2324, 1987.

Larsen, E., Celi, A., Gilbert, G,E., Furie, B.C., Erban, J.K., Bonfanti, R., Wagner, D.D. and Furie, B. PADGEM Protein: A receptor that mediates the interaction of activated platelets with neutrophils and monocytes. *Cell* 59:305-312, 1989.

Le, J. and Vilcek., J. Biology of Disease: Tumor necrosis factor and interleukin 1: Cytokines with multiple overlapping biological activities. *Laboratory Invest.* 56:234-248, 1987.

Le, J.M. and Vilcek, J. Interleukin 6: a multifunctional cytokine regulating immune reactions and the acute phase protein response. *Lab. Invest.* 61:588-602, 1989.

Libby, P. Regulation of smooth muscle proliferation in hyperplastic arterial lesions. *J. Vasc. Surgery* (In Press):l989.

Libby, P., Friedman, G.B. and Salomon, R.N. Cytokines as modulators of cell proliferation in fibrotic diseases. *Am. Rev. Respir. Dis.* 140:1114-1117, 1989a.

Libby, P., Loppnow, H. and Girard, J. The roles of endothelial cells in septic shock, p.25-31. In Beaufils, F., and Mercier, J.C. (eds.), *Chocs septiques en pediatrie*, Arnette, Paris, 1989b.

Libby, P., Salomon, R.N., Payne, D.D., Schoen, F.J. and Pober, J.S. Functions of vascular wall cells related to the development of transplantation-associated coronary arteriosclerosis. *Transplantation Proc.* 21:3677-3684, 1989c.

Libby, P., Warner, S.J.C. and Friedman, G.B. Interleukin-1: a mitogen for human vascular smooth muscle cells that induces the release of growth-inhibitory prostanoids. *J. Clin. Invest.* 88:487-498, 1988a.

Libby, P,, Warner, S.J.C., Salomon, R.N. and Birinyi, L.K. Production of platelet-derived growth factor-like mitogen by smooth-muscle cells from human atheromata. *New England J. Med.* 318:1493-1498, 1988b.

Loppnow, H., Brade, H., Durrbaum, I., Dinarello, C.A., Kusumoto, S., Rietschel, E.T. and Flad, H.-D. ILI induction capacity of defined lipopolysaccharide partial structures. *J. Immunol.* 142:3229-3238, 1989.

Loppnow, H. and Libby, P. Adult human vascular endothelial cells express the IL6 gene differentially in response to LPS or IL1. *Cell Immunol.* 122:493-503, 1989a.

Loppnow, H. and Libby, P. Comparative analysis of cytokine induction in human vascular endothelial and smooth muscle cells. *Lymph. Res.* 8:293-299, 1989b.

Loppnow, H. and Libby, P. Proliferating human smooth muscle cells express the gene for the multipotent cytokine interleukin 1 (IL6). *J. Clin. Invest.* 85:731-738, 1990.

Majack, R.A., Majesky, M.W. and Goodman, L.V. Role of PDGF-A expression in the control of vascular smooth muscle cell growth by transforming growth factor-beta. *J. Cell Biol.* 111:239-247, 1990.

Majno, G., Shea, S.M. and Leventhal, M. Endothelial contraction induced by histamine-type mediators: an electron microscopic study. *J. Cell Biol.* 42:647-72, 1969.

Mathison, J,C., Wolfson, E. and Ulevitch, R.J. Participation of tumor necrosis factor in the mediation of gram negative bacterial lipopolysaccharide-induced injury in rabbits. *J. Clin. Invest.* 81:1925-1937, 1988.

McEver, R.P., Beckstead, J.H., Moore, K.L., Marshall-Carlson, L. and Bainton, D.F. GMP--140, a platelet alpha-granule membrane protein, is also synthesized by vascular endothelial cells and is localized in Weibel-Palade bodies. *J. Clin. Invest.* 84:92-99, 1989.

Nawroth, P.P., Bank, I., Handley, D., Cassimeris, J., Chess, L. and Stern, D. Tumor necrosis factor/cachectin interacts with endothelial cell receptors to induce release of interleukin 1. *J. Exp. Med.* 163:1363-1375, 1986.

Nawroth, P.P, and Stern, D.M. Modulation of endothelial cell hemostatic properties by tumor necrosis factor. *J. Exp. Med.* 163:740-745, 1986.

Nossal, G.J.V. The basic components of the immune system. *N. Engl. J. Med.* 320-1325, 1987.

Okusawa, S., Gelfand, J.A., Ikejima, T., Connolly, R.J. and Dinarello, C.A. Interleukin 1 induces a shock-like State in rabbits. Synergism with tumor necrosis factor and the effect of cyclooxygenase inhibition. *J. Clin. Invest.* 81:1162-1172, 1988.

Pober, J.S., Collins, T., Gimbrone, M.A., Jr., Libby, P. and Reiss, C.S. Inducible expression of class II major histocompatibility complex antigens and the immunogenicity of vascular endothelium. *Transplantation* 41:141-146, 1986.

Pober, J.S. and Cotran, R.S. Cytokines and endothelial cell biology. *Physiol. Rev.* 70:427, 1990.

Raines, E.W., Dower, S.K. and Ross, R. Interleukin-1 mitogenic activity for fibroblasts and smooth muscle cells is due to PDGF-AA. *Science* 243:393-396, 1989.

Ross, R. The pathogenesis of atherosclerosis - an update. *New Engl. J. Med.* 314:488-500, 1986.

Ross, R. and Glomset, J.A. Atherosclerosis and the arterial smooth muscle cells. *Science.* 180:1332-1339, 1973.

Ryan, G.B, and Majno, G. Acute inflammation. A review. *Am. J. Pathol.* 86:183-276, 1977.

Samuelsson, B. Leukotrienes: mediators of immediate hypersensitivity reactions and inflammation. *Science* 220:568-75, 1983.

Sejersen, T., Betsholtz, C,, Sjolund, M., Heldin, C.H., Westermark, B, and Thyberg, J. Rat skeletal myoblasts and arterial smooth muscle cells express the gene for the A chain but not the gene for the B chain (C-sis) of platelet-derived growth factor (PDGF) and produce a PDGF-like protein. *Proc. Natl. Acad. Sci. U.S.A.* 83:6844-6848, 1986.

Sironi, M., Breviario, F., Proserpio, P., Biondi, A., Vecchi, A., Van Damme, J., Dejana, E. and Mantovani, A. IL1 stimulates IL6 production in endothelial cells. *J. Immunol.* 142:549-553, 1989.

Sjolund, M., Hedin, U., Sejersen, T,, Heldin, C.-H. and Thyberg, J. Arterial smooth muscle cells express platelet-derived growth factor (PDGF). A chain mRNA, secrete a PDGF-like mitogen, and bind exogenous PDGF in a phenotype- and growth state-dependent manner. *J. Cell Biol.* 106:403-413, 1988.

Stern, D.M., Bank, I., Nawroth, P.P., Cassimeris, J., Kisiel, W., Fenton, J.W.2., Dinarello, C., Chess, L. and Jaffe, E.A. Self-regulation of procoagulant events on the endothelial cell surface. *J. Exp. Med.* 162:1223-1235, 1985.

Thomas, K.A., Rios, C.M., Gimenez, G.G., DiSalvo, J., Bennett, C., Rodkey, J. and Fitzpatrick, S. Pure brain-derived acidic fibroblast growth factor is a potent angiogenic vascular endothelial cell mitogen with sequence homology to interleukin 1. *Proc. Natl. Acad. Sci. U.S.A.* 82:6409-6413, 1985.

Tracey, J.J., Beutler, B., Lowry, F., Merryweather, J., Wolpe, S., Milsark, W., Hariri, R.J., Fahey III, T.J., Zentella, A., Albert, J.D., Shires, G.T. and Cerami, A. Shock and tissue injury induced by recombinant human cachectin. *Science* 234:470-474, 1986.

Tracey, K.J., Wei, H., Manogue, K.R., Fong, Y., Hesse, D.G., Nguyen, H.T., Kuo, C.G., Beutler, B., Cotran, R.S., Cerami, A, and et al. Cachectin/tumor necrosis factor induces cachexia, anemia, and inflammation. *J. Exp. Med.* 167:1211-1127, 1988.

Wagner, C.R., Vetto, R.M. and Burger, D.R. Expression of I-region-associated antigen (Ia) and interleukin 1 by subcultured human endothelial cells. *Cell Immunol.* 93:91-104, 1985.

Warner, S.J.C,, Friedman, G.B. and Libby, P. Immune interferon inhibits proliferation and induces 2'-5'-oligoadenylate synthetase gene expression in human vascular smooth muscle cells. *J. Clin. Invest.* 83:1174-1182, 1989a.

Warner, S.J.C., Friedman, G.B. and Libby, P. Regulation of major histocompatibility gene expression in cultured human vascular smooth muscle cells. *Arteriosclerosis* 9:279-288, 1989b.

THE BIOCHEMISTRY, CELL AND MOLECULAR BIOLOGY OF TYPE 1 PLASMINOGEN ACTIVATOR INHIBITOR

David J. Loskutoff

Research Institute of Scripps Clinic
La Jolla, California 92037, U.S.A.

INTRODUCTION

Plasminogen activation (Figure 1) provides an important source of localized proteolytic activity in the blood during fibrinolysis, and in the tissues during a variety of normal and pathologic processes that in general involve invasive or degradative events (*Bachmann and Kruithof, 1984; Astrup, 1978; Hekman and Loskutoff, 1987*). Precise regulation of plasminogen activator (PA) activity thus constitutes a critical feature of many biological processes, and abnormalities in this regulation may lead to clinical problems (*Collen and Lijnen, 1986*). This control is achieved at many levels (Figure 2). For example, the synthesis of tissue-type PA (t-PA) and/or urokinase-like PA (u-PA) by endothelial cells of the vessel wall, is regulated by hormones, cytokines and growth factors (*Bachmann and Kruithof, 1984; Medcalf et al., 1988; Bevilacqua et al., 1986; Saksela et al., 1987*). Once these PAs are released from the cells, their activity is further modulated through interactions with cell surface receptors (*Roldan et al., 1990*), with components of the extracellular matrix (*Salonen et al., 1985; Silverstein et al., 1986*), and with plasma cofactors (*Hoylaerts et al., 1982*). However, the primary regulator of plasminogen activation may well be type 1 plasminogen activator inhibitor, or PAI-1 (*Sprengers and Kluft, 1987; Loskutoff et al., 1988*). PAI-1 appears to be the physiological inhibitor of t-PA, at least in the blood. This conclusion is based on the observation that naturally occurring complexes between single-chain t-PA and PAI-1, but not between t-PA and other inhibitors (*Hanss and Collen, 1987*), can be detected in the blood. In addition, the second order rate constant for the interaction of single-chain t-PA with PAI-1 is approximately 20,000 times higher than the constant derived for the interaction of t-PA with PAI-2 and protease nexin (*Colucci et al., 1986; Hekman and Loskutoff, 1988a*), two other molecules with PAI activity. Finally, a number of recent clinical observations relate alterations in plasma levels of PAI-1 with the pathogenesis of vascular disease. Thus, it is now clear that patients with an elevated level of PAI-1 in their plasma have, or are at risk to develop, thrombotic disease. For example, the levels of PAI-1 increase 10 to 20-fold during gram negative infections (*Colucci et al., 1985*), and disseminated intravascular coagulation is one of the major clinical manifestations of this condition. Elevated PAI activity also appears to be associated with deep vein thrombosis (*Paramo et al., 1985*), myocardial infarction (*Almer and Ohlin, 1987*) and pregnancy (*Kruithof et al., 1987b*), and has been detected in critically ill patients with unrelated diseases such as pancreatitis, malignancy, and liver disease (*Juhan-Vague et al., 1984*). Diseases of the liver may be expected to lead to altered levels of plasma PAI-1 since PAI-1 is cleared from the circulation by the liver (*Emeis, 1985*) and may be produced by hepatocytes (*Sprengers et al., 1985; Risberg et al., 1987*). High PAI activity also was detected in the blood of obese (*Vague et al., 1987*) and elderly (*Kruithof et al., 1987a; Aillaud et al., 1986*) individuals. PAI-1 activity in plasma frequently increases after major surgery in response to acute trauma (*Kluft et al., 1985*), suggesting that PAI-1 is an acute-phase reactant. The importance of PAI-1 regulation is further emphasized by the finding that patients with decreased PAI-1 have bleeding disorders (*Schleef et al., 1989; Francis et al.,*

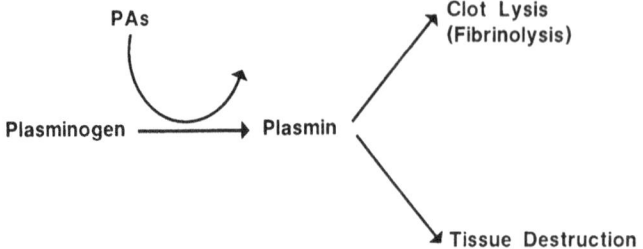

FIGURE 1: Schematic Representation of Plasminogen Activation

TABLE I: PAI-1 Concentration in Various Biological Samples

SAMPLE	PAI-1 CONCENTRATION (ng/ml)	PROTEIN CONCENTRATION (mg/ml)
Plasma (Normal)	5 - 20	70
Plasma (Sepsis)	50 - 200	
Platelets (10^9)	100 - 300	
Endothelial cells*	1000 - 5000	0.05

*Also human fibrosarcoma (HT - 1080) and melanoma (MJZJ) cells

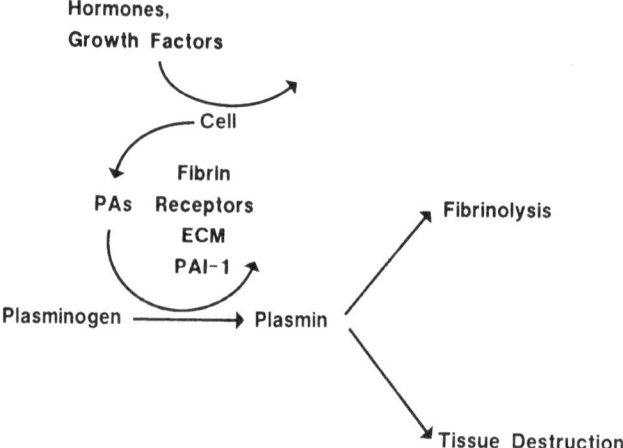

FIGURE 2: Regulation of Plasminogen Activation

1986). Finally, activated protein C stimulates the fibrinolytic activity of endothelial cells and plasma quite dramatically, and does so not by increasing t-PA or u-PA production, or by dissociating preexisting t-PA/PAI-1 complexes, but rather by decreasing PAI-1 itself (*van Hinsbergh et al., 1985; Sakata et al., 1986; Loskutoff et al., 1988*).

These observations thus suggest that the net fibrinolytic activity of blood or cells is a reflection of the balance between PAs and PAI-1, and that changes in any of these molecules may lead to thrombotic problems or to a bleeding diathesis. It follows that understanding the nature of signals that regulate PAI-1 itself, and delineating their mechanism of action, will provide new insights into the fibrinolytic system and its regulation in health and disease. Moreover, this information may lead to the development of new thrombolytic therapies based entirely on PAI-1 (i.e., on the identification of drugs that chronically lower PAI-1 and thus effectively elevate t-PA). For these reasons, we have made a major effort to understand PAI-1 biosynthesis, both at the cellular and molecular level, and much of this review will be devoted to summarizing our results using cultured bovine aortic endothelial cells (BAEs) as a model. However, before I begin this discussion on regulation, I would like to briefly summarize some of the more relevant and interesting properties of PAI-1 itself (*Sprengers and Kluft, 1987; Loskutoff et al., 1988*).

PROPERTIES OF PAI-1

The explosive growth of information on the biochemistry, cell and molecular biology of PAI-1 can be traced back to the decision by a number of the early workers to study PAI-1 production by cultured cells, including endothelial cells. This was a critical decision because while PAI-1 is a trace protein in plasma (Table I), it is a major biosynthetic product of endothelial cells, and may represent 10-15% of the protein secreted by these cells into the serum-free conditioned medium. The high rate of production of PAI-1 by these cells, together with the relative simplicity of conditioned medium compared to plasma, made the purification (*van Mourik et al., 1984; Wagner and Binder, 1986; Andreasen et al., 1986a*) and cloning (*Ny et al., 1986; Pannekoek et al., 1986; Ginsburg et al., 1986; Andreasen et al., 1986b*) of PAI-1 relatively straight forward and led to the development of monoclonal and polyclonal antibodies that could be used in assays to quantitate PAI-1 not only in cells, but also in plasma.

The structure of PAI-1 has been investigated using both the purified protein and its cDNA (*Loskutoff et al., 1988*). These studies reveal that PAI-1 is a single-chain glycoprotein with an approximate molecular weight of 50,000 and an isoelectric point of 4.5-5.0. More precisely, the cDNA revealed that the mature human protein consists of 379 amino acids, three of which represent potential sites for the attachment of n-linked carbohydrate side chains. Carbohydrates constitute approximately 13% of the mass of the molecule. Comparison of the sequence of the PAI-1 cDNA with that of α_1-antitrypsin indicates that its reactive center is located at the carboxyterminal end of the molecule, at Arg_{346}-Met_{347} (*Andreasen et al., 1986b*), and that it is a member of the serine proteinase inhibitor (Serpin) gene family. It is noteworthy that PAI-1 lacks cysteine residues, a property that may account for its stability under reducing conditions (*van Mourik et al., 1984*).

Even at the early times during the initial characterization of PAI-1, it was obvious that this was an unusual protease inhibitor. For example, it appeared to migrate with β-mobility when analyzed by agarose zone electrophoresis (*Erickson et al., 1986*), while most plasma protease inhibitors displayed α-mobility when analyzed in this way. Moreover, its activity was still apparent after SDS-PAGE and reduction (*van Mourik et al., 1984*), treatments that irreversibly inactivate most other protease inhibitors. In spite of this unusual stability, the molecule was rapidly and efficiently inactivated by oxidants (*Lawrence and Loskutoff, 1986*). This sensitivity to oxidants may represent a means by which activated neutrophils locally inactivate PAI-1 during inflammatory processes when large amounts of oxidants are liberated. Finally, an unexpected but rather consistent finding is that a latent form of PAI-1 exists in the conditioned medium of a large variety of cells (*reviewed in Loskutoff et al., 1988*) and also may be present in platelets (*Booth et al., 1988*). The basis for this conclusion is that while PAI-1 antigen is readily detected in these samples, it frequently has less than 1% of its theoretical activity. However, the inactive form can be converted into the active inhibitor by treatment with denaturants (*Hekman and Loskutoff, 1985*), heat (*Katagiri et al., 1988*), negatively charged phospholipids (*Lambers et al., 1987*) and vitronectin (*Wun et al., 1989*). These observations have led to the hypothesis that PAI-1 may be produced and stored in this latent (pro-inhibitor) form, and as such, may represent a large potential reservoir of inhibitory activity.

This hypothesis lacks experimental support since biologically relevant activators of latent PAI-1 in plasma and cells have not yet been identified. In addition, the time required for vitronectin to convert latent PAI-1 into its active form is many days (*Wun et al., 1989*), but the clearance time for PAI-1 *in vivo* is only minutes (*Emeis, 1985*). The available data argue against this hypothesis of a pro-inhibitor. Thus, all detectable intracellular PAI-1 appears to be active (*Levin and Santell, 1987b; Kooistra et al., 1986b*), suggesting that it is synthesized in the active, not latent form. Moreover, intracellular PAI-1 is quite labile, decaying into the latent form with a half-life of 2-3 hours after it is secreted (*Hekman and Loskutoff, 1988b*). These observations are most consistent with the simple idea (*Loskutoff et al., 1988*) that PAI-1 is produced in the active form but is inherently unstable and rapidly decays into the inactive form once secreted, perhaps because of conformational changes in the molecule. According to this hypothesis, activating agents alter the 3-dimensional structure of the molecule, re-exposing its reactive center. Interestingly, PAI-1 is present in the extracellular matrix (ECM) of a variety of cells (*Laiho et al., 1986; Knudsen et al., 1987; Levin and Santell, 1987a; Pollanen et al., 1987; Mimuro et al., 1987*), where it appears to be distributed as a rather homogeneous carpet under the cells (*Pollanen et al., 1987*). In contrast to the situation in conditioned medium, the majority of PAI-1 in extracellular matrix (*Mimuro et al., 1987b*) and plasma (*Chmielewska et al., 1987*) is active, not latent. These samples contain molecules that specifically bind to PAI-1 and protect it from the spontaneous loss of activity that occurs in solution, presumably by stabilizing it in the active configuration.

These results indicate that the PAI-1 binding protein is an important regulator of PAI-1 activity, and we thus set out to purify it. We used bovine plasma as our starting material, both, because it was difficult to obtain enough protein from matrix and because PAI-1 in plasma is also bound to a high molecular weight protein (*Wiman et al., 1988; Declerck et al., 1988*). We employed affinity chromatography using PAI-1 immobilized on Sepharose to isolate a polypeptide greatly enriched in PAI-1 binding protein activity (*Mimuro and Loskutoff, 1989*) and showed that it was vitronectin, an adhesive glycoprotein also known as serum-spreading factor (*Hayman et al., 1983*) and the S-protein of the complement system (*Dahlback and Podack, 1985*). A PAI-1 binding protein was purified recently from human plasma and also shown to be Vn (vitronectin) (*Declerck et al., 1988; Wiman et al., 1988*). The primary PAI-1 binding protein in ECM is also Vn (*Seiffert & Loskutoff, 1990*). Interestingly, this Vn is not synthesized by these cells but rather is derived from the serum in the growth medium (*Seiffert & Loskutoff, 1990*). The binding of PAI-1 to Vn is relatively specific since PAI-1 does not bind to laminin, fibronectin, type IV collagen, or other matrix proteins (*Seiffert & Loskutoff, 1990*).

The biological significance of the interaction between PAI-1 and vitronectin in plasma remains unclear. For example, it was reported that plasma vitronectin stabilizes PAI-1 in solution (*Declerck et al., 1988*), increasing its half-life from 2 hours to 4 hours, a significant prolongation. However, since the clearance time for PAI-1 *in vivo* is only 5-10 minutes (*Emeis, 1985*), the increase does not appear to be biologically relevant. Vn has been detected in tissues (*Hayman et al., 1983*) and the subendothelium of the vessel wall (*Guettier et al., 1989; Niculescu et al., 1989*) suggesting that the major function of the Vn/PAI-1 interaction may be to localize PAI-1 in tissues, thus protecting them from cellular proteases.

The human PAI-1 gene is located on chromosome number 7 (*Ginsburg et al., 1986*), close to the locus for cystic fibrosis (*Klinger et al., 1987*). The entire PAI-1 gene was isolated (Figure 3) and shown to be 12.2 Kb in length and to be organized into 8 introns and 9 exons (*Loskutoff et al., 1987; Bosma et al., 1988*). It specifies two distinct transcripts of 3.2 and 2.3 Kb which are co-linear from their 5´-end and appear to be formed by alternative polyadenylation. The 3.2 Kb cDNA contains a small 5´ non-translated region, a region that codes for a 23 amino acid signal peptide, the coding region, and a rather large, 3´ non-translated region which accounts for over 50% of the cDNA (*Ny et al., 1986; Pannekoek et al., 1986; Ginsburg et al., 1986; Andreasen et al., 1986b*).

REGULATION OF PAI-1 BIOSYNTHESIS

PAI-1 is interesting, not only because it is the primary inhibitor of t-PA in the blood, but also because it is a major biosynthetic product of endothelial cells and its synthesis is highly regulated (*Loskutoff et al., 1988*). For these reasons, a number of investigators have been

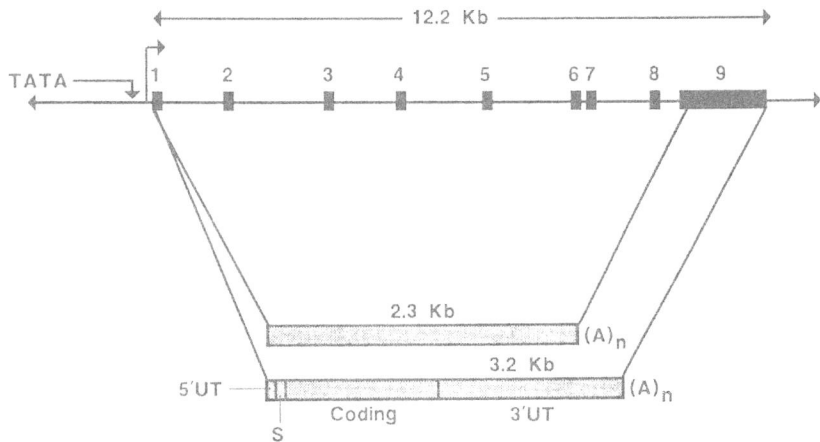

FIGURE 3: Structure of the Human PAI-1 Gene

trying to define the signals that modulate PAI-1 production by various cells and to understand their mechanism of action. We have focused on molecules likely to be released by activated leukocytes and platelets. In initial experiments, we compared the effects of the immune monokines interleukin 1 (IL-1) and tumor necrosis factor (TNF) on PAI-1 synthesis in human umbilical vein endothelial cells (*Schleef et al., 1988*). Both molecules increased PAI-1 antigen levels by 5 to 6-fold as compared to untreated controls, and both also appeared to suppress t-PA production by these cells. Northern blot analysis showed that the cytokine-mediated increases in PAI-1 antigen occur as a result of similar increases in the steady state level of PAI-1 mRNA. These agents also induce tissue factor, a protein with procoagulant activity (*Bevilacqua et al., 1986*). The elaboration of cytokines at inflammatory sites thus may suppress the fibrinolytic system and activate the coagulation system, and in the process, convert the endothelium from a tissue that maintains vessel patency, to one which promotes clot formation. Fibrin deposition is one of the hallmarks of the inflammatory process *in vivo*.

The endothelium *in vivo* is a non-growing, quiescent tissue. However, it can be induced to undergo explosive growth at sites of vascular injury, presumably in response to growth factors elaborated from platelets and leukocytes. Thus, growth factors represent another group of molecules that may modulate PAI-1 biosynthesis. We examined the effect of a variety of growth factors on PAI-1 synthesis and secretion using cultured bovine aortic endothelial cells as a model (*Mimuro and Loskutoff, 1987a*). Of those tested, only transforming growth factor beta (TGFβ), a growth modulator from platelets, induced PAI-1. Platelet-derived growth factor (PDGF) and epidermal growth factor were without effect in this system. The TGFβ effect was dose-dependent with a half-maximal response at 0.5-1.0 ng/ml, and increased secreted PAI-1 by over 100-fold. These results with TGFβ thus raised the possibility that activated platelets may be able to suppress the fibrinolytic system of the vessel wall by releasing TGFβ which in turn stimulates PAI-1 biosynthesis in endothelial cells. To test this possibility, we added thrombin-induced platelet releasates to endothelial cells in the presence of ^{35}S-methionine, and looked for changes in PAI-1 biosynthesis (*Slivka and Loskutoff, 1991a*). We found that the platelets increased the rate of PAI-1 biosynthesis and did so in a dose and time dependent manner. This increase could be blocked by pre-treating the platelet releasates with antisera to TGFβ but not with pre-immune sera. Thus, platelets can communicate with endothelial cells to modulate their fibrinolytic system. This result suggests that platelet-rich thrombi may have high local concentrations of PAI-1, both as a direct consequence of its release from platelet α-granules, and because its production by surrounding endothelial cells is stimulated by platelet-derived TGFβ. These considerations may account for the recent observation that platelet-rich thrombi are frequently resistant to thrombolytic therapy with t-PA (*Yasuda et al., 1990*).

The induction of PAI-1 antigen by TGFβ is preceded by an increase in the steady state level of PAI-1 mRNA (*Sawdey et al., 1989*), and the kinetics of induction are similar to those obtained with TNF and endotoxin (or LPS), a molecule shown previously (*Colucci et al., 1985*) to stimulate PAI-1 synthesis by endothelial cells. This induction of PAI-1 mRNA does not appear to require new protein synthesis since each agent induces PAI-1 mRNA in the

TABLE II: Agents that Modify PAI-1 Production by Various Cells

Dexamethasone	FTO2B; HT-1080
TGF	BAEs; HEP 3B; LTK; A549
IL-1	HUVECs
TNF	BAEs; HUVECs; HT-1080
LPS	BAEs; HUVECs; FT02B
IL-6	FTO2B; HEP 3B
Insulin	FTO2B; HEP 3B; HEP G2
Estradiol	MCF-7
Serum	BAEs; HUVECs, LTK
PMA	BAEs; FTO2B
Forskolin	BAEs

presence of cycloheximide (*Sawdey et al., 1989*), an inhibitor of protein synthesis. Although the half-life of PAI-1 mRNA in BAEs was short (approximately 2 1/2 hours), it was not altered by any of these agents, suggesting that they do not act by increasing the half-life of PAI-1 mRNA. And finally, nuclear run-on experiments (*Sawdey et al., 1989*) indicate that the increases in PAI-1 mRNA caused by these agents result primarily from an increase in the rate of transcription of the PAI-1 gene. Based on these observations, our working hypothesis is that these agents act directly on the gene to increase the rate of transcription of PAI-1 mRNA.

The intracellular mechanism of PAI-1 induction by these various agents is largely unknown. The observation that the tumor promoter, phorbol myristate acetate (PMA), stimulates PAI-1 production by some cells (*Thalacker and Nilsen-Hamilton, 1987*) suggests that the protein kinase C pathway may be involved since PMA activates protein kinase C. However, our preliminary results (*Slivka and Loskutoff, 1990b*) suggest that these effects are not mediated by activation of the protein kinase C pathway or by decreases in cyclic AMP.

The large diversity of molecules that alter PAI-1 biosynthesis (*Loskutoff et al., 1988; Table II*) implies that the regulatory region of the PAI-1 gene must be unusually complex, containing DNA elements responsive to all of these molecules. Thus, in order to study PAI-1 gene expression in detail, we have begun to characterize the 5'-flanking region (*van Zonneveld et al., 1988*) since it is likely to contain regulatory information. Initially, we focused on an 874 bp EcoRI-Hind III fragment immediately upstream of the cDNA. This fragment contained the transcription initiation (cap) site at an "A" nucleotide located 145 bp upstream of the methionine initiation codon, and a perfect TATA box was located at -23 to -28, the consensus distance from the cap site. These results suggest that this region contains the PAI-1 promoter (*Ptashne, 1988*). Transfection experiments were performed to verify this conclusion experimentally, and at the same time, to begin to identify those DNA elements in it that might be important for regulation by some of the above agents. In these experiments, we used restriction fragments from this region that were fused to the promoterless firefly luciferase gene as a reporter gene (Figure 4). We asked whether the cloned fragments provide the information to activate the reporter gene as indicated by the generation of luciferase activity and light. We found (*van Zonneveld et al., 1988*) that fragments 187 bp and 1500 bp in length exhibited high promoter activity in mouse fibroblasts and in BAEs, consistent with the observations that these cells also express PAI-1 in culture. HeLa cells do not express significant amounts of PAI-1 under normal conditions, and very little promoter activity was detected in transfection experiments using these cells. On the other hand, rat hepatoma cells, which normally do not make PAI-1, can be induced to synthesize high amounts of PAI-1 by the synthetic glucocorticoid, dexamethasone, and dexamethasone induced the PAI-1 promoter activity of both fragments transfected into the rat hepatoma cell line, FT02B. Taken together, these results indicate the presence of a strong, and relatively tissue-specific promoter within the first 187 bp of the 5´ flanking region of the PAI-1 gene. This promoter is inducible by dexamethasone in rat FT02B hepatoma cells.

There appear to be two regions in the promoter that mediate a dexamethasone dependent rise in promoter efficiency (*van Zonneveld et al., 1988*). The first, located in the region between nucleotides -90 and +75, raises promoter activity about ten fold. The second is

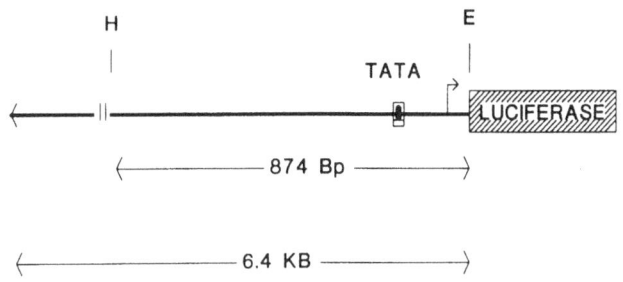

FIGURE 4: Analysis of the 5´-Flanking Region of the PAI-1 Gene

located in the sequence -800 to -549 and mediates an additional 4-fold increase in promoter efficiency. We have been able to establish that the proximal element functions in both the normal and reversed orientation suggesting that it is an authentic glucocorticoid responsive enhancer. Finally, we have made a series of overlapping substitution mutants through the proximal region and will employ them to perform transfection experiments as well as DNA "footprinting" and gel shift experiments to more precisely define the relevant sequences. We are taking similar approaches to localize the CIS-acting elements responsible for the TGFβ and TNF effects. Such experiments should provide important clues concerning the mechanism of PAI-1 regulation *in vivo*.

REFERENCES

Aillaud, M.F., Pignol, F., Alessi, M.C., Harle, J.R., Escande, M., Mongin, M., Juhan-Vague, I. Increase in plasma concentration of plasminogen activator inhibitor, fibrinogen, von Willebrand Factor, Factor VIIIc, and in erythrocyte sedimentation rate with age. *Thromb. Haemost.* 55:330-332, 1986.

Almer, L., Ohlin, H. Elevated levels of the rapid inhibitor of plasminogen activator (t-PAI) in acute myocardial infarction. *Thromb. Res.* 47:335-339, 1987.

Andreasen, P.A., Nielsen, L.S., Kristensen, P., Grondahl-Hansen, J., Skriver, L., Dano, K. Plasminogen activator inhibitor from human fibrosarcoma cells binds urokinase-type plasminogen activator, but not its proenzyme. *J. Biol. Chem.* 261:7644-7651, 1986a.

Andreasen, P.A., Riccio, A., Welinder, K.G., Sartorio, R., Nielsen, L.S., Oppenheimer, C., Blasi, F., Dano, K. Plasminogen activator inhibitor type-I: Reactive center and amino terminal heterogeneity determined by protein and cDNA sequencing. *FEBS Lett.* 209:213-218, 1986b.

Astrup, T. Fibrinolysis. An overview. In: Davidson, J.F., Rowan, R.M., Samama, M.M., Desnoyers, P.C. (eds), Progress in *Chemical Fibrinolysis and Thrombolysis*. New York, Raven Press Vol 3:1-89, 1978.

Bachmann, F., Kruithof, E.K.O. Tissue plasminogen activator: Chemical and physiological aspects. *Sem. Thromb. Hemostas.* 10: 6-17, 1984.

Bevilacqua, M.P., Schleef, R.R., Gimbrone, M.A., Jr., Loskutoff, D.J. Regulation of the fibrinolytic system of cultured human vascular endothelium by interleukin 1. *J. Clin. Invest.* 78:587-591, 1986.

Booth, N.A., Simpson, A.J., Croll, A., Bennett, B. and MacGregor, I.R. Plasminogen activator inhibitor (PAI-1) in plasma and platelets. *Br. J. Haematol.* 70:327-333, 1988.

Bosma, P.J., van den Berg, E.V., and Kooistra, T. Human plasminogen activator inhibitor-1 gene. *J. Biol. Chem.* 263:9129-9141, 1988.

Chmielewska, J., Carlsson, T., Urden, G., Wiman, B. On the relationship between different molecular forms of the fast inhibitor of tissue plasminogen activator. *Fibrinolysis* 1:67-73, 1987.

Collen, D., Lijnen, H.R. In: Davis, S. (Ed, CRC Critical Reviews in *Oncology/Hematology*). Boca Raton, CRC Press Vol 4 (3), p 249-301, 1986.

Colucci, M., Paramo, J.A., Collen, D. Generation in plasma of a fast-acting inhibitor of plasminogen activator in response to endotoxin stimulation. *J. Clin. Invest.* 75:818-824, 1985.

Colucci, M., Paramo, J.A., Collen, D. Inhibition of one-chain and two-chain forms of human tissue-type plasminogen activator by the fast-acting inhibitor of plasminogen activator *in vitro* and *in vivo*. *J. Lab. Clin. Med.* 108:53-59, 1986.

Dahlback, B., Podack, E.R. Characterization of human S protein, an inhibitor of the membrane attack complex of complement. Demonstration of a free reactive thiol group. *Biochem.* 24:2368-2374, 1985.

Declerck, P.J., De Mol, M., Alessi, M-C., Baudner, S., Paques, E-P., Preissner, K.T., Muller-Berghaus, G., Collen, D. Purification and characterization of a plasminogen activator inhibitor 1 binding protein from human plasma. *J. Biol. Chem.* 263:15454-15461, 1988.

Emeis, J.J. Fast hepatic clearance of plasminogen activator inhibitor. *Thromb. Haemost.* 54:230, 1985.

Erickson, L.A., Hekman, C.M., Loskutoff, D.J. Denaturant-induced stimulation of the β-migrating plasminogen activator inhibitor in endothelial cells and serum. *Blood* 68:1298-1305, 1986.

Francis, R.B. Jr., Liebman, H., Koehler, S., Feinstein, D.I. Accelerated fibrinolysis in amyloidosis. Specific binding of tissue plasminogen activator inhibitor by an amyloidogenic monoclonal IgG. *Blood* 68:333a, 1986.

Ginsburg, D., Zeheb, R., Yang, A.Y., Rafferty, U.M., Andreasen, P.A., Nielsen, L., Dano, K., Lebo, R.V., Gelehrter, T.D. cDNA cloning of human plasminogen activator inhibitor from endothelial cells. *J. Clin. Invest.* 78:1673-1680, 1986.

Guettier, C., Hinglais, N., Bruneval, P., Kazatchkine, M., Bariety, J., and Camilleri, J-P. Immunohistochemical localization of S protein/vitronectin in human atherosclerotic versus arteriosclerotic arteries. *Virchows Archiv. A. Pathol. Anat.*, 414, 309-313, 1989.

Hanss, M., Collen, D. Secretion of tissue type plasminogen activator and plasminogen activator inhibitor by cultured human endothelial cells; modulation by thrombin, endotoxin, and histamine. *J. Lab. Clin. Med.* 109:97-104, 1987.

Hayman, E.G., Pierschbacher, M.D., Ohgren, Y., and Ruoslahti, E. Serum spreading factor (vitronectin) is present at the cell surface and in tissues. *Proc. Natl. Acad. Sci. USA* 80:4003-4007, 1983.

Hekman, C.M., Loskutoff, D.J. Endothelial cells produce a latent inhibitor of plasminogen activators that can be activated by denaturants. *J. Biol. Chem.* 260:11581-11587, 1985.

Hekman, C.M., Loskutoff, D.J. Fibrinolytic pathways and the endothelium. *Semin. Thromb. Hemost.* 13:514-527, 1987.

Hekman, C.M., Loskutoff, D.J. Kinetic analysis of the interactions between plasminogen activator inhibitor 1 and both urokinase and tissue plasminogen activator. *Arch. Biochem. Biophys.* 262:199-210, 1988a.

Hekman, C.M., Loskutoff, D.J. Bovine plasminogen activator inhibitor 1: Specificity determinations and comparison of the active, latent and guanidine-activated forms. *Biochem.* 27:2911-2918, 1988b.

Hoylaerts, M., Rijken, D.C., Lijnen, H.R., Collen, D. Kinetics of the activation of plasminogen by human tissue plasminogen activator. Role of fibrin. *J. Biol. Chem.* 257:2912-2919, 1982.

Juhan-Vague, I., Moerman, B., De Cock, F., Aillaud, M.F., Collen, D. Plasma levels of a specific inhibitor of tissue-type plasminogen activator (and urokinase) in normal and pathological conditions. *Thromb. Res.* 33:523-530, 1984.

Katagiri, K., Okada, K., Hattori, H., and Yano, M. Bovine endothelial cell plasminogen activator inhibitor. *Eur. J. Biochem.* 176:81-87, 1988.

Klinger, K.W., Winquist, R., Riccio, A., Andreasen, P.A., Sartorio, R., Nielsen, L.S., Stuart, N., Stanislovitis, P., Watkins, P., Douglas, R., Grzeschik, K-H., Alitalo, K., Blasi, F., Dano, K. Plasminogen activator inhibitor type 1 gene is located at region q21.3-q22 of chromosome 7 and genetically linked with cystic fibrosis. *Proc. Natl. Acad. Sci. U.S.A.* 84:8548-8552, 1987.

Kluft, C., Verheijen, J.H., Jie, A.F.H., Rijken, D.C., Preston, F.E., Sue-Ling, H.M., Jespersen, J., Aasen, A.D. The postoperative fibrinolytic shutdown: a rapidly reverting acute phase pattern for the fast-acting inhibitor of tissue-type plasminogen activator after trauma. *Scand. J. Clin. Lab. Invest.* 45:605-610, 1985.

Knudsen, B.S., Harpel, P.C., Nachman, R.L. Plasminogen activator inhibitor is associated with the extracellular matrix of cultured bovine smooth muscle cells. *J. Clin. Invest.* 80:1082-1089, 1987.

Kooistra, T., Sprengers, E.D., van Hinsbergh, V.W.M. Rapid inactivation of the plasminogen-activator inhibitor upon secretion from cultured human endothelial cells. *Biochem. J.* 239:497-503, 1986.

Kruithof, E.K.O., Nicolosa, G., Bachmann, F. Plasminogen activator inhibitor 1: Development of a radioimmunoassay and observations on its plasma concentration during venous occlusion and after platelet aggregation. *Blood* 70: 1645-1653, 1987a.

Kruithof, E.K.O., Tran-Thang, C., Gudinchet, A., Hauert, J., Nicolosa, G., Genton, C., Welti, H., Bachmann, F. Fibrinolysis in pregnancy. A study of plasminogen activator inhibitors. *Blood* 69:460-466, 1987b.

Laiho, M., Saksela, O., Andreasen, P.A., Keski-Oja, J. Enhanced production and extracellular deposition of the endothelial-type plasminogen activator inhibitor in cultured human lung fibroblasts by transforming growth factor-β. *J. Cell. Biol.* 103:2403-2410, 1986.

Lambers, J.W.J., Cammenga, M., Konig, B., Pannekoek, H., van Mourik, J.A. Activation of human endothelial type plasminogen activator inhibitor (PAI-1) by negatively charged phospholipids. *J. Biol. Chem.* 262:17492-17496, 1987.

Lawrence, D.A., Loskutoff, D.J. Inactivation of plasminogen activator inhibitor by oxidants. *Biochem.* 25:6351-6355, 1986.

Levin, E.G., Santell, L. Association of plasminogen activator inhibitor (PAI-1) with the growth substratum and membrane of human endothelial cells. *J. Cell Biol.* 105:2543-2549, 1987a.

Levin, E.G., Santell, L. Conversion of active to latent plasminogen activator inhibitor from human endothelial cells. *Blood* 70:1090-1098, 1987b.

Loskutoff, D.J., Linders, M., Keijer, J., Veerman, H., van Heerikhuizen, H., Pannekoek, H. The structure of the human plasminogen activator inhibitor 1 gene: Non-random distribution of introns. *Biochem.* 26:3763-3768, 1987.

Loskutoff, D.J., Sawdey, M., Mimuro, J. Type 1 plasminogen activator. In: *Progress in Hemostas Thromb*, Coller, B.S. (ed., WB Saunders Company, Philadelphia) vol. 9, p 87-115, 1988.

Medcalf, R.L., Van den Berg, E., and Schleuning, W-D. Glucocorticoid-modulated gene expression of tissue-and urinary-type plasminogen activator and plasminogen activator inhibitor 1 and 2. *J. Cell Biol.* 106:971-978, 1988.

Mimuro, J., Loskutoff, D.J. Effect of transforming growth factor-β (TGFβ) on the fibrinolytic system of cultured bovine aortic endothelial cells (BAEs). *Thromb. Haemost.* 58:1647, 1987a.

Mimuro, J., Schleef, R. R., Loskutoff, D. J. The extracellular matrix of cultured bovine aortic endothelial cells contains functionally active type 1 plasminogen activator inhibitor. *Blood* 70:721-728, 1987b.

Mimuro, J., Loskutoff, D. Purification of a protein from bovine plasma that binds to type 1 plasminogen activator and prevents its interaction with extracellular matrix. Evidence that the protein is vitronectin. *J. Biol. Chem.*, 264:936-939, 1989.

Niculescu, F., Rus, H. G., Porutiu, D., Ghiurca, V., and Vlaicu, R. Immunoelectron-microscopic localization of S-protein/vitronectin in human atherosclerotic wall. *Atherosclerosis*, 78:197-203, 1989.

Ny, T., Sawdey, M., Lawrence, D.A., Millan, J.L., Loskutoff, D.J. Cloning and sequence of a cDNA coding for the human β-migrating endothelial-cell-type plasminogen activator inhibitor. *Proc. Natl. Acad. Sci. USA* 83:6776-6780, 1986.

Pannekoek, H., Veerman, H., Lambers, H., Diergaarde, P., Verweij, C.L., van Zonneveld, A.J., van Mourik, J.A. Endothelial plasminogen activator inhibitor (PAI): A new member of serpin gene family. *EMBO J.* 5:2539-2544, 1986.

Paramo, J.A., DeBoer, A., Colucci, M., Jonker, J.J.C. Plasminogen activator inhibitor (PA-inhibitor) activity in the blood of patients with deep vein thrombosis. *Thromb. Haemost.* 54:725, 1985.

Pollanen, J., Saksela, O., Salonen, E.M., Andreason, P., Nielsen, L., Dano, K., Vaheri, A. Distinct localizations of urokinase-type plasminogen activator and its type 1 inhibitor under cultured human fibroblast and sarcoma cells. *J. Cell Biol.* 104:1085-1096, 1987.

Ptashne, M. How eukaryotic transcriptional activators work. *Nature* 335:683-689, 1988.

Risberg, B., Hansson, G.K., Eriksson, E., Wiman, B. Immunohistochemical localization of plasminogen activator inhibitor (PAI) in tissue. *Thromb. Haemost.* 58:446, 1987.

Roldan, A. L., Cubellis, M. V., Masucci, M. T., Behrendt, N., Lund, L. R., Dano, K., Appella, E., and Blasi, F. Cloning and expression of the receptor for human urokinase plasminogen activator, a central molecule in cell surface, plasmin dependent proteolysis. *EMBO J.* 9:467-474, 1990.

Sakata, Y., Loskutoff, D.J., Gladson, C.L., Hekman, C.M., Griffin, J.H. Mechanism of protein C-dependent clot lysis: Role of plasminogen activator inhibitor. *Blood* 68:1218-1223, 1986.

Saksela, O., Moscatelli, D., Rifkin, D.B. The opposing effects of basic fibroblast growth factor and transforming growth factor beta on the regulation of plasminogen activator activity in capillary endothelial cells. *J. Cell Biol.* 105:957-963, 1987.

Salonen, E., Saksela, O., Vartio, T., Vaheri, A., Nielsen, L., Zeuthen, J. Plasminogen and tissue-type plasminogen activator bind to immobilized fibronectin. *J. Biol. Chem.* 260:12302-12307, 1985.

Sawdey, M., Podor, T.J., Loskutoff, D.J. Regulation of type 1 plasminogen activator inhibitor gene expression in cultured bovine aortic endothelial cells: Induction by transforming growth factor-β, lipopolysaccharide, and tumor necrosis factor-α. *J. Biol. Chem.* 264:10396-10401, 1989.

Schleef, R.R., Bevilacqua, M.P., Sawdey, M., Gimbrone, M.A. Jr., Loskutoff, D.J. Interleukin 1 (IL-1) and tumor necrosis factor (TNF) activation of vascular endothelium: Effects on plasminogen activator inhibitor (PAI-1) and tissue type plasminogen activator (tPA). *J. Biol. Chem.* 263:5797-5803, 1988.

Schleef, R.R., Higgins, D.L., Pillemer, E., and Levitt, L.J. Bleeding diathesis due to decreased functional activity of type 1 plasminogen activator inhibitor. *J. Clin. Invest.* 83:1747-1752, 1989.

Seiffert, D. and Loskutoff, D.J. Serum-derived vitronectin influences the pericellular distribution of type 1 plasminogen activator inhibitor. *J. Cell Biol.* 111:1283-1291, 1990.

Silverstein, R., Harpel, P., Nachman, R. Tissue plasminogen activator and urokinase enhance the binding of plasminogen to thrombospondin. *J. Biol. Chem.* 261:9959-9965, 1986.

Slivka, S., and Loskutoff, D.J. Platelets stimulate endothelial cells to synthesize type 1 plasminogen activator inhibitor. *Blood* 77(in press), 1991a.

Slivka, S., and Loskutoff, D.J. Evaluation of the roles of protein kinase C and cAMP in the regulation of type I plasminogen activator inhibitor synthesis. Submitted, 1991b.

Sprengers, E.D., Kluft, C. Plasminogen activator inhibitors. *Blood* 69:381-387, 1987.

Sprengers, E.D., Princen, H.M.B., Kooistra, T., van Hinsbergh, V.W.N. Inhibition of plasminogen activators by conditioned medium of human hepatocytes and hepatoma cell line hep G2. *J. Lab. Clin. Med.* 105:751-758, 1985.

Thalacker, F.W., Nilsen-Hamilton, M. Specific induction of secreted proteins by transforming growth factor-β and 12-0-tetradecanoylphorbol-13-acetate. *J. Biol. Chem.* 262:2283-2290, 1987.

Vague, P.H., Juhan-Vague, I., Alessi, M.C., Badier, C., Valadier, J. Metformin decreases the high plasminogen activator inhibition capacity, plasma insulin and triglyceride levels in non-diabetic obese subjects. *Thromb. Haemost.* 58:326-328, 1987.

van Hinsbergh, V.W.N., Bertina, R.M., van Wijngaarden, A., van Tilburg, N.H., Emeis, J.J., Haverkate, F. Activated protein C decreases plasminogen activator-inhibitor activity in endothelial cell-conditioned medium. *Blood* 65:444-451, 1985.

van Mourik, J.A., Lawrence, D.A., Loskutoff, D.J. Purification of an inhibitor of plasminogen activator (antiactivator) synthesized by endothelial cells. *J. Biol. Chem.* 259:14914-14921, 1984.

van Zonneveld, A-J., Curriden, S.A., Loskutoff, D.J. Type 1 plasminogen activator inhibitor gene: Functional analysis and glucocorticoid regulation of its promoter. *Proc. Natl. Acad. Sci. U.S.A.* 85:5525-5529, 1988.

Wagner, O.F., Binder, B. Purification of an active plasminogen activator inhibitor immunologically related to the endothelial type plasminogen activator inhibitor from the-conditioned media of a human melanoma cell line. *J. Biol. Chem.* 261:14474-14481, 1986.

Wiman, B., Lindahl, T., and Almqvist, A. Evidence for a discrete plasminogen activator inhibitor binding protein in plasma. *Thromb. Haemostas.* 59:392-395, 1988.

Wun, T.C., Palmier, M.O., Siegel, N.R., Smith, C.E. Affinity purification of active plasminogen activator inhibitor-1 (PAI-1) using immobilized anhydrourokinase. *J. Biol. Chem.* 264:7862-7868

Yasuda, T., Gold, H. K., Leinbach, R. C., Saito, T., Guerrero, J. L., Jang, I-K., Holt, R., Fallon, J. T. and Collen D. Lysis of rt-PA resistant platelet-rich coronary arterial thrombus with combined bolus injection of recombinant tissue-type plasminogen activator (rt-PA) and anti-platelet GPIIb/IIIa antibody. (In press), 1990.

IV. ENDOTHELIAL CELL INTERACTION WITH BLOOD COMPONENTS

PURINE REGULATION OF ENDOTHELIAL CELLS: RELEVANCE TO PATHOPHYSIOLOGY

John L. Gordon

British Bio-technology
Watlington Road, Cowley,
Oxford, OX4 5LY, United Kingdom

INTRODUCTION

The purines that endothelial cells are likely to encounter in biologically significant concentrations are adenosine, ATP and ADP.

The interactions between these purines and endothelial cells take several forms:
- o Uptake into cells
- o Intracellular metabolism
- o Biological responses of endothelial cells
- o Metabolism of extracellular purines by endothelial ectoenzymes
- o Release of purines from endothelium.

Underlying these interactions is the question of how purines arise extracellularly and can thus interact with the endothelial surface - in other words, what are the cellular sources of the extracellular purines?

SOURCES OF EXTRACELLULAR PURINES

The vascular endothelium is the interface between the blood and the rest of the body, and therefore endothelial cells are the first tissue which any purines in the plasma encounter. Endothelium can also interact with purines released from extravascular tissue.

ATP and ADP are released in high concentrations when platelets aggregate and degranulate - concentrations in the high micromolar range are readily achieved in the plasma following platelet activation.

ATP and ADP are also released from other cells, both those which circulate in the blood and those fixed in tissues. A common stimulus for such adenine nucleotide release is physical or chemical damage - for example, Born and Kratzer (1984) demonstrated that ATP was released into the plasma flowing from a cut blood vessel, and they identified this ATP as coming from two sources: red blood cells and the vessel wall itself. It had already been established that ATP could be released from viable vascular endothelial cells (and smooth muscle cells) in culture (*Pearson and Gordon, 1979; Lollar and Owen, 1981*).

The concentration of ATP in the cytoplasm of most cell types is around 1-10 mM, and therefore, even a modest percentage release can result in concentrations in the micromolar range in the pericellular environment.

However, physical damage is not essential for inducing the release of ATP from tissues; for example, ischaemia or muscular work is sufficient (*Forrester, 1972*). In addition, ATP and ADP are released extracellularly at the neuromuscular junction.

As well as releasing ATP and ADP, the working heart can also release adenosine from coronary microvascular endothelial cells and from cardiomyocytes (*Nees, 1989*).

Adenosine and ATP in the plasma can both have profound biological effects on the circulatory system. The endothelium plays a significant role in these cardiovascular effects of purines, and endothelial cells are also critically important in removing purines from the circulating plasma.

STIMULATION OF ENDOTHELIAL CELLS BY ADENOSINE

Endothelial cells have P_1 receptors for adenosine on their surface, of the A_2 subclass, through which adenosine induces an increase in the intracellular level of cyclic AMP; adenosine also promotes the export of cyclic AMP from endothelial cells and this effect is potentiated by ATP (*Goldman et al, 1983*). This demonstration that the selective release of cyclic AMP from mammalian cells can be stimulated raises the question of whether it can act as an intercellular messenger in mammals, as is the case with some lower organisms such as Dictyostelium discoideum.

UPTAKE AND METABOLISM OF ADENOSINE BY ENDOTHELIAL CELLS

Adenosine in plasma is removed rapidly - not by deamination through the action of plasma enzymes, as was originally thought (*Clarke et al, 1952*), but through uptake and subsequent intracellular metabolism (*Kolassa et al, 1971; Pearson et al, 1978*). The cell types mainly responsible are vascular endothelium and red blood cells. Both of these are very numerous - more than 10^{12} endothelial cells in the human microcirculation, and more than 10^{12} red blood cells per litre of human blood. Consequently, their relative contributions to adenosine uptake depend on the geometry - uptake by red blood cells will predominate in large blood vessels and uptake by endothelial cells will predominate in the microcirculation. The uptake site for adenosine on both cell types is blocked by dipyridamole and related compounds, and it is a distinct entity from the P_1 receptor for adenosine on the endothelial cell surface.

After uptake, adenosine is metabolized intracellularly in endothelium, mainly by rephosphorylation (predominantly to ATP) if the substrate concentration is low, and mainly by deamination, with subsequent release of the products from the cells, if the substrate concentration is much more than about $10^{-5}M$ (*Pearson et al, 1978*).

STIMULATION OF ENDOTHELIAL CELLS BY ATP

In most blood vessels with resting tone, exposure to extracellular ATP induces vasodilation. This response is mainly because of the action of ATP on P2 receptors on the endothelial cells, which results in the release of nitric oxide (NO) derived from L-arginine, and release of prostacyclin, derived from plasma membrane phospholipids through the action of phospholipase. The phospholipase releases arachidonate, which is then subsequently converted to prostacyclin and related products through the action of cyclo-oxygenase and the other enzymes in the prostaglandin synthesis pathways.

Nitric oxide, which induces vasodilation by increasing cyclic GMP (through stimulation of guanylate cyclase) in vascular smooth muscle cells, was suggested to be the compound responsible for the activity previously known as Endothelium-Derived Relaxing Factor (*Palmer et al, 1987*). Subsequent observations (*Myers et al, 1990*) indicated that EDRF is more likely to be a compound such as S-nitrosocysteine, which contains NO within its structure but is a more potent vasodilator than NO itself. Prostacyclin, which induces vasodilation by increasing cyclic AMP (through stimulation of adenylate cyclase), can act synergistically with NO (*see Moncada et al, 1988 for review*) but it appears that vasodilation by NO is functionally the more important.

The P_2 receptor on endothelial cells is of the P_{2Y} subclass (*Burnstock and Kennedy, 1985*). Studies with analogues of ATP have shown that substitutions in the 2 position increase potency, that the stereochemistry of the sugar moiety is critical, and that the actions of ATP on this receptor are unrelated to its metabolism: analogues resistant to metabolic degradation show equivalent potency to those that are easily broken down (*See Gordon, 1986, for review*).

The production of both vasodilator substances (prostacyclin and NO) is very rapid and, in the case of prostacyclin, is also transient - the prostacyclin production shows desensitization on prolonged exposure (or repeated exposure) to the stimulus. (*See Needham et al, 1987 and Pearson and Gordon, 1989*).

It has also been found that endothelium can induce a vasoconstrictor response in certain blood vessels through P_2 receptor activation. The mechanisms and mediators involved have not been unequivocally established, although in other examples of endothelium-dependent vasoconstriction, mediators such as superoxide anion, thromboxane A_2 and endothelin have been suggested (*Shirahase et al, 1985; Vanhoutte and Katusic, 1988*). Likewise, the classification of the P_2 receptor responsible has not been reported, but it may well be different from the receptor designations in current use. These are P_{2X}, which stimulates smooth muscle contraction; P_{2Y}, which stimulates prostaglandin production (and NO production - see above); P_{2Z}, which induces permeability of cell membranes; P_{2T}, which stimulates thrombocytes, platelets and megacaryocytes (and responds to ADP, with ATP acting as an antagonist); and, possibly, a subclass of receptor which produces similar effects to the P_{2Y} receptor but at which UTP is equi-active with ATP. It should be emphasized that no entirely satisfactory antagonists are yet available, and attempts at cloning these receptors are only now beginning; consequently, details of P_2 receptor classification remain to be resolved.

The minimum effective concentration for ATP on endothelial cells in culture is around 1 µM or a little less, although 2-chloro-ATP can stimulate prostaglandin production at concentrations around 5 nM. However, it appears that brain microvascular endothelial cells may be particularly susceptible to the effects of ATP; such vessels responded to ATP at concentrations of 1 nM or less (*Forrester et al, 1979; Hardebo et al, 1987*).

In most experimental conditions, ADP is almost equi-active with ATP at stimulating endothelial responses; where ADP is significantly less active, this is probably because there is significant metabolism by ectonucleotidases, allowing ADP to be converted to the inactive AMP. (See below for discussion of endothelial ectonucleotidases.)

INTRACELLULAR SIGNALLING PATHWAYS COUPLED TO ENDOTHELIAL P2 RECEPTORS

Stimulation of the endothelial P_{2Y} receptor by ATP results in dose-dependent increases in free cytoplasmic calcium and in inositol triphosphate; the latter event suggests that the P_{2Y} receptor is linked to phospholipase C (*Hallam and Pearson, 1986; Luckoff and Busse, 1986, Pirotton et al, 1987, Carter et al, 1988*). The first, transient increase in intracellular calcium results from its release from intracellular stores; the second, more sustained calcium increase results from the opening of receptor-operated channels in the plasma membrane that allow the ingress of ionized calcium from the extracellular fluid (*Carter et al, 1988*). It appears that this secondary increase in calcium level may result from a summation of spikes in single cells in the population, giving a mean steady state (*Jacob et al, 1988*). Additional intracellular changes are also noted following P_{2Y} receptor occupation, probably again as a consequence of the increased intracellular calcium. These include K+ efflux (*Gordon and Martin, 1983; Sauve et al, 1988*), and a decrease followed by an increase in intracellular pH, resulting from Na/H exchange (*Kitazono et al, 1988*). P_{2Y} receptor activation also results in the phosphorylation of several proteins, some (but not all) resulting from the activity of protein kinase C (*Demolle et al, 1988*).

METABOLISM OF ATP BY ENDOTHELIAL CELLS

Extracellular ATP exerts a range of important biological effects, some of them mediated through its actions on endothelial cells, as indicated above. Because of the biological importance of the effects of extracellular ATP, coupled with the fact that ATP is an almost ubiquitous intracellular constituent of mammalian cells, present in high concentrations and readily released, the mechanisms that exist for its removal from the extracellular compartment (especially from the plasma, in the context of the present discussion) are of particular interest.

ATP is not metabolized rapidly in plasma *in vitro*: the t½ is around 30 mins at 37°C. However, ATP (and other adenine nucleotides), when injected into the circulation, are metabolized very rapidly - the t½ is, for example, less than 1 s in the pulmonary microvascular bed. The fate of ATP in the circulation is determined mainly by ectonucleotidases on the surface of endothelial cells (*for review see Pearson and Gordon, 1985*). Not only are these enzymes very efficient, but the huge area of vascular endothelium (greater than 2,000 m^2)

ensures very rapid degradation of nucleotides. ATP is sequentially converted by these enzymes to ADP, AMP and adenosine (which is then taken up by the endothelium or by red blood cells - see above).

The characteristics and kinetics of adenine nucleotide metabolism by endothelial ectonucleotidases have been studied in detail, especially using perfused columns of cultured endothelial cells on microcarrier beads (*Gordon et al, 1986*). Such studies have revealed complex interactive controls in this system of ectonucleotidases - notably, that the conversion of AMP to adenosine is powerfully inhibited by ADP, thus increasing the time interval between the metabolism of the P_2 receptor agonists (ATP and ADP) to the inactive AMP, and the subsequent production of adenosine. In the circulation, this delay results in an anatomical separation, so that P_2 receptor-mediated responses occur first, and at a different site from those responses mediated by the P_1 receptor, which is activated by the adenosine eventually formed.

Ectonucleotidases are present on several other cell types, but their properties and organization are not identical to those on the endothelium. For example, on vascular smooth muscle cells, unlike vascular endothelial cells, the products of the ATPase and ADPase are preferentially delivered to the next enzyme down the chain - in other words, the substrate for the subsequent enzyme does not equilibrate with the bulk phase before being metabolized.

CONCLUSION

It is now widely accepted that several classes of receptors for extracellular ATP and ADP exist - not just on vascular endothelial cells, as summarized here, but on many other cell types too. These receptors, when activated, result in powerful biological effects: in the circulatory system, the main effect is usually vasodilation (often mediated through an action on the endothelial cells), although vasoconstriction can also occur, particularly through a direct action on P_2 receptors on the vascular smooth muscle.

Effects on vascular endothelial cells also include secretory responses, resulting in the release of biochemical messengers such as prostacyclin and NO, which themselves have a number of different biological actions.

Although ATP and ADP are ubiquitous and abundant intracellular constituents, they can also be released from several cell types (including the endothelial cells themselves), reaching pericellular concentrations well in excess of those needed to induce biological responses.

The nucleotides that are released into the plasma are metabolized mainly by ectonucleotidases on the endothelial cell surface, resulting in very rapid, sequential conversion of ATP to ADP, AMP and adenosine. The adenosine thus formed - or itself released into the plasma from cells of the vessel wall or surrounding tissues, including the myocardium - exerts potent cardiovascular effects via the endothelium, the smooth muscle of the vessel wall and the heart cells. Adenosine in the plasma can be removed from the circulation by uptake into the endothelial cells and subsequent intracellular metabolism (usually phosphorylation to form ATP).

Vascular endothelial cells thus have the capacity to release adenine nucleotides from intracellular pools; to respond to these nucleotides via P_2 receptors on their surface; to metabolize the nucleotides through ectonucleotidases, thus forming adenosine; to respond to this adenosine by P_1 receptors of the A_2 class on their surface; and to remove the adenosine through high affinity uptake sites on the cell surface. Finally, they can metabolize the adenosine intracellularly, forming ATP once more.

In summary, the vascular endothelium exerts a central controlling influence on the production and metabolism of extracellular purines at the blood-tissue interface and regulates the cardiovascular responses to these powerful biochemical mediators.

REFERENCES

Born, G.V.R. and Kratzer, M.A.A. Source and concentration of extracellular adenosine triphosphate during hemostasis in rats, rabbits and man. *J. Physiol.* 354:419-429, 1984.

Burnstock, G. and Kennedy, C. Is there a basis for distinguishing two types of P_2-purinoceptor? *Gen. Pharmacol.* 16:433-440, 1985.

Carter, T.D., Hallam, T.J., Cusack, N.J. and Pearson, J.D. Regulation of P_{2Y} purinoceptor-mediated prostacyclin release from human endothelial cells by cytoplasmic calcium concentration. *Br. J. Pharmacol.* 94:1181-1190, 1988.

Clarke, D.A., Davoli, J., Philips, F.S. and Brown, G.B. Enzymatic deamination and vasodepressor effects of adenosine analogues. *J. Pharmacol. Exp. Ther.* 106:291-302, 1952.

Demolle, D., Lecomte, M. and Boeynaems, J-M. Pattern of protein phosphorylation in aortic endothelial cells. Modulation by adenine nucleotides and bradykinin. *J. Biol. Chem.* 263:18459-18465, 1988.

Forrester, T. An estimate of ATP release into the venous effluent from exercising human forearm muscle. *J. Physiol.* 224:611-628, 1972.

Forrester, T., Harper, A.M., Mackenzie, E.T., and Thomson, E.M. Effect of adenosine triphosphate and some derivatives on cerebral blood flow and metabolism. *J. Physiol. (Lond)* 296:343-355, 1979.

Goldman, S.J., Dickinson, E.S. and Slakey, L.L. Effect of adenosine on synthesis and release of cyclic AMP by cultured vascular cells from swine. *J. Cyclic Nucleotide Protein Phosphor. Res.* 9:69-78, 1983.

Gordon, E.L., Pearson, J.D. and Slakey, L.L. The hydrolysis of extracellular adenine nucleotides by cultured endothelial cells from pig aorta: feed-forward inhibition of adenosine production at cell surface. *J. Biol. Chem.* 261:15496-15504, 1986.

Gordon, J.L. Review Article: Extracellular ATP: effects, sources and fate. *Biochem. J.* 233:309-319, 1986.

Gordon, J.L., Martin, W. Endothelium-dependent relaxation of the pig aorta: relationship to stimulation of ^{86}Rb efflux from isolated endothelial cells. *Br. J. Pharmacol.* 79:531-541, 1983.

Hallam, T.J. and Pearson, J.D. Exogenous ATP raises cytoplasmic free calcium in fura-2 loaded piglet aortic endothelial cells. *FEBS Lett* 207:95-99, 1986

Hardebo, J.E., Kahrstrom, J. and Owman, C. P1- and P2- purine receptors in brain circulation. *Eur. J. Pharmacol.* 144:343-352, 1987.

Jacob, R., Merritt, J.E., Hallam, T.J., and Rink, T.J. Repetitive spikes in cytoplasmic calcium evoked by histamine in human endothelial cells. *Nature* 335:40-45, 1988.

Kitazono, T., Takeshige, K., Cragoe, E.J. Jr. and Minakami, S. Intracellular pH changes of cultured bovine aortic endothelial cells in response to ATP addition. *Biochem. Biophys. Res. Commun.* 152:1304-1309, 1988.

Kolassa, N., Pfleger, K. and Tram, M. Species differences in action and elimination of adenosine after dipyridamole and hexobendine. *Eur. J. Pharmacol.* 13:320-325, 1971.

Lollar, P. and Owen, W.G. Active-site dependent thrombin-induced release of nucleotides from cultured human endothelial cells. *Annals of the New York Academy of Science* 370:51-56, 1981.

Luckoff, A. and Busse, R. Increased free calcium in endothelial cells under stimulation with adenine nucleotides. *J. Cell Physiol.* 126:414-420, 1986.

Moncada, S., Palmer, R.M.J. and Higgs, E.A. The discovery of nitric oxide as the endogenous nitrovasodilator. *Hypertension*, 12:365-372.

Myers, P.R., Minor, Jr, R.L., Guerra, Jr, R., Bates, J.N., Harrison, D.G. Vasorelaxant properties of the endothelium-derived relaxing factor more closely resemble S-nitrosocysteine than nitric oxide. *Nature* 345:161-163.

Needham, L., Cusack, N.J., Pearson, J.D., and Gordon, J.L. Characteristics of the P2 purinoceptor that mediates prostacyclin production by pig aortic endothelial cells. *Eur. J. Pharmacol.* 134:199-209, 1987.

Nees, S. The adenosine hypothesis of metabolic regulation of coronary flow in the light of newly recognized properties of the coronary endothelium. *Z. Kardiol.* 78:suppl.6, 42-49, 1989.

Palmer, R.M., Ferrige, A.G. and Moncada, S. Nitric Oxide release accounts for the biological activity of endothelium-derived relaxing factor. *Nature* 327:524-526, 1987.

Pearson, J.D., Carleton, S., Hutchings, A. and Gordon, J.L. Uptake and metabolism of adenosine by pig aortic endothelial and smooth-muscle cells in culture. *Biochem. J.* 170:265-271, 1978.

Pearson, J.D. and Gordon, J.L. Vascular endothelial and smooth muscle cells in culture selectively release adenine nucleotides. *Nature* 281:384-386, 1979.

Pearson, J.D. and Gordon, J.L. Nucleotide metabolism by endothelium. *Ann. Rev. Physiol.* 47:617-627, 1985.

Pearson, J.D. and Gordon, J.L. P2 purinoceptors in the blood vessel wall. *Biochem. Pharmacol.* Vol.38,No.23:4157-4163, 1989.

Pirotton, S., Raspe, E., Demolle, D., Erneux, C. and Boeynaems, J-M. Involvement of inositol 1,4,5-trisphosphate and calcium in the action of adenine nucleotides on aortic endothelial cells. *J. Biol. Chem.* 262:17461-17466, 1987.

Sauve, R., Parent, L., Simoneau, C. and Roy, G. External ATP triggers a biphasic activation process of a calcium-dependent K+ channel in cultured bovine aortic endothelial cells. *Pflugers Arch.* 412:469-481, 1988.

Shirahase, H., Ushi, H., Manabe, K., Kurahashi, K. and Fujiwara, M. Endothelium-dependent contraction and -independent relaxation induced by adenine nucleotides and nucleoside in the canine basilar artery. *J. Pharmacol. Exp. Ther.* 247:1152-1157, 1985.

Vanhoutte, P.M. and Katusic, Z.S. Endothelium-derived contracting factors: endothelin and/or superoxide anion? *Trends Pharmacol. Sci.* 9:229-230, 1988.

AUTOANTIBODIES TO ENDOTHELIAL CELLS

Jeremy D. Pearson

Section of Vascular Biology
MRC Clinical Research Centre
Watford Road, Harrow,
Middlesex HA1 3UJ, United Kingdom

DETECTION OF ANTI-ENDOTHELIAL CELL ANTIBODIES IN AUTOIMMUNE DISEASE

It has become apparent over the last 10 years that anti-endothelial cell antibodies (AECA) are present in a proportion of patients with a wide variety of autoimmune diseases that exhibit vascular pathology. These antibodies have, in general, been detected in one of two ways: by their ability to kill endothelial cells (EC) *in vitro* in the presence of complement or by their ability to bind to cultured EC monolayers and are therefore presumed to recognize antigens at the surface of EC.

One of the earliest suggestions that AECA might contribute causally to vascular pathology was by Shingu & Hurd (1981), who noted selective binding of Ig from the sera of patients with systemic lupus erythematous (SLE) to cytospin preparations of cultured umbilical vein EC, detected by immunofluorescence. The antibodies did not bind to fibroblasts and were interpreted as staining cytoplasmic components of the endothelium. Normal sera, and sera from patients with rheumatoid arthritis, did not stain endothelial cells.

Subsequently, AECA (both IgG and IgM) that are present in SLE sera but not normal sera, and which bind to endothelial cells in cell-based ELISA methods, have been reported several times (*Cines et al. 1984; Hashemi et al. 1987; Rosenbaum et al. 1988*), and the list of disease states in which similar AECA have been detected has expanded considerably. It includes Wegener's granulomatosis or microscopic polyarteritis (*Brasile et al. 1989; Ferraro et al. 1990*); rheumatoid vasculitis (*Heurkens et al. 1989*); peripheral vascular disease (*Cerilli et al. 1985a; D'Anastasio et al. 1988*); systemic sclerosis and mixed connective tissue disease (*Hashemi et al. 1987; Rosenbaum et al. 1988; Bodolay et al. 1989*); thrombotic thrombocytopenic purpura and hemolytic uremic syndrome (*Burns et al. 1982; Leung et al. 1988*); heparin-associated thrombocytopenia (*Cines et al. 1987*); Kawasaki disease (*Leung et al. 1986a, 1986b*); pre-eclampsia (*Rappaport et al. 1990*); IgA nephropathy (*Yap et al. 1988*); and experimental autoimmune encephalitis and multiple sclerosis (*Tsukuda et al. 1986; Tanaka et al. 1987*).

In most of these diseases, however, there is a plethora of autoantibodies of multiple specificities, and the relevance of AECA is poorly understood. To date, a causal role for AECA in pathogenesis remains unproven, and the molecular nature of the EC antigenic determinants has rarely been identified. This review summarizes what is currently known of the relationship between AECA and other autoantibodies, candidate EC target antigens and their regulation (including the special case of AECA in transplant rejection), and the possible cytotoxic or other dysfunctional effects of AECA.

RELATIONSHIP OF AECA TO OTHER AUTOANTIBODIES

Some studies have noted the presence within EC of antigenic determinants directly related to those found in other cell types and diagnostically characteristic of the disease process, eg. the common anti-Scl-70 Ig in systemic sclerosis reacts with an analogous component (but of higher molecular mass) in EC (*Alderuccio et al. 1986*). However, in most studies where it has been examined, it seems that although patients often possess AECA together with other autoantibodies characteristic of their disease, these cannot be identical, since patients with only AECA or only other autoantibodies are also found. Furthermore, by direct adsorption of other Ig (anticardiolipin in SLE, anti DNA in systemic sclerosis) Rosenbaum et al. (1988) were able to show that the levels of AECA were unaffected. Similarly, Ferraro et al., (1990) and Savage et al., (1991) demonstrated that AECA in patients with Wegener's granulomatosis were not related to the diagnostic antineutrophil cytoplasmic antigen (ANCA) found in many of these patients.

The cellular specificity of AECA seems to differ amongst disease states. In general, AECA do not cross-react with determinants on red or white blood cells or platelets [see eg. in SLE, *Rosenbaum et al. (1988), Vismara et al. (1988)*; in pre-eclampsia, *Rappaport et al. (1990)*; in mixed connective tissue disease, *Bodolay et al. (1989)*], though it should be noted that one subset of AECA, particularly associated with transplant rejection, does crossreact with monocytes or granulocytes (see below). However, whereas Leung et al. (1986a) found no binding of AECA from Kawasaki disease patients to fibroblasts or smooth muscle cells (or blood cells), AECA that crossreact equally well with fibroblasts have been found in SLE, systemic sclerosis and rheumatoid vasculitis (*Rosenbaum et al. 1988; Vismara et al, 1988; Heurkens et al, 1989*).

It thus appears that the specificities of AECA are heterogeneous between different disease states. It is likely that they exhibit multiple specificities within individual patients, though immunoblotting experiments have rarely been reported. When purified IgG from a series of SLE patients with AECA were tested for their ability to block the binding of radiolabelled AECA IgG from one patient to human umbilical vein EC in culture, it was found that some individuals had IgG that could inhibit binding, whereas others did not, showing that some (but not all) AECA from different individuals share specificity (*Rosenbaum et al. 1988*).

AECA IN TRANSPLANT REJECTION

The importance of microvascular damage in the pathogenesis of accelerated or acute graft rejection when the donor tissue expresses identical major histocompatibility antigens has been recognized for over 20 years (see refs. in *Paul et al, 1985*). Several groups have now demonstrated the presence in such cases of autoantibodies, pre-existing in cases of hyperacute rejection, which are directly cytotoxic to cultured EC in the presence of complement. As noted above, these AECA are also cytotoxic to monocytes and sometimes to granulocytes, in distinction to many AECA found in autoimmune vasculitic disease (*Moraes & Stasny, 1977; Cerilli et al. 1985b; Paul et al. 1985*). The antigen(s) recognized by these AECA are present on human umbilical vein EC, and on microvascular EC in renal and coronary beds. Although not MHC class I or II molecules, family studies have indicated that the ability of AECA to kill monocytes and EC segregates with HLA haplotype (*Brasile et al. 1985; Cerilli et al. 1985b*), suggesting that they may recognize antigens closely genetically linked to HLA.

It is plausible that the AECA in transplant rejection are directed at one or a small number of allotypically polymorphic determinants, which seems less likely to be the case for AECA in autoimmune disease. These findings have prompted studies to produce monoclonal antibodies (MAb) that are cytotoxic to both EC and monocytes, with the aim of identifying the antigens involved (*Schook et al. 1987; Russ et al. 1987*). Interestingly, the former group found that their 3 MAb reacted with 3 different EC glycoprotein determinants (*Wood et al. 1988*), but in each case the target was more strongly expressed on Ia$^+$ cell lines, and the latter group noted their 2 MAb killed EC in culture better when the cells were pretreated with mixed cytokines derived from stimulated T cell supernatants (known to upregulate Ia-related antigens on EC). While these findings were interpreted as supporting the idea that MHC-linked EC and monocyte surface proteins are recognized by the AECA found in transplant rejection, they are also relevant to the independent demonstration that AECA in certain autoimmune diseases can recognize EC determinants that are modulated by cytokine treatment (see below), and indicate possible mechanistic links between the two processes.

AECA DIRECTED AT MODULATABLE EC SURFACE DETERMINANTS IN AUTOIMMUNE VASCULITIDES

Kawasaki disease is characterized by diffuse vasculitis and lymphocyte activation. Leung et al. (1986a) first reported the interesting finding that although sera from patients in the acute phase of the disease did not possess cytotoxic AECA that could damage EC cultured under routine conditions, these sera did contain AECA that were cytotoxic when EC were pretreated with interferon (IFN)τ. This cytokine, produced by stimulated lymphocytes, was already known to induce or upregulate a variety of EC surface molecules, including class II MHC molecules (*Pober et al. 1983*). Since these AECA did not damage similarly pretreated fibroblasts or smooth muscle cells, which also increased class II MHC expression, the authors concluded that the EC-specific determinants recognized by the AECA were not class II MHC but were molecules upregulated specifically on EC in a similar way. Subsequently, the same group demonstrated that pretreatment of EC with interleukin (IL)1 or tumor necrosis factor (TNF)α also led to killing by AECA in Kawasaki disease, though cross-adsorption studies indicated that the target antigen specificity differed according to which cytokine treatment was used (*Leung et al. 1986b*). Production of TNFα and IL1 has been shown to be stimulated in monocytes from patients with acute Kawasaki disease (*Lang et al. 1989; Leung et al. 1989*).

Leung and colleagues did not use an ELISA method to detect AECA bound to EC. It therefore remains uncertain whether non-lytic (but perhaps functionally important) AECA, which recognize unstimulated EC, are present in Kawasaki sera as they are in other autoimmune vasculitic diseases. Similarly, few studies have yet examined whether the binding of AECA detected by ELISA in other vasculitic diseases is altered by pretreatment of EC with cytokines. Rappaport et al. (1990) found the binding of AECA from pre-eclamptic patients to EC was not modulated by IFNτ treatment. We have, however, found that cytokine treatment of EC enhanced AECA binding in sera from patients with Wegener's granulomatosis (*Savage et al., 1991*).

More recently, Leung et al. (1988) showed that cytotoxic AECA in sera from patients with hemolytic uremic syndrome lysed nontreated EC in the presence of complement, but this ability was lost when EC were pretreated with IFNτ, suggesting that, in this instance, a class of antigen on EC that is downregulated by IFNτ is recognized.

AECA AND EC CYTOTOXICITY

Unfortunately, few studies have used both AECA binding and complement-dependent cytotoxicity assays, so further work is needed to clarify the relative incidence of, and relationship between, AECA detected in these different ways. As noted above, cytotoxic AECA have been found in Kawasaki disease and hemolytic uremic syndrome. They have also been found in mixed connective tissue disease (*Bodolay et al. 1989*), some patients with other vasculitic disorders, including Wegener's granulomatosis (*Brasile et al. 1989*), and noted in one patient with Raynaud's phenomenon (*Baguley & Hughes, 1988*). However, others have failed to find cytotoxic AECA in Wegener's granulomatosis, SLE or systemic sclerosis (*Summers et al. 1984; Rosenbaum et al. 1988; Ferraro et al. 1990*).

Although several further studies have reported that sera from vasculitic patients could be cytotoxic to EC, in these cases it was either not clear or excluded that this was Ig-mediated (eg. *Kahaleh et al. 1979; Meyer et al. 1983; Ianaccone et al. 1984; Drenk et al. 1988*). Together with reports that storage of sera from such patients can generate EC cytotoxic activity (*Blake et al. 1985*), and the significant technical differences between laboratories in the methods used to detect cytotoxicity, it is obvious that there is a need for additional careful studies to assess the incidence of cytotoxic Ig (or other factors) that damage EC in vasculitic diseases.

Another related area, which requires further study in the light of recent advances in our understanding of antigen presentation and lymphocyte classification, is the ability of specific subclasses of lymphocytes from vasculitic patients to directly, or in an antibody-dependent manner, damage EC, since several groups have reported cell-mediated cytotoxicity to EC, usually antibody-dependent with mononuclear cell preparations from SLE or systemic sclerosis patients (*Penning et al. 1984, 1985; Marks et al. 1988; Holt et al. 1989*).

FUNCTIONAL CHANGES IN EC INDUCED BY AECA

Significant changes in endothelial reactivity, i.e. sublethal and reversible forms of injury, occur in response to a variety of damaging stimuli (*Gordon & Pearson, 1982*). Though not all AECA seem to be directly cytotoxic to EC, the hypothesis that they play a causal role in the genesis of the vascular pathology found in the connective tissue and other autoimmune diseases would be strengthened substantially if they induced selective alterations of EC functions, particularly those related to the control of hemostasis or vascular permeability, which is often disturbed in these diseases.

One of the first suggestions that this could occur was by Carreras et al. (1981a), who found that IgG from the serum of one SLE patient inhibited EC prostacyclin (PGI_2) synthesis by comparison with control IgG. A series of subsequent papers, however, provided conflicting results. In general, no consistent alteration of PGI_2 synthesis has been found using sera from patients with SLE. Purified Ig have yielded increased or decreased PGI_2 production in different studies or even between samples; and attempts to correlate responses with the presence of AECA or other autoantibodies (in particular, antiphospholipid or anticardiolipin Ig) have not been fruitful (eg. *Cines et al. 1984; Walker et al. 1988; Austin et al. 1988; Coade et al. 1989*). Similarly, suggestions that PGI_2 synthesis is decreased in other diseases, including systemic sclerosis and pre-eclampsia, have been disputed and are not clearly attributable to AECA (*Carreras et al. 1981b; Marks et al. 1988; Hasselaar et al. 1988*).

There is general agreement that the plasma levels of von Willebrand factor (vWf) are elevated in patients with SLE or systemic sclerosis, which almost certainly reflects alteration or damage to EC function since this cell type is the source of circulating vWf (*Kahaleh et al. 1981; Byron et al. 1987; Gordon et al. 1987*). In addition, the ability of EC to release active tissue plasminogen activator (tPA) in response to a challenge (eg. venous stasis or infusion of a vasopressin analogue) appears to be diminished in SLE patients, though not all studies agree (*Belch et al. 1987; Byron et al. 1987; Awada et al. 1988*). Both of these alterations in EC functions could contribute to the procoagulant complications experienced in SLE, but there has been no published work to indicate that these processes can be modified by AECA acting on EC *in vitro*, and in our own laboratory we have failed to find convincing *in vitro* effects of IgG fractions from patients with a variety of autoimmune vasculitic diseases on vWf release (Coade & Pearson, unpublished data).

The most clearcut example of a relevant pathological alteration in EC function in response to AECA was published by Tannenbaum et al. (1986a). These authors demonstrated that IgG from SLE patients induced the synthesis of procoagulant Tissue Factor (TF) activity by EC *in vitro*, whereas IgG from normal sera were essentially inactive. Tannenbaum et al. (1986a,b) found in addition that incubation of EC with heat-aggregated normal IgG induced TF and inhibited EC fibrinolytic activity (due to increased secretion of PAI-1, the inhibitor of tPA). They suggested that natural immune complexes might therefore alter EC functions in a procoagulant manner. Uchmann et al. (1988), however, failed to induce TF activity in EC with insulin-antiinsulin immune complexes, though these complexes did stimulate monocytes to produce TF and soluble factors (presumably cytokines), which in turn indirectly induced EC TF.

Cariou et al. (1988) studied the effects on EC of purified IgG from patients with circulating lupus anticoagulant (ie. anticardiolipin and/or antiphospholipid Ig). They found that these IgG, though without effect on the release of PGI_2, tPA or PAI-1 from EC, significantly inhibited the ability of EC to activate protein C in the presence of thrombin, an anticoagulant function attributable to the binding of thrombin by EC thrombomodulin. This effect was antagonized by phospholipids and was thus attributed to antiphospholipid Ig. Rosenbaum et al. (1988) showed that the majority of AECA in SLE serum samples was not anticardiolipin Ig, but the results of Cariou et al. (1988) indicate that significant effects on EC coagulant activity may nonetheless occur in response to a particular subfraction of AECA.

A further demonstration that functional changes in EC can be brought about by this mechanism was the finding that Ig from patients with heparin-induced thrombocytopenia, where autoantibodies to heparin are generated, consistently induce EC TF activity (*Cines et al. 1987*), presumably as a consequence of binding to the abundant heparin-like molecules on the EC surface. Another example of how cross-reacting autoantibodies could have functional effects on EC is suggested by the finding that sera from three patients with autoimmune thrombotic thrombocytopenic purpura inhibited binding of a MAb to the vitronectin receptor on EC, which is homologous with platelet GPIIb/IIIa (*Nakajima et al. 1987*). The authors did not prove that the effects were due to Ig, but the results are consistent with the hypothesis that autoantibodies to platelet GPIIb/IIIa were present (total platelet associated Ig was markedly raised) and could in addition bind to the related EC molecule. Platelet alloantigens have been implicated in neonatal and post-transfusion thrombotic disorders, where autoantibodies are generated that recognize epitopes on a variety of integrins present on platelets and EC, and therefore may also act through effects on EC function (*Van Mourik et al, 1990*).

Recently, McCarty et al (1990) reported that autoantibodies recognizing the glycoprotein GMP140 were present in patients with systemic vasculitis. This antigen, rapidly inducible on the surface of platelets and EC in response to agonists such as thrombin, has been shown to be an adhesive ligand for neutrophil leukocytes (*Larsen et al, 1989; Toothill et al, 1990*). AECA binding GMP140 could therefore be expected to modulate neutrophil-EC interactions in vasculitis, though *in vitro* tests of the effects of these AECA have not yet been reported.

CONCLUSIONS AND FUTURE STUDIES

The evidence that AECA are directly or indirectly involved in the pathogenesis of vascular disease is increasing. In the special case of transplants there is no doubt that cytotoxic AECA, perhaps directed to a very limited repertoire of EC antigens, can be responsible for the vascular damage that causes rejection. Libby et al. (1989) have suggested that similar AECA, in conjunction with enhanced MHC Class II expression, could be involved in recruiting lymphocytes and monocytes in the pathogenesis of the accelerated arteriosclerosis that is a consistent feature of transplanted blood vessels. Other experimental studies have shown the production of vascular damage in response to the infusion of antibodies directed against specific EC antigens, though it is likely that this involves cell-mediated mechanisms in addition to or instead of direct cytotoxicity (*Matsuo et al. 1987; Camussi et al. 1987*).

Future studies are needed to determine the range of antigens recognized by AECA in autoimmune diseases. If, as seems likely, in diseases such as SLE and systemic sclerosis, this range is wide and heterogenous between and within individuals, it reduces the likelihood that AECA play a direct or primary role, though they perhaps remain a sensitive indicator of altered EC structure in response to other stimuli. These AECA are commonly detected in patient sera by binding of Ig to monolayers of cultured EC that have not been damaged or stimulated in any specific manner. This suggests either that AECA bind to normal EC surface components merely because of crossreactivity with foreign or neoantigens primarily expressed elsewhere, or that EC in culture in fact modulate their antigenic expression in such a way as to mimic EC damaged or specifically stimulated *in vivo*.

It is now appreciated that EC, in response to cytokines and bacterial lipopolysaccharides, modulate their surface components (and their ability to secrete cytokines that affect leukocyte function) in a variety of ways leading to distinct but overlapping patterns of expression of antigens, notably cell adhesion molecules and MHC Class II molecules (*Pober & Cotran, 1990*). In addition, viral infection of EC induces the surface expression of neoantigens, including Fc and C3 receptors and molecules involved in leukocyte adhesion and the coagulation process (*Ryan et al. 1981; Cines et al. 1982; Etingin et al. 1990*). The known increase in circulating cytokines and the suspected viral etiology in autoimmune diseases such as Kawasaki disease are thus entirely consistent with the generation of EC neoantigens, perhaps of a limited number, which subsequently induce the generation of cytotoxic AECA with a direct role in pathogenesis.

It is less clear whether similar pathogenic mechanisms occur in the diseases such as SLE or systemic sclerosis, where AECA are not the major or distinguishing diagnostic class of autoantibody. There is, however, the potential for indirectly pathogenic roles for such AECA, as in the apparent antibody-dependent T cell cytotoxicity in SLE and systemic sclerosis (*Penning et al. 1984, 1985; Marks et al. 1988*), or in the enhanced ability of neutrophils to disrupt EC monolayers *in vitro* in the presence of immune complexes from patients with rheumatoid arthritis (*Breedveld et al. 1988*). More studies are therefore needed to determine the role of AECA in leukocyte-mediated injury to EC.

EC can present antigen to T cells in the context of MHC Class II molecules (*Pober et al. 1983*). Aberrant Class II expression and antigen presentation have been proposed as a necessary mechanism involved in the induction of autoimmunity, in view of the well-documented association between specific HLA allotypes and autoimmune disease (*Bottazzo et al. 1983*). New investigations are therefore required to discover whether EC neoantigens (whether intrinsic to EC or acquired by other means) can play a role in the induction of cellular immunity in autoimmune vasculitic disease, and to investigate the relationship of such neoantigens to AECA.

In summary, AECA occur in a wide variety of autoimmune diseases with a vasculitic component, but the target antigens they recognize and the consequences for vascular function are still poorly defined. At the least, the measurement of AECA may provide a useful indicator of EC damage, perhaps predictive of clinical outcome, in these disorders and thus an aid to the monitoring of therapeutic efficacy. More speculatively, the elucidation of the targets for AECA, and of the mechanisms of direct or leukocyte-mediated EC dysfunction or cytotoxicity in response to AECA, could not only increase our understanding of vascular damage in autoimmune vasculitic disease but also be of general relevance to pathogenic mechanisms in other autoimmune diseases.

REFERENCES

Alderuccio, F., Barnett, A.J., Campbell, J.H., Pedersen, J.S. and Toh, B.H. Scl-95/100; doublet of endothelial marker autoantigens in progressive systemic sclerosis. *Clin. Exp. Immunol.* 64:94-100, 1986.

Awada, H., Barlowatz-Meimon, G., Dougados, M., Maisonneuve, P., Sultan, Y. and Amor, B. Fibrinolysis abnormalities in systemic lupus erythematosus and their relation to vasculitis. *J. Lab. Clin. Med.* 111:229-236, 1988.

Baguley, E. and Hughes, G.R.V. Lytic IgG anti-endothelial cell antibodies in vasculitis. *Lancet* ii:907, 1988.

Belch, J.J.F., Zoma, A., McLaughlin, K., Curran, L., Capell, H.A., Forbes, C.D. and Sturrock, R.D. Fibrinolysis in systemic lupus erythematosus: effect of desamino D-arginine vasopressin infusion. *Br. J. Rheumatol.* 26:262-266, 1987.

Blake, D.R., Winyard, P., Scott, D.G.I., Brailsford, S., Blann, A. and Lunec, J. Endothelial cell cytotoxicity in inflammatory vascular diseases - the possible role of oxidized lipoproteins. *Ann. Rheum. Dis.* 44:176-182, 1985.

Bodolay, E., Bojan, F., Szegedi, G., Stenszky, V. and Farid, N.R. Cytotoxic endothelial cell antibodies in mixed connective tissue disease. *Immunol. Lett.* 20:163-168, 1989.

Botazzo, G.F., Pujol-Borrell, R., Hanafusa, T. and Feldmann, M. Role of aberrant HLA-DR expression and antigen presentation in induction of endocrine autoimmunity. *Lancet* ii:1115-1119, 1983.

Brasile, L., Galouzis, T., Clarke, J. and Cerilli, J. Clinical significance of the vascular endothelial cell antigen system - evidence for genetic linkage between the endothelial cell antigen system and the major histocompatibility complex. *Transpl. Proc.* 17:2318-2321, 1985.

Brasile, L., Kremer, J.M., Clarke, J. and Cerilli, J. Identification of an autoantibody to vascular endothelial cell-specific antigens in patients with systemic vasculitis. *Am. J. Med.* 87:74-80, 1989.

Breedveld, F.C., Heurkens, A.H.M., Lafeber, J.M., Van Hinsbergh, V.W.M. and Cats, A. Immune complexes in sera from patients with rheumatoid vasculitis induce polymorphonuclear cell-mediated injury to endothelial cells. *Clin. Immunol. Immunopathol.* 48:202-213, 1988.

Burns, E.R. and Zucker-Franklin, D. Pathologic effects of plasma from patients with thrombotic thrombocytopenia purpura on platelets and cultured vascular endothelial cells. *Blood* 60:1030-1037, 1982.

Byron, M.A., Allington, M.J., Chapel, H.M., Mowat, A.G. and Cederholm-Williams, S.A. Indications of vascular endothelial cell dysfunction in systemic lupus erythematosus. *Ann. Rheum. Dis.* 46:741-745, 1987.

Camussi, G., Caldwell, B., Andres, G. and Brentjens, J.R. Lung injury mediated by antibodies to endothelium. *Am. J. Pathol.* 127:216-228, 1987.

Cariou, R., Tobelem, G., Bellucci, S., Soria, J., Soria, C., Maclouf, J. and Caen, J. Effects of lupus anticoagulant on antithrombogenic properties of endothelial cells - inhibition of thrombomodulin-dependent Protein C activation. *Thrombos. Haemostas.* 60:54-58, 1988.

Carreras, L.O., Defreyn, G., Machin, S.J., Vermylen, J., Deman, R., Spitz, B. and Van Assche, A. Arterial thrombosis, intrauterine death and lupus anticoagulant: detection of immunoglobulin interfering with prostacyclin formation. *Lancet* i:244-246, 1981a.

Carreras, L.O., Defreyn, G., Van Houtte, E., Vermylen, J. and Van Assche, A. Prostacyclin and pre-eclampsia. *Lancet i*:442, 1981b.

Cerilli, J., Brasile, L. and Karmody, A. Role of the vascular endothelial cell antigen system in the etiology of atherosclerosis. *Ann. Surg.* 202:329-334, 1985a.

Cerilli, J., Brasile, L., Galouzis, T., Lempert, N. and Clarke, J. The vascular endothelial cell antigen system. *Transplantation* 39:286-289, 1985b.

Cines, D.B., Lyss, A.P., Bina, M., Corkey, R., Kefalides, N.A. and Friedman, H.M. Fc and C3 receptors induced by Herpes simplex virus on cultured human endothelial cells. *J. Clin. Invest.* 69:123-128, 1982.

Cines, D.B., Lyss, A.P., Reeber, M., Bina, M. and DeHoratius, R.J. Presence of complement-fixing anti-endothelial cell antibodies in systemic lupus erythematosus. *J. Clin. Invest.* 73:611-625, 1984.

Cines, D.B, Tomaski, A. and Tannenbaum, S. Immune endothelial cell injury in heparin-associated thrombo-cytopenia. *N. Eng. J. Med.* 316:581-589, 1987.

Coade, S.B., Van Haaren, E., Loizou, S., Walport, M.J., Denman, A.M. and Pearson, J.D. Endothelial prostacyclin release in systemic lupus erythematosus. *Thrombos. Haemostas.* 61:97-100, 1989.

D'Anastasio, C., Impallomeni, M., McPherson, G.A.D., Clements, W.G., Howells, G.L., Brooks, P.A. and Batchelor, J.R. Antibodies against monocytes and endothelial cells in the sera of patients with atherosclerotic peripheral arterial disease. *Atherosclerosis* 74:99-105, 1988.

Drenk, F. and Deicher, H.R.G. Pathophysiological effects of endothelial cytotoxic activity derived from sera of patients with progressive systemic sclerosis. *J. Rheumatol.* 15:468-474, 1988.

Etingin, D.R., Silverstein, R.L., Friedman, H.M. and Hajjar, D.P. Viral activation of the coagulation cascade: molecular interactions at the surface of infected endothelial cells. *Cell* 61:657-662, 1990.

Ferraro, G., Meroni, P.L., Tincani, A., Sinico, A., Barcellini, W., Radice, A., Gregorini, G., Froldi, M., Borghi, M.O. and Balestrieri, G. Anti-endothelial cell antibodies in patients with Wegener's granulomatosis and micropolyarteritis. *Clin. Exp. Immunol.* 79:47-53, 1990.

Gordon, J.L. and Pearson, J.D. Responses of endothelial cells to injury. In: Nossel, H.L. and Vogel, H.J. (eds); *Pathobiology of the endothelial cell.* Academic Press, New York, pp. 433-454, 1982.

Gordon, J.L., Pottinger, B.E., Woo, P., Rosenbaum, J. and Black, C.M. Plasma von Willebrand factor in connective tissue disease. *Ann. Rheum. Dis.* 46:491-492, 1987.

Hashemi, S., Smith, C.D. and Izaguirre, C.A. Anti-endothelial cell antibodies: detection and characterization using a cellular enzyme-linked immunosorbent assay. *J. Lab. Clin. Med.* 109:434-440, 1987.

Hasselaar, P., Derksen, R.H.W.M., Blokzijl, L. and de Groot, P.G. Thrombosis associated with antiphospholipid antibodies cannot be explained by effects on endothelial and platelet prostanoid synthesis. *Thromb. Haemostas.* 59:80-85, 1988.

Heurkens, A.H.M., Hiemstra, P.S., Lafeber, G.J.M., Daha, M.R. and Breedveld, F.C. Anti-endothelial cell antibodies in patients with rheumatoid arthritis complicated by vasculitis. *Clin. Exp. Immunol.* 78:7-12, 1989.

Holt, C.M, Lindsey, N., Moult, J., Malia, R.G., Greaves, M., Hume, A., Rowell, N.R., Hughes, P. Antibody-dependent cellular cytotoxicity of vascular endothelium: characterization and pathogenic associations in systemic sclerosis. *Clin. Exp. Immunol.* 78:359-365, 1989.

Ianaccone, S.T., Bowed, D., Yarom, A. and Ciraolo, G. *In vitro* study of cytotoxic factors against endothelium in childhood dermatomyositis. *Arch. Neurol.* 41:862-864, 1984.

Kahaleh, M.B., Sherer, G.K. and LeRoy, E.C. Endothelial injury in scleroderma. *J. Exp. Med.* 149:1326-1335, 1979.

Kahaleh, M.B., Osborn, I. and LeRoy, E.C. Increased factor VIII/von Willebrand factor antigen and von Willebrand factor activity in scleroderma and Raynaud's phenomenon. *Ann. Intern. Med.* 94:482-484, 1981.

Lang, B.A., Silverman, E.D., Laxer, R.M. and Lau, A.S. Spontaneous tumor necrosis factor production in Kawasaki disease. *J. Pediatr.* 115:939-943, 1989.

Larsen, E., Celi, A., Gilbert, G.E., Furie, B.C., Erban, J.K., Bonfanti, R. and Wagner, D.D. PADGEM protein: a receptor that mediates the interaction of activated platelets with neutrophils and monocytes. *Cell* 59:305-312, 1989.

Leung, D.Y.M., Collins, T., Lapierre, L.A., Geha, R.S. and Pober, J.S. Immunoglobulin M antibodies present in the acute phase of Kawasaki syndrome lyse cultured vascular endothelial cells stimulated by gamma interferon. *J. Clin. Invest.* 77:1428-1435, 1986a.

Leung, D.Y.M., Geha, R.F., Newburger, J.W., Burns, J.C., Fiers, W., Lapierre, L.A. and Pober, J.S. Two monokines, interleukin 1 and tumor necrosis factor render cultured vascular endothelial cells susceptible to lysis by antibodies circulating during Kawasaki syndrome. *J. Exp. Med.* 164:1958-1972, 1986b.

Leung, D.Y.M., Moake, J.L., Havens, P.L., Kim, M. and Pober, J.S. Lytic anti-endothelial cell antibodies in hemolytic-uraemic syndrome. *Lancet ii*:183-186, 1988.

Leung, D.Y.M., Cotran, R.S., Kurt-Jones, E., Burns, J.C., Newburger, J.W. and Pober, J.S. Endothelial cell activation and high interleukin-1 secretion in the pathogenesis of acute Kawasaki disease. *Lancet ii*:1298-1302, 1989.

Libby, P., Salomom, R.N., Payne, D.D., Schoen, F.J. and Pober, J.S. Functions of vascular walls cells related to the development of transplantation-associated atherosclerosis. *Transpl. Proc.* 21:3677-3684, 1989.

Marks, R.M., Czerniecki, M., Andrews, B.S. and Penny, R. The effects of scleroderma serum on human microvascular endothelial cells. *Arthr. Rheum.* 31:1524-1534, 1988.

Matsuo, S., Fukatsu, A., Taub, M.A., Caldwell, P.R.B., Brentjens, J.R. and Andres, G. Glomerulonephritis induced in rabbits by antiendothelial antibodies. *J. Clin. Invest.* 79:17-98-1811, 1987.

McCarty, G.A., Lister, K.A., Reichlin, M. and McEver, R. Autoantibody to a novel endothelial cell granule membrane protein antigen in vasculitis. *Clin. Res.* 38:316A, 1990.

Moraes, J.R. and Stasny, P. A new antigen system expressed in human endothelial cells. *J. Clin. Invest.* 60:449-454, 1977.

Nakajima, T., Koyama, T., Kakishita, E. and Nagai, K. Inhibitory effects of TTP sera on binding of anti-platelet glycoprotein IIb-IIIa monoclonal antibodies to human vascular endothelial cells. *Am. J. Hematol.* 25:115-118, 1987.

Paul, L.C., Baldwin, W.M. and Van Es, L.A. Vascular endothelial alloantigens in renal transplantation. *Transplantation* 40:117-123, 1985.

Penning, C.A., Cunningham, J., French, M.A.H., Harrison, G., Rowell, N.R. and Hughes, P. Antibody-dependent cellular cytotoxicity of human vascular endothelium in systemic sclerosis. *Clin. Exp. Immunol.* 57:548-556, 1984.

Penning, C.A., French, M.A.H., Rowell, N.R. and Hughes, P. Antibody-dependent cellular cytotoxicity of human vascular endothelium in systemic lupus erythematosus. *J. Clin. Lab. Immunol.* 17:125-130, 1985.

Pober, J.S., Collins, T., Gimbrone, M.A., Cotran, R.S., Gitlin, J.D., Fiers, W., Clayberger, C., Kreusky, A.M., Burakoff, S.J. and Reiss, C.S. Lymphocytes recognize human vascular endothelial and dermal fibroblast Ia antigens induced by recombinant immune interferon. *Nature* 305:726-729, 1983.

Pober, J.S. and Cotran, R.S. Cytokines and endothelial cell biology. *Physiological Reviews* 70:427-451, 1990.

Rappaport, V.J., Hirata, G., Yap, H.K. and Jordan, S.C. Anti-vascular endothelial cell antibodies in severe preeclampsia. *Am. J. Obstet. Gynecol.* 168:138-146, 1990.

Rosenbaum, J.R., Pottinger, B.E., Woo, P., Black, C.M., Loizou, S., Byron, M.A., Pearson, J.D. Measurement and characterization of circulating anti-endothelial cell IgG in connective tissue diseases. *Clin. Exp. Immunol.* 72:450-456, 1988.

Russ, G.R., Starr, R., Nicholls, C., Johnson, P. and Day, A. Monoclonal antibodies against antigens on human monocytes. *Transpl. Proc.* 19:2882-2884, 1987.

Ryan, U.S., Schultz, D.R. and Ryan, J.W. Fc and C3b receptors on pulmonary endothelial cells. Induction by injury. *Science* 214:557-558, 1981.

Savage, C.O.S., Pottinger, B.E., Gaskin, G., Wickwood, C.M., Pusey, C.D. and Pearson, J.D. Endothelial cell damage in Wegener's granulomatosis and microscopic polyarteritis: presence of anti-endothelial cell antibodies and their relation to anti-neutrophil cytoplasm antibodies. *Clin. Exp. Immunol.* (in press), 1991.

Schook, L.B., Wood, N. and Mohanakumar, T. Identification of human vascular endothelial cell/monocyte antigenic system using monoclonal antibodies. *Transplantation* 44:412-416, 1987.

Shingu, M. and Hurd, E.R. Sera from patients with systemic lupus erythematosus reactive with human endothelial cells. *J. Rheumatol.* 8:581-586, 1981.

Summers, G.D., Weiss, J.B. and Jayson, M.I.V. Failure of sera from patients with scleroderma to exhibit cytotoxicity towards human umbilical vein endothelial cells. *Rheumatol. Int.* 5:9-13, 1984.

Tanaka, N., Tsukada, N., Kar, C-S. and Yanagisawa, N. Antiendothelial cell antibodies and circulating immune complexes in sera of patients with multiple sclerosis. *J. Neuroimmunol.* 17:49-59, 1987.

Tannenbaum, S.H., Finko, R. and Cines, D.B. Antibody and immune complexes induce tissue factor production by human endothelial cells. *J. Immunol.* 137:1532-1537, 1986a.

Tannenbaum, S.H., Barnathan, E., Van der Keyl, H., Kuo, A., Murray, S. and Cines, D.B. Aggregated IgG stimulates the fibrinolytic inhibitory activity of human endothelial cells. *Blood* 68:(Suppl):235, 1986b.

Toothill, V.J., Van Mourik, J.A., Niewenhuis, H.K., Metzelaar, M.J. and Pearson, J.D. Characterization of the enhanced adhesion of neutrophil leukocytes to thrombin-stimulated endothelial cells. *J. Immunol.* 145:283-291, 1990.

Tsukada, N., Inoue, A., Yanagisawa, N., Behan, W.M.H. and Behan, P.O. Anti-endothelial cell antibody and immune complexes in the sera of animals with acute experimental allergic encephalomyelitis and chronic relapsing experimental allergic encephalomyelitis. *J. Neuroimmunol.* 12:89-97, 1986.

Uchmann, B., Bang, N.U., Rathbun, M.J., Fineberg, N.S., Davidson, J.K. and Fineberg, S.E. Effect of insulin immune complexes on human blood monocyte and endothelial cell procoagulant activity. *J. Lab. Clin. Med.* 112:652-659, 1988.

Van Mourik, J.A., Von dem Borne, A.E.G.K. and Giltay, J.G. Pathophysiological significance of integrin expression by vascular endothelial cells. *Biochem. Pharmacol.* 39:233-239, 1990.

Vismara A, Meroni PL, Tincani A, Harris, E.N., Barcellini, W., Brucato, A., Khamashta, M., Hughes, G.R.V., Zanussi, C. and Balestrieri, G. Relationship between anti-cardiolipin and anti-endothelial cell antibodies in systemic lupus erythematosus. *Clin. Exp. Immunol.* 74:247-253, 1988.

Walker, T.S., Triplett, D.A., Javed, N. and Musgrave, K. Evaluation of lupus anticoagulants: antiphospholipid antibodies, endothelium associated immunoglobulin, endothelial prostacyclin secretion, and antigenic Protein S levels. *Thrombos. Res.* 51:267-281, 1988.

Wood, N.L., Schook, L.B., Studer, E.J. and Monaakumar, T. Biochemical characterization of human vascular endothelial cell-monocyte antigens defined by monoclonal antibodies. *Transplantation* 45:787-792, 1988.

Yap, H.K., Sakai, R.S., Bahn, L., Rappaport, V., Woo, K.T., Ananthuraman, V., Lim, C.H., Chiang, G.S.C. and Jordan, S.C. Anti-vascular endothelial cell antibodies in patients with IgA nephropathy: frequency and clinical significance. *Clin. Immunol. Immunopathol.* 49:450-462, 1988.

LEUKOCYTE-ENDOTHELIAL CELL INTERACTIONS

James Varani

Department of Pathology
University of Michigan Medical School
Ann Arbor, Michigan 48109, U.S.A.

INTRODUCTION

Neutrophils normally circulate within the vasculature as unstimulated cells and do not, even when present on the endothelial cell surface as marginated cells, damage the vascular endothelium. However, these cells can become activated in response to a variety of stimuli produced within or outside of the vascular compartment. Once activated, neutrophils produce potent reactive oxygen metabolites and release the contents of their granules, which contain proteolytic enzymes as well as other hydrolyses and a large amount of cationic peptides. Products of activated neutrophils are capable of damaging the vascular endothelium and are responsible for the severe endothelial damage which is a hallmark of acute inflammatory disease. The active agents in the neutrophil armamentarium have been well-described in previous reviews (*Fantone and Ward, 1983; Weiss, 1989*) and this information will not be repeated here. Rather, this review will focus on two topics related to the interaction of neutrophils with the endothelium. These include 1) mechanisms of adhesive interactions between vascular endothelial cells and neutrophils and 2) events which occur within the endothelial cell under attack by neutrophils.

MECHANISMS OF NEUTROPHIL-ENDOTHELIAL CELL ADHESIVE INTERACTIONS

Neutrophil injury to the endothelium requires a priori that there be close contact between the effector cell and the target. Neutrophil oxidants, which play a major role in neutrophil-mediated injury, are for the most part short-lived species. Furthermore, even stable products such as hydrogen peroxide (H_2O_2) are present in small enough quantities that if they were diluted in the entire volume of the vascular space, they would be ineffective in producing measurable damage to the vascular lining cells. Additionally, plasma constituents are capable of scavenging oxygen radicals and inhibiting the activity of neutrophil proteolytic enzymes. The absolute requirement for a specific adhesive interaction between target and effector cell has prompted the search for the molecules responsible for mediating the binding of neutrophils to the endothelium. As indicated above, non-activated neutrophils within the vasculature do not bind tightly to normal vascular endothelial cells. In contrast, a variety of inflammatory mediators activate neutrophil adhesive functions concomitant with activation of the respiratory burst and granule secretion (*Varani et al., 1985*). Stimuli which are effective include immune complexes, zymosan particles, chemotactic peptides, such as C5a, and, to a lesser extent, formylated peptides and platelet-activating factor. A family of three adhesion molecules, termed leukocyte-function associated molecule-1 (LFA-1; CD11a/CD18), Mo1 (CD11b/CD18), and p150,95 (LeuM5; CD11c/CD18) have been identified by a number of investigators as being the major adhesion molecules up-regulated on the neutrophil surface

TABLE 1: Inhibition of Neutrophil Adhesion to Rat Pulmonary Artery Endothelial Cells with Antibodies to CD11/CD18-related molecules

TREATMENT GROUP	PERCENT ADHERENCE
Unstimulated human neutrophils	8 ± 2
Phorbol ester-stimulated human neutrophils	
NO antibody	72 ± 6
IgG1 (control)	76 ± 10
IgG2a (control)	80 ± 4
Anti-CD11a	64 ± 15
Anti-CD11b	13 ± 1
Anti-CD11c	49 ± 6
Anti-CD18	0 ± 3
Unstimulated rat neutrophils	15 ± 4
Phorbol ester-stimulated neutrophils	
No antibody	58 ± 2
IgM (control)	56 ± 10
Anti-CD11b	31 ± 11

Values shown represent means and standard deviations based on triplicate samples in a single experiment. Each antibody was examined on at least two separate occasions with similar results.

during activation. (*Trowbridge and Omary, 1981; Sanchez-Madrid et al., 1983; Todd et al., 1984; Lanier et al., 1985; Arnaout, Lanier and Faller, 1988*) These molecules are members of the integrin supergene family. Each of the three molecules contains a common ß subunit (i.e., CD18) linked non-covalently to a unique α subunit (i.e., CD11a, b, or c). CD11b and CD11c are present on neutrophils and monocytes as well as lymphocytes with NK activity while CD11a is present on T and B lymphocytes as well. The relative contribution of these adhesion proteins to leukocyte-endothelial cell adhesion has been well-studied using human umbilical vein endothelial cells as targets. With neutrophils activated in the presence of phorbol ester, nearly all of the adhesion to untreated human umbilical vein endothelial cells can be blocked with antibodies to the common β chain (anti-CD18) or with antibodies to Mo1 (anti-CD11b). Very little inhibition is seen with antibodies to the other two α chains, suggesting that Mo1 is primarily responsible for mediating this adhesion interaction. It should be noted that up-regulation of CD18-dependent neutrophil adhesion is associated with activation for cytotoxicity. The same stimuli which up-regulate adhesion also activate the respiratory burst (*Varani, et al., 1985*). Thus, the stimuli which promote neutrophil-endothelial cell adhesion also activate the neutrophils to injure the target cells.

Most studies documenting the role of CD11/CD18-related molecules in neutrophil-endothelial cell interactions have been conducted using human umbilical vein endothelial cells as the target. In contrast, our laboratory has made use of cells derived from rat pulmonary artery as the target in similar studies. Since the rat has been used extensively to elucidate mechanisms of inflammatory lung disease, it is important to identify the adhesive mechanisms that are relevant to these cells. In both autologous and heterologous systems, a role for similar moieties has been identified. That is, with both human and rat neutrophils, monoclonal antibodies to CD11/CD18-related molecules have been shown to provide significant inhibition of adhesion. With human neutrophils, anti-CD11b and anti-CD11c (but not anti-CD11a) antibodies provide inhibition (Table 1). Not surprising, inhibition of adhesion with monoclonal antibodies to CD11/CD18-related ligands results in a corresponding inhibition of cytotoxicity (*Wencel et al., 1989*). Finding a possible role for CD11c in the rat model is of interest because similar studies employing human umbilical vein endothelial cells as target identified a role for CD11b but not CD11a or CD11c (*Arnaout et al., 1988*). The rat pulmonary artery endothelial cells may, therefore, provide a good model in which to probe for the as yet unidentified ligand on the endothelial cell for the CD11c adhesion receptor on the neutrophil. The rat may also provide a useful model for probing the involvement of CD11c in leukocyte trafficking *in vivo*.

A full complement of monoclonal antibodies to CD11/CD18-related molecules on rat cells are not available and monoclonal antibodies to the human counterparts do not cross-react. Based on findings with a single antibody to CD11b, however, we found that adhesion of phorbol ester-stimulated rat peritoneal neutrophils to rat pulmonary artery endothelial cells could be blocked (Table 1).

These data suggest that CD18-related molecules are responsible for the majority of the binding of phorbol ester-activated neutrophils to untreated endothelial cells. However, this is not the only pathway for stimulation of neutrophil adhesion. We recently showed that human neutrophils treated with plasma and lipopolysaccharide demonstrated increased adhesion to endothelial cells that was not Mo1 inhibitable (*Wencel et al., 1989*). Functional significance for binding through this mechanism was suggested by the finding in the same study that neutrophils activated in this manner were able to injure endothelial cells *in vitro* and produce lung injury in a perfusion model.

In addition to the adhesion-promoting molecules on the neutrophil, there are endothelial cell surface moieties that can be up-regulated to promote binding of various types of leukocytes including neutrophils. Studies conducted during the early 1980s demonstrated that cytokines such as tumor necrosis factor-a (TNF-α), interleukin-1 (IL-1) and interferon-γ (IFN-g) stimulated the expression on the surface of human umbilical vein endothelial cells of molecules that promoted leukocyte adhesion (summarized in: *Cotran and Pober, 1989*). Endothelial cell expression of leukocyte adhesion molecules following cytokine stimulation occurs with a lag period of several hours and is blocked with inhibitors of transcription or protein synthesis. One of the moieties stimulated by TNF-α or IL-1 is intercellular adhesion molecule-1 (ICAM-1) (*Dustin et al., 1986*). This moiety is induced slowly (peaks at approximately 24 hours) and remains expressed as long as the cytokine is present. Marlin and Springer (1987) have provided evidence that ICAM-1 is the receptor for CD11a (LFA-1). A second neutrophil adhesion factor induced on cytokine-activated endothelial cells is referred to as endothelial leukocyte adhesion factor-1 (ELAM-1) (*Bevilacqua et al., 1986; 1987; Pober et al., 1986*). This receptor is not found on the surface of unstimulated human umbilical vein endothelial cells but is up-regulated following treatment with TNF-α or IL-1. It appears rapidly after cytokine treatment (peak expression is within 4-6 hours) and declines thereafter. This occurs regardless of whether the cytokine is continually present or not. ELAM-1 has structural homology with a group of adhesion factors that have been referred to as LEC-CAMs (*Stoolman, 1989*). This group of factors is characterized by the presence of an N-terminal sequence homologous to a variety of Ca2+ - dependent animal lectins in the cell-binding region of the molecule and includes MEL-14 (mouse), Leu8/TQ1/Lam-1 (human homologue of MEL-14) and GMP-140 (human). At present the corresponding neutrophil ligand for ELAM-1 has not been identified.

In addition to this small group of adhesion moieties that have already been identified, there are undoubtedly a number of other molecules that participate in the adhesion of various leukocytic populations with the vascular endothelium. Since virtually all of the work that has been conducted so far has made use of human umbilical vein endothelial cells as target, a major question that needs to be addressed is the role of these previously-identified adhesion factors in leukocyte adhesion to endothelial cells from other sources. Such studies are, for the most part, in their infancy. Studies have been conducted in our laboratory utilizing endothelial cells from rat pulmonary artery as a target of leukocyte adhesion. While our interest has been mainly on the intra (endothelial) cellular events that occur during injury mediated by human neutrophils, we have also probed the molecular basis of the adhesive interaction between these target cells and human neutrophils. As indicated above, we have found that the adhesion of phorbol ester-activated human neutrophils to rat pulmonary artery endothelial cells is mediated through the CD18-dependent pathway with a significant portion of the adhesion being due to CD11b and CD11c. In their ability to support adhesion of activated human neutrophils, therefore, the rat pulmonary artery endothelial cells appear to be functionally equivalent to human umbilical vein endothelial cells. We have also observed

TABLE 2: Inhibition of Neutrophil-Mediated Killing of Rat Pulmonary Artery Endothelial Cells

TREATMENT GROUP	PERCENT KILLED
Unstimulated human neutrophils	9 ± 1
Phorbol ester-stimulated human neutrophils	
+ No inhibitor	36 ± 4
SOD (280 units)	47 ± 3
+ Catalase (1800 units)	18 ± 3
+ SBT1 (100 μg)	41 ± 4
+ Desferol (750 μM)	14 ± 3
+ Sodium azide (1 mM)	54 ± 2

The cytotoxicity assay was carried out using human neutrophils stimulated with PMA (100 ng/ml); the effector to target radio was 20:1 and the inhibitors were added immediately before the leukocytes. Values shown are means and standard deviations based on triplicate samples in a single experiment. Each inhibitor was examined on two or more occasions with similar results. See: *Varani et al., 1985* for additional information.

functional equivalence between these two endothelial cell populations in their ability to support CD18-dependent lymphocyte and monocyte adhesion (unpublished).

With regard to the adhesion factors that can be up-regulated on the rat pulmonary artery endothelial cell surface, very little is known. We have found, however, that both TNF-α and IL-1 induce changes in these cells that increase their adhesion for unstimulated neutrophils (*Varani et al., 1988*). With TNF-α, responsiveness is observed within 4-6 hours and declines after 24 hours regardless of cytokine presence. This is consistent with induced expression of an ELAM-1 - like molecule, but there is no definitive evidence that it is the same molecule. If it turns out that rat pulmonary artery endothelial cells utilize similar molecules as do human umbilical vein endothelial cells in their support of neutrophil adhesion, this will provide a powerful approach to investigate the importance of these moieties in acute and chronic inflammatory disease conditions.

INTRACELLULAR EVENTS IN THE INJURY OF ENDOTHELIAL CELLS BY HUMAN NEUTROPHILS

Human peripheral blood neutrophils have a variety of potent effector mechanisms capable of injuring the vascular endothelium. These include a series of reactive oxygen metabolites, a number of potent proteolytic enzymes, as well as other hydrolyses (for example, phospholipase A2) and cationic proteins. The neutrophil hydrolyses and cationic proteins are contained within storage granules and are released by an exocytotic process during activation. The reactive oxygen metabolites are synthesized *de novo* from oxygen via activation of the membrane NADPH oxidase. The superoxide anion (O_2^-) generated in this reaction can be converted spontaneously or by the action of superoxide dismutase (SOD) to hydrogen peroxide (H_2O_2). H_2O_2 can be converted to the highly reactive hydroxyl radical (HO•) in a Fenton type reaction or converted to hypochlorous acid and chloramines through the action of the neutrophilic myeloperoxidase. A number of independent studies have demonstrated the central role of H_2O_2 in the killing of endothelial cells by activated neutrophils (*Sacks et al., 1978; Weiss et al., 1981; Martin, 1984; Varani et al., 1985*). This is based primarily on the strong protective effects seen with catalase and the ineffectiveness of SOD or proteolytic enzyme inhibitors under the same conditions. Further studies have shown that low molecular weight iron chelators, such as desferol and HO• scavengers, such as dimethylthiourea and

TABLE 3: Protection of Endothelial Cells from Neutrophil-Mediated Killing by Pretreatment/Washing with Desferol and Xanthine Oxidase Inhibitors

	Percent Protection	
Inhibitor	Neutrophils Treated[a]	Endothelial Cells Treated[b]
Desferol	-2 ± 3	88 ± 3
Allopurinol	-3 ± 1	64 ± 4
Oxypurinol	-4 ± 1	28 ± 2
Lodoxamide	0 ± 1	13 ± 1

[a]Human peripheral blood neutrophils were incubated for 30 minutes with 1.0 mM of the desired inhibitor. Following this, the neutrophils were washed tow times and added to monolayers of endothelial cells. Cytotoxicity was determined in the normal manner. The values shown are expressed as percent protection and represent triplicate values in a single experiment.

[b]Monolayers of rat pulmonary artery endothelial cells were incubated for 4 hours with 1.0 mM of the desired inhibitor. Following this, the monolayers were washed two times and neutrophils added. Cytotoxicity was determined in the normal manner. Values shown are expressed as percent protection and represent triplicate values in a single experiment. See: *Gannon et al., 1987* and *Phan et al., 1989*, for additional information.

mannitol, also provide substantial protection of endothelial cells against killing by activated neutrophils while myeloperoxidase inhibitors do not (*Varani et al., 1985*). This suggests that the toxic radical directly responsible for cell injury is the HO• (or perhaps some other iron-oxygen complex) rather than a product of the myeloperoxidase pathway such as hypochlorous acid or chloramines. Data from our own experiments are shown in Table 2.

Using rat pulmonary artery endothelial cells as the target and activated human peripheral blood neutrophils as the effector cells, our laboratory has made an effort to understand the biochemical events that occur during the injury process. The findings from these studies strongly suggest that the target endothelial cells are not passive "bystander" cells but actively participate in their own injury.

Our data suggest that the generation of the toxic radical occurs within the target cell, utilizing reactants generated by the endothelial cells themselves.

a. Source of iron in neutrophil-mediated killing of rat pulmonary artery endothelial cells: Because of the critical importance of iron to the cytotoxic process, studies were conducted to identify the source of the iron used in the process. Our studies showed that treatment of endothelial cells with the low molecular weight iron chelator desferol, results in the chelator entering the endothelial cell. A complex of iron and chelator is formed within the target cell. Although iron can be removed from the cell in the presence of desferol, the process occurs slowly (over the course of days) and is too slow to account for the protective effects. Thus, we suspect that the presence of desferol within the target cells prevents the iron from participating in HO• generating reactions or alters the distribution of iron within the cell such that the HO• is not produced in close proximity to critical target structures. In contrast to its protective effect within the target cells, treatment of neutrophils with the same iron chelator does not hinder their ability to injure rat pulmonary artery endothelial cells upon subsequent activation. Results obtained with desferol are summarized in Table 3 and are described in detail in a past report (*Gannon et al., 1987*).

b. Oxidant source within target endothelial cells: Iron must be present within the reduced form (Fe^{2+}) to participate in HO• generation from H_2O_2. Yet nearly all of the iron within

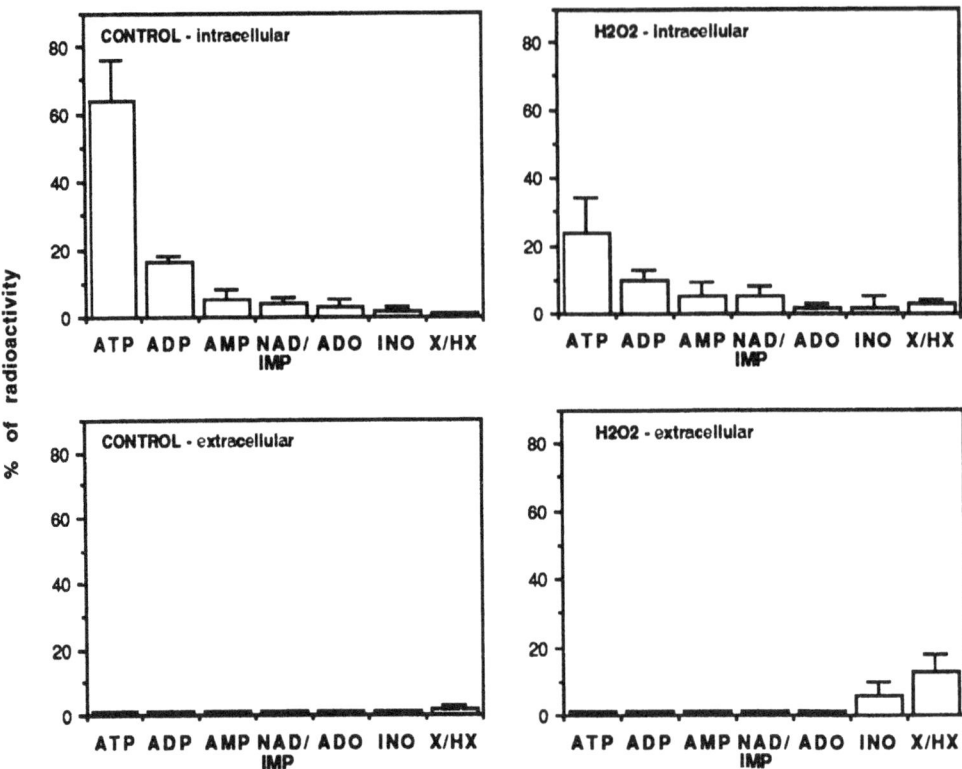

FIGURE 1: Distribution of radioactivity following incorporation of ^{14}C-adenosine into rat pulmonary artery endothelial cells and exposure of the cells for 30 minutes to buffer alone or to 500 nmols/ml of H_2O_2. Metabolites were separated by HPLC, quantitated by on-line 14C detection and compared to authentic standards. Values shown represent average percentages of total radioactivity co-migrating with authentic standards + standard deviations based on six separate experiments, each with duplicate samples. See: Varani et al., 1990 for additional information.

cells is in the Fe^{3+} form. It can be assumed, therefore, that a source of reducing equivalents must be present within susceptible cells to reduce iron from the Fe^{3+} form to the Fe^{2+} form. The enzyme xanthine oxidase is a good candidate for such a source in the rat pulmonary artery endothelial cells since rat tissues are rich in this enzyme. This enzyme can exist in two functional forms. In the dehydrogenase form it uses NAD as the electron acceptor, but in the oxidase form it uses O_2 as the electron acceptor, generating O_2^- in the process. That the xanthine oxidase-generated O_2^- may be involved in the cytotoxicity process is suggested by the finding that exposure of the endothelial cells to factors that result in cytotoxicity causes a conversion of the enzyme from the dehydrogenase form to the oxidase form and by the finding that inhibitors of xanthine oxidase protect endothelial cells against killing by activated neutrophils or reagent H_2O_2. These data, which have been described in recent reports (*Phan et al., 1989; Varani et al., 1990*) are summarized in Table 3.

As part of our effort to delineate the role of endothelial cell xanthine oxidase in the cytotoxicity process, we examined cells for the presence of substrate (xanthine and hypoxanthine) for this enzyme (*Varani et al, 1990*). Cells were labeled for 45 minutes with ^{14}C-adenosine. Following this, they were washed two times to remove unincorporated label and then incubated for varying periods of time in control buffer or in the presence of H_2O_2. At the end of the incubation periods, extracts were prepared and analyzed by HPLC for ^{14}C-labeled metabolites of adenosine. As shown in Figure 1, approximately 80% of the incorporated label was found in control cells in a fraction co-migrating with authentic ATP. Approximately 15% of the label co-migrated with ADP with small amounts associating with AMP, adenosine and NAD/inosine monophosphate. There was virtually no radioactivity associating with xanthine or hypoxanthine. In contrast, radioactivity co-migrating with ATP dropped precipitously within 5 minutes of exposure to H_2O_2. Simultaneously, there was a build-up of radioactivity in the xanthine/hypoxanthine fractions. By 30 minutes after exposure to H_2O_2, up to 20% of the incorporated radioactivity was found in this fraction. It is of interest that although ATP levels fell rapidly after H_2O_2 exposure, this was, of itself, not indicative of cytotoxicity. Addition of catalase to the buffer 30-60 minutes after treatment with H_2O_2 resulted in substantial protection from killing, even though the greatest changes in ATP levels had already taken place. In addition, agents which provided significant protection against killing, including desferol and allopurinol, did not inhibit the loss of ATP from the H_2O_2-exposed cells or the build-up of xanthine and hypoxanthine. These data are, therefore, consistent with the suggestion that the loss of ATP per se is not responsible for oxidant injury but may participate by providing substrate for the generation of oxidant species within the target cell.

The model of endothelial cell injury developed on the basis of these findings can be summarized as follows: It suggests that neutrophil-mediated killing of rat pulmonary artery endothelial cells requires intimate contact between target and effector cells; requires activation of surface HADPH oxidase, resulting in the generation of O_2^- and subsequently H_2O_2; requires availability of ferrous iron and an oxidant (presumably derived from xanthine oxidase activity) within the target cell and ultimately involves the generation within the target cell of HO• by ferrous iron reduction of H_2O_2.

OTHER MECHANISMS OF NEUTROPHIL-MEDIATED ENDOTHELIAL CELL INJURY

It should not be implied that the scenario described above is the only way in which activated neutrophils can injure the vascular endothelium. It has been shown by Harlan et al. (1981; 1985) that activated neutrophils can cause disruption of intact endothelial cell monolayers by an oxygen radical-independent, protease-dependent mechanism. While this is not associated with lethal injury to the endothelial cells, disruption of the vascular lining could have important consequences for the establishment or amplification of inflammatory processes.

In addition to this model of non-lethal endothelial cell injury, studies by other investigators have shown that microvascular endothelial cells can be lethally injured by neutrophil-derived proteolytic enzymes (*Smedly et al., 1986*). Neutrophil elastase and a membrane-bound neutrophil protease have been implicated as the effector moieties responsible for the injury in these models (*Smedly et al., 1986; Pontremoli et al., 1986*). Perhaps the most interesting model, in light of the findings presented in this report, is one described by Rodell et al. (1987) showing that inhibitors of xanthine oxidase protect bovine pulmonary artery endothelial cells from elastase-mediated injury. The implication of this is that even in protease-mediated endothelial cell injury, there is a role for events which occur within the target cell and that these events may depend on the generation of an oxidant.

SUMMARY

It should be obvious from this short review that neutrophil-mediated injury of vascular endothelial cells is a complex event and may involve multiple independent and/or overlapping processes. In all of these, the target endothelial cell is not a passive bystander but rather, actively participates in its own injury. Thus, understanding the molecular events that occur during inflammatory vascular disease requires understanding the biochemical processes that occur in the target cell as well as the biochemical processes of the effector cell.

REFERENCES

Arnaout, M.A., Lanier, L.L. and Faller, D.V. Relative contribution of the leukocyte molecules Mol, LFA-1 and P150, 95 (Leu M5) in adhesion of granulocytes and monocytes to vascular endothelium is tissue and stimulus-specific. *J. Cell. Physiol.* 137:305-309, 1988.

Bevilacqua, M.P., Pober, J.S., Mendrick, D.L., Cotran, R.S. and Gimbrone, M.A. Jr. Identification of an inducible endothelial-leukocyte adhesion molecule. *Proc. Natl. Acad. Sci. USA* 84:9238-9242, 1987.

Bevilacqua, M.P., Pober, J.S., Wheeler, M.E., Fiers, W., Mendrick, D., Cotran, R.S. and Gimbrone, M.A. Jr. Endothelial-dependent mechanisms of leukocyte adhesion: Regulation by IL-1 and TNF, In: *Leukocyte Emigration and its Sequelae*, Basel, S Karger, 1987.

Cotran, R.S. and Pober, J.S. Effects of cytokines on vascular endothelium: Their role in vascular and immune injury. *Kidney International* 35:969-975, 1989.

Dustin, M.L., Rothlein, R., Bhan, A.K., Dinarello, C.A. and Springer, T.A. A human intercellular adhesion molecule (ICAM-1) distinct from LFA-1. *J. Immunol.* 137:1270-1274, 1986.

Fantone, J.C. and Ward, P.A. Mechanism of neutrophil-dependent lung injury. In: *Immunology of Inflammation:Handbook of Inflammation.* Vol IV, edited by P.A. Ward p 89. Amsterdam, Elsevier, 1983.

Gannon, D.E., Varani, J., Phan, S.H., Ward, J.H., Kaplan, J., Till, G.O., Simon, R.H., Ryan, U.S. and Ward, P.A. Source of iron in neutrophil-mediated killing of endothelial cells. *Lab. Invest.* 57:37-44, 1987.

Lanier, L.L., Arnaout, M.A., Schwarting, R., Warner, N.L. and Ross, G.D. p 150/95, third member of the LFA-1/CR3 polypeptide family identified by anti-Leu M5 monoclonal antibody. *Eur. J. Immunol.* 15:713-718, 1985.

Marlin, S.D. and Springer, T.A. Purified intercellular adhesion molecule-1 (ICAM-1) is a ligand for lymphocyte function-associated antigen 1 (LFA-1). *Cell* 51:813-819, 1987.

Martin, W.J. Neutrophils kill pulmonary endothelial cells by a hydrogen peroxide-dependent pathway: An *in vitro* model of neutrophil-mediated lung injury. *Am. Rev. Res. Dis.* 130:209-213, 1984.

Phan, S.H., Gannon, D.E., Varani, J., Ryan, U.S. and Ward, P.A. Xanthine oxidase activity in rat pulmonary artery endothelial cells and its modulation by activated neutrophils. *Am. J. Pathol.* 134:1201-1211, 1989.

Pober, J.S., Bevilacqua, M.P., Mendrick, D.L., Lapierre, L.A., Fiers, W. and Gimbrone, M.A. Jr. Two distinct monokines, interleukin 1 and tumor necrosis factor, each independently induce biosynthesis and transient expression of the same antigen on the surface of cultured human vascular endothelial cells. *J. Immunol.* 136:1680-1687, 1986.

Pontremoli, S., Melloni, E., Michetti, M., Sacco, O., Sparatore, B., Salamino, F., Damiani, G. and Horecker, B.L. Cytolytic effects of neutrophils: Role for a membrane-bound neutral proteinase. *Proc. Natl. Acad. Sci. USA* 83:1685-1689, 1986.

Rodell, T.C., Cheronis, J.C., Ohnemus, C.L., Piemattei, D.J. and Repine, J.E. Xanthine oxidase mediates elastase induced injury in isolated lungs and endothelium. *J. Appl. Physiol.* 63:2159-2163, 1987.

Sanchez-Madrid, F., Nagy, J.A., Robins, E., Simon, P. and Springer, T.A. A human leukocyte differentiation family with distinct alpha subunit and a common beta subunit: The lymphocyte function associated antigen (LFA-1), the C3bi complement receptor (OKM 1/Mac 1) and the p 150,95 molecule. *J. Exp. Med.* 58:1785-1803, 1983.

Sacks, T., Moldow, C.F., Craddock, P.R., Boweres, T.K. and Jacob, H.A. Oxygen radicals mediate endothelial cell damage by complement-stimulated granulocytes: An *in vitro* model of immune complex vasculitis. *J. Clin. Invest.* 61:1161-1167, 1978.

Smedly, L.A., Tonnesen, M.G., Sandhaus, R.A., Haslett, C., Guthrie, L.A., Johnson, R.B., Henson, P.M. and Worthen, G.S. Neutrophil-mediated injury to endothelial cells: Enhancement by endotoxin and essential role of neutrophil elastase. *J. Clin. Invest.* 77:1233-1243, 1986.

Stoolman, L.M. Adhesion molecules controlling lymphocyte migration. *Cell* 56:907-910, 1989.

Todd, R.F. III, Bahn, A.K., Kabawat, S.E. and Schlossman, S.F. Human myelomonocytic differentiation antigens defined by monoclonal antibodies. In BA Bounswell, J Dausset, CF Milstein, and SF Schlossman (eds): *Leukocyte Typing* New York: Springer-Verlag, pp 424-433, 1984.

Trowbridge, I.S. and Omary, M.B. Molecular complexity of leukocyte surface glycoproteins related to the macrophage differentiation antigen Mac-1. *J. Exp. Med.* 154:1517-1524, 1981.

Varani, J., Bendelow, M.J., Sealey, D.E., Kunkel, S.L., Gannon, D.E., Ryan, U.S. and Ward, P.A. Tumor necrosis factor enhances susceptibility of vascular endothelial cells to neutrophil-mediated killing. *Lab. Invest.* 59:292-295, 1988.

Varani, J., Fligiel, S.E.G., Till, G.O., Kunkel, R.G., Ryan, U.S. and Ward, P.A. Pulmonary endothelial cell killing by human neutrophils: Possible involvement of hydroxyl radical. *Lab. Invest.* 53:656-663, 1985.

Varani, J., Phan, S.H., Gibbs, D.F., Ryan, U.S. and Ward, P.A. H_2O_2-mediated cytotoxicity of rat pulmonary endothelial cells: Changes in purine products and effects of protective intervention. *Lab. Invest.* 63:683-689, 1990.

Weiss, S.J. Tissue destruction by neutrophils. *New Eng. J. Med.* 320:365-375, 1989.

Weiss, S.J., Young, J., LoBuglio, A.F., Slivka, A. and Nimeh, N.F. Role of hydrogen peroxide in neutrophil-mediated destruction of cultured endothelial cells. *J. Clin. Invest.* 68:714-720, 1981.

Wencel, M.L., Morganroth, M.L., Schoeneich, S.O., Gannon, D.E., Varani, J., Todd, R.F., Ryan, U.S. and Ward, P.A. Cytoplasts and Mol-deficient neutrophils pretreated with plasma and LPS induce lung injury. *Am. J. Physiol.* 256:H751-H759, 1989.

V. VASCULAR RESPONSES TO ENDOTHELIAL CELL INJURY

ENDOTHELIUM FUNCTION IN HUMAN CORONARY BYPASS GRAFTS

Thomas F. Lüscher and Z. Yang

Department of Internal Medicine
Division of Cardiology, and Department of Research
Vascular Research
University Hospital
Basel, Switzerland

ARTERIAL AND VENOUS CORONARY BYPASS GRAFTS

Coronary bypass grafting is an established method in the surgical management of patients with coronary artery disease (*Lüscher and Vanhoutte, 1990*). Most commonly, the saphenous vein and the internal mammary artery are used as coronary bypass vessels (*Effler et al., 1963; Green et al., 1970; Grondin et al., 1984; Miller, Jr. et al., 1981; Okies et al., 1984; Tector et al., 1984*). More recently, the right gastroepiploic artery has been introduced as an alternative coronary bypass graft (*Verkkala et al., 1989*).

Several clinical studies have demonstrated that with mammary artery grafts the long-term patient morbidity and mortality is lower and the graft patency is higher as compared to venous grafts (Figure 1; *Grodin et al., 1984; Loop et al., 1986; Lytle et al., 1985; Tector et al., 1981*). Since this difference persists even when arterial and venous grafts supplying the same epicardial coronary artery (for instance the left anterior descending coronary artery) are compared, it has been suggested that different biological properties of arteries and veins play an important role for graft function.

CAUSES OF GRAFT FAILURE

Coronary bypass graft failure has many causes, and their importance differs at different time points after implantation (*Fuster and Chesebro, 1986; Lüscher et al., 1988; Yang and Lüscher, 1989*). Perfusion pressure through the graft, run-off in the native coronary vascular bed, platelet-vessel wall interaction and the physiological behavior of the graft itself are important determinants of graft function and patency (*Lüscher et al., 1988*). This review will focus on the latter two aspects.

Blood flow to the native coronary circulation through a bypass graft can be impaired due to vasospasm (*D'Souza et al., 1984; Sarabu et al., 1987*), platelet aggregation and thrombus formation (*Fuster and Chesebro, 1986; Lüscher et al., 1988; Yang and Lüscher, 1989*), and/or due to atherosclerotic plaques and intimal hyperplasia (*Grondin et al., 1984; Campeau et al., 1984; Spencer, 1986; Singh, 1983*). The development of such alterations is greatly influenced by biological properties of the grafts, in particular the function of the endothelium and vascular smooth muscle cells. Both cell types importantly contribute to the regulation of interactions of blood cells (i.e. platelets, monocytes) with the vessel wall (*Lüscher and Vanhoutte, 1990*) and to proliferative processes (*Garg and Hassid, 1989*).

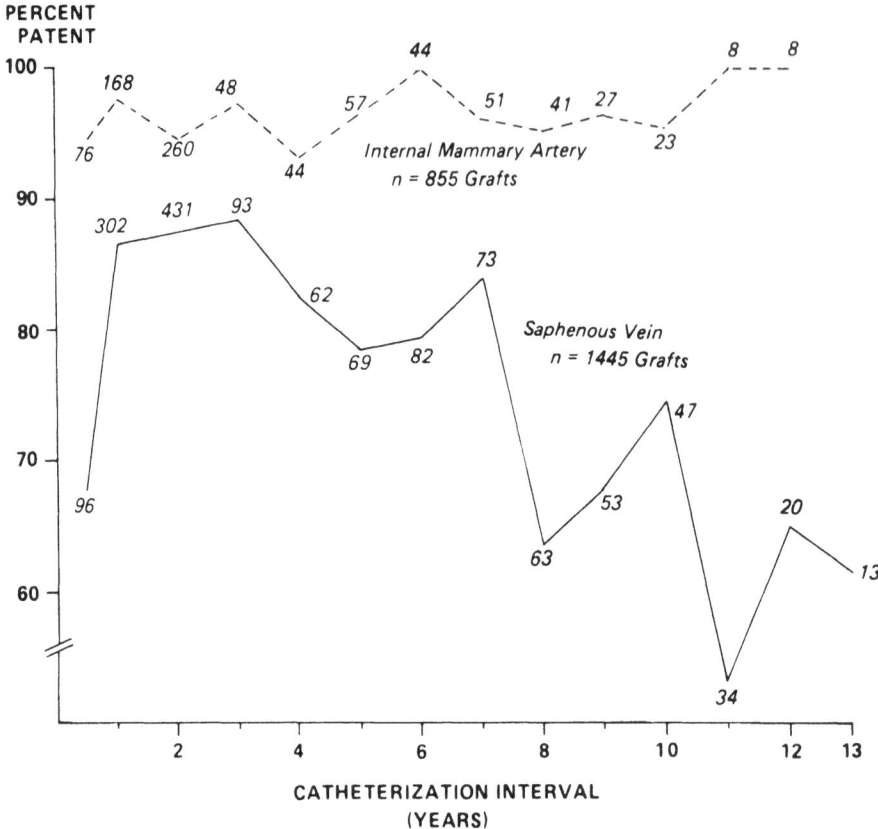

FIGURE 1: Patency rate of internal mammary artery grafts (dashed line) and saphenous vein grafts (solid line) at different catheterization intervals in patients who underwent coronary bypass grafting to the left anterior descending coronary artery. Note the remarkably higher patency rate of arterial as compared to venous grafts (*Loop, et al., 1986, by permission*).

THE ENDOTHELIUM AND VASCULAR FUNCTION

The vascular endothelium has many roles in the circulation (*Lüscher and Vanhoutte, 1990*). Particularly important are its antiadhesive properties which prevent the adherence of platelets (Figure 2: *Azuma et al., 1986; Radomski et al., 1987a,b; Busse et al., 1987*), leukocytes and monocytes. In addition, the cells produce factors involved in blood clotting and play a crucial role in capillary transport. Endothelial cells profoundly affect local vascular tone by activating (i.e. angiotensin I) and inactivating (i.e. norepinephrine, serotonin, bradykinin and adenosine 5′-diphosphate) local and circulating hormones and by releasing vasodilator (i.e. endothelium-derived relaxing factor, prostacyclin) and vasoconstrictor substances (i.e. endothelin, angiotensin, endothelium-derived constricting factors, histamine). Finally, the endothelium must participate in the regulation of vascular growth. The cells are a source of growth inhibitors (such as heparin and heparin sulfates and nitric oxide; *Garg and Hassid, 1989; DiCorletto and Fox, 1987*) and growth factors (platelet-derived growth factor; *DiCorletto and Fox, 1987*).

ENDOTHELIUM-DERIVED NITRIC OXIDE IN HUMAN BLOOD VESSELS

In 1980, Furchgott and Zawadzki demonstrated that in the isolated aorta of the rabbit, relaxations induced by acetylcholine are endothelium-dependent (Figure 3; *Furchgott and Zawadski, 1980*). The mediator of the response was named endothelium-derived relaxing factor (EDRF) and has recently been identified as nitric oxide (*Palmer et al., 1987; Palmer et al., 1988; Ignarro et al., 1987*). Nitric oxide is formed from the precursor aminoacid L-argi-

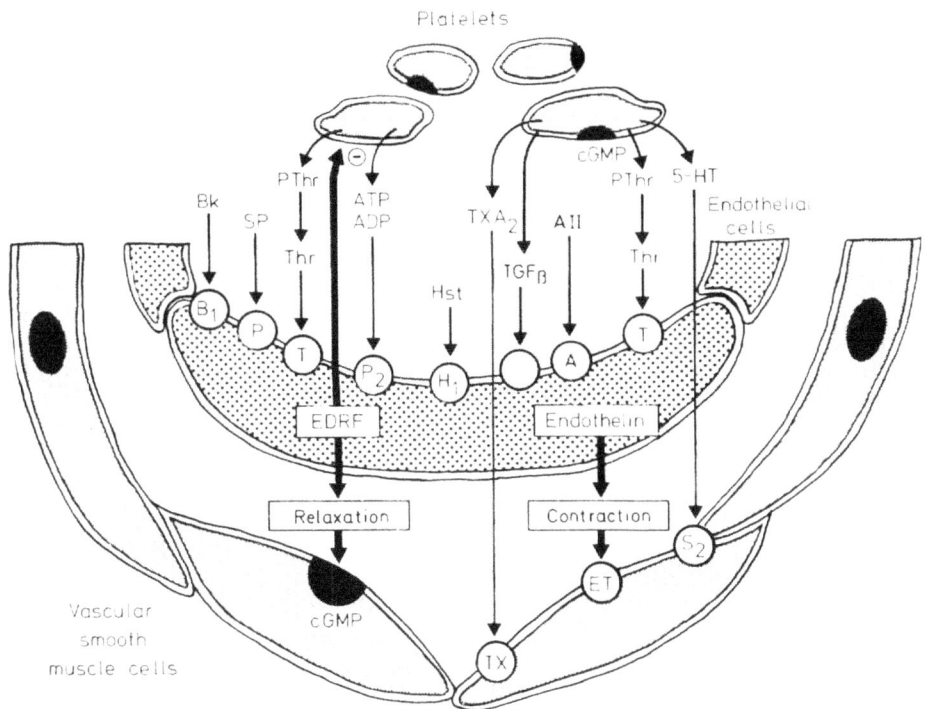

FIGURE 2: Release of endothelium-derived relaxing factor (EDRF) and endothelin (GT) from the blood vessel wall. EDRF causes relaxation and inhibition of platelet function while endothelin is a potent vasoconstrictor substance. The substances are released after activation of specific receptors. A = angiotensinergic receptor; AII = angiotensin II; ADP/ATP = adenosine diphosphate/triphosphate receptor; Bk = bradykinin; 5-HT = serotonin (5-hydroxytryptamine); H_1 = histaminergic receptor; HST = histamine; P = substance P receptor; P_1 = P_1-purinergic receptor; SP = substance P; S_2 = serotonergic receptor; Thr = thrombin; TGF = transforming growth factor ; TXA_2 = thromboxane A_2; TX = thromboxane receptor.

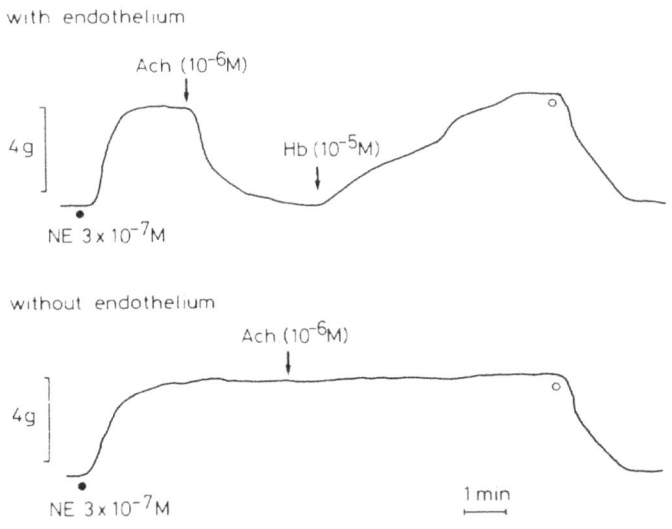

FIGURE 3: Endothelium-dependent relaxations to acetylcholine; (Ach) in the human internal mammary artery. NE = Norepinephrine, Hb = hemoglobin (Original recording from *Lüscher et al., 1988*, by permission).

nine by the action of specific cytosolic enzymes (*Palmer et al., 1988*). Debate continues whether the endogenous nitrate is liberated as nitric oxide or bound to a carrier molecule such as nitrosocysteine (*Myers et al., 1990*).

Early attempts to demonstrate endothelium-dependent relaxations in human arteries obtained at autopsy have failed, indicating that the endothelium, soon after death, looses its capacity to form and/or release nitric oxide (*Kalsner, 1985*). In human blood vessels obtained intraoperatively, however, endothelium-dependent responses do occur (*Lüscher et al., 1987; Lüscher et al., 1988; Lüscher and Vanhoutte, 1988; Yang et al., 1989; Forstermann et al., 1988*). In human arteries of various anatomical origin, acetylcholine induces complete inhibition of norepinephrine-induced tone in rings with, but not in those without, endothelium (Figure 3; *Lüscher et al., 1988*). The response is unaffected by cyclooxygenase inhibitors (which excludes prostacyclin as the mediator) but it can be markedly reduced (with acetylcholine as agonist) or prevented (in the case of histamine) by the inhibitor of nitric oxide formation from L-arginine L-N^G-monomethyl arginine (L-NMMA; Figure 4; *Yang et al., 1991a*). Hemoglobin (which binds nitric oxide in its ferrous ring; *Martin et al., 1985*) can reverse the response (*Lüscher et al., 1988*). The endothelium-dependent relaxations are associated with a rise in

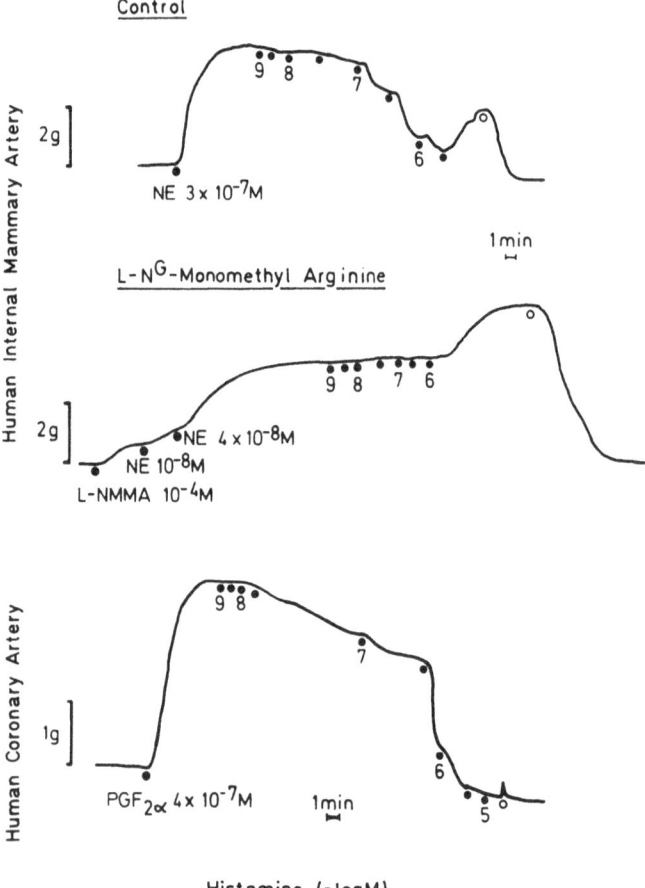

FIGURE 4: Endothelium-dependent relaxations to histamine in the human internal mammary artery (top panel) and human left anterior descending coronary artery (lower panel). In both blood vessels, histamine evokes potent relaxations in the presence of the endothelium. Since the inhibitor of nitric oxide formation from L-arginine L-N^G-monomethyl arginine (L-NMMA) prevents the response, nitric oxide must be the mediator. NE = Norepinephrine; PGF$_{2\alpha}$ = Prostaglandin F$_{2\alpha}$ (Unpublished observation).

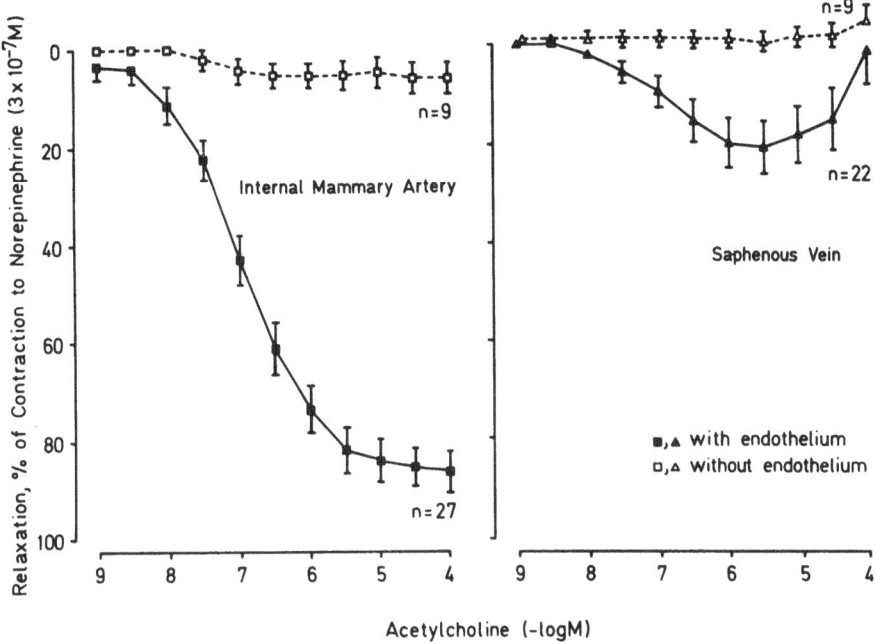

FIGURE 5: Endothelium-dependent relaxations to acetylcholine in the human internal mammary artery and saphenous vein. Note a much more pronounced response in the artery as compared to the vein (from *Lüscher et al., 1988*, by permission).

intracellular cyclic GMP in vascular smooth muscle cells (*Rapoport and Murad, 1983*). Since the inhibitor of soluble guanylate cyclase methylene blue inhibits the relaxation, the formation of cGMP must be essential for the response.

ENDOTHELIUM-DEPENDENT RESPONSES IN HUMAN BYPASS VESSELS

(a) Basal release of endothelium-derived nitric oxide: In quiescent human internal mammary arteries, L-NMMA causes endothelium-dependent contractions demonstrating the basal release of endothelium-derived nitric oxide (*Yang et al., 1991a*). As the response to L-NMMA is much weaker in the saphenous and mammary vein than in the mammary artery, the basal formation of the endogenous nitrate is much less pronounced in veins than in arteries. In line with this observation, removal of the endothelium augments the contractile response to norepinephrine and serotonin in the mammary artery, but not in the saphenous vein (*Yang and Lüscher, 1990*). As L-NMMA has similar effects on the response to norepinephrine and serotonin in preparations with, but not in those without endothelium, endothelium-derived nitric oxide must be involved.

(b) Stimulated release of endothelium-derived nitric oxide: Acetylcholine causes much more pronounced endothelium-dependent relaxations in the mammary artery than in the saphenous vein (Figure 5; *Lüscher et al., 1988*). In the right gastroepiploic artery, endothelium-dependent relaxations are comparable to those seen in the mammary artery (Yang et al., submitted). Although acetylcholine is much less effective in the saphenous vein, the blood vessel responds well both to nitric oxide donors such as sodium nitroprusside, SIN-1 (the active metabolite of molsidomine; Figure 6; *Lüscher et al., 1989*), exogenous nitric oxide and to arterial EDRF transferred from an arterial segment with endothelium (*Lüscher et al., 1988*). Thus, the different endothelium-dependent relaxations in arteries and veins are not due to an inability of venous vascular smooth muscle to relax, but rather due to a decreased release of endothelium-derived relaxing factor(s) and/or due to differences in the release of contracting factors (*Yang et al., 1991a*).

In the mammary artery, nitric oxide is the primary mediator of endothelium-dependent relaxations (Figure 7; *Yang et al., 1991a*). Similarly, in the saphenous vein, L-NMMA fully prevents the endothelium-dependent relaxations to acetylcholine indicating that, in contrast to the dog (*Miller and Vanhoutte, 1989*), nitric oxide mediates endothelium-dependent relaxations in human veins. Unlike the mammary artery, indomethacin and the inhibitors of thromboxane synthetase CGS-13080 augment the relaxations to acetylcholine in the saphenous vein, suggesting that the venous endothelium produces both nitric oxide and an cyclo-oxygenase-dependent endothelium-derived contracting factor (EDCF$_2$), most likely thromboxane A$_2$, after muscarinic stimulation (Figure 7; *Yang et al., 1991a*).

Histamine causes potent endothelium-dependent relaxations in mammary arteries (via an endothelial H$_1$-histaminergic receptor) as well as in the right gastroepiploic artery, but contractions in the saphenous vein (*Yang et al., 1989; Yang et al., 1991a*; Yang et al., submitted). Indomethacin and the thromboxane/endoperoxide receptor antagonist SQ 30741 unmask moderate endothelium-dependent relaxations to histamine in the veins (Yang and Lüscher, unpublished). As the thromboxane synthetase inhibitor CGS-13080 is ineffective under these conditions, this suggests that histamine releases both nitric oxide and endoperoxides (most likely prostaglandin H$_2$; *Kato et al., 1990*) from the venous endothelium (Figure 7).

(c) Endothelin-1: In human bypass vessels, endothelin-1 evokes potent contractions in a concentration range about two orders of magnitude lower than that with norepinephrine (Figure 8; *Lüscher et al., 1990*). The endothelium plays a minor regulatory role as removal of the intima only augments the maximal contraction in response to endothelin-1 (but not the sensitivity to the peptide) in the artery and leaves the response of the saphenous vein unaffected (*Lüscher et al., 1990*).

EDRF is a potent inhibitor of endothelin-1-induced contractions (Figure 9; *Lüscher et al., 1990*). Indeed, in the mammary artery, acetylcholine fully reverses the contractions to the peptide to a similar degree as those evoked by norepinephrine. In contrast, in the saphenous vein, endothelin-1 reduces the capacity of EDRF (released by bradykinin) to relax the blood vessels (Figure 9; *Lüscher et al., 1990*). As nitric oxide and sodium nitroprusside also are less effective to relax the saphenous vein contracted with endothelin-1 than that contracted with norepinephrine, the peptide must reduce the capacity of venous vascular smooth muscle to relax via a cyclic GMP-dependent mechanism (*Lüscher et al., 1990*). Most likely this is related

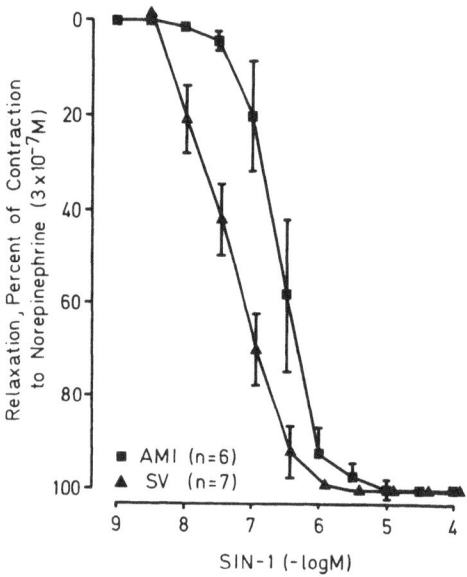

FIGURE 6: Relaxations induced by the nitric oxide donor 8-morpholino-sydnomine (SIN-1) in the internal mammary artery (IMA) and saphenous vein (SV) contracted with norepinephrine. Note that the vein exhibits an augmented sensitivity, but a comparable maximal response to the nitrovasodilator as compared to the artery (*from Lüscher et al., 1989 by permission*).

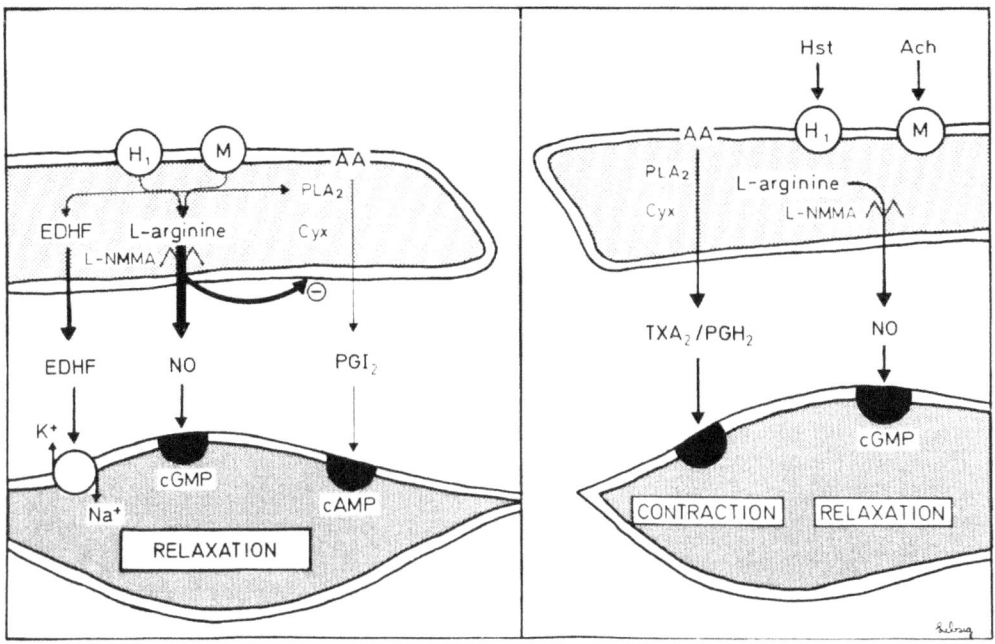

FIGURE 7: Endothelium-derived mediators in the human internal mammary artery (left panel) and saphenous vein (right panel). In the artery, nitric oxide (NO) formed from L-arginine is by far the most important mediator in response to acetylcholine (Ach) and histamine (Hst). Although the arteries formed prostacyclin (PGI_2) after activation of phospholipase A_2 (PLA_2) and cyclooxygenase (CYX), PGI_2 does not significantly contribute to endothelium-dependent relaxations. It remains possible, however, that an endothelium-derived hyperpolarizing factor (EDHF) is released after stimulation of muscarinic receptors. In the saphenous vein, less NO is formed and its effects are in part inhibited by endothelium-derived contracting factors originating from the cyclooxygenase pathway such as thromboxane A_2 (TXA_2) and prostaglandin H_2 (PGH_2). AA = Arachidonic acid; cAMP/cGMP = cyclic Adenosine/Guanosine monophosphate; H_1 = Histamine receptor; L-NMMA = L-N^G-monomethyl arginine; M = muscarine receptor.

to a more pronounced depolarization of venous as compared to arterial vascular smooth muscle (*Miller et al., 1989*).

PLATELET-VESSEL WALL INTERACTIONS

Increased platelet-vessel wall interactions (see Figure 2) with platelet activation and adhesion to the vessel wall have been implicated in venous graft occlusion (*Fuster and Chesebro, 1986; Lüscher et al., 1988; Yang and Lüscher, 1989*). Thus, if EDRF were involved in short-term or long-term patency of bypass grafts, the factor should be released differently in arterial and venous grafts in response to aggregating platelets, platelet-derived substances and coagulation products.

(a) Role of the endothelium: In the mammary artery, aggregating platelets evoke endothelium-dependent relaxations (*Yang and Lüscher, 1990*). As the response is prevented by apyrase (which breaks down adenine nucleotide; *Houston et al., 1986*) and L-NMMA, adenosine tri- and diphosphate (ATP/ADP) released from platelets must activate endothelial purinergic receptors linked to the release of endothelium-derived nitric oxide. Indeed, ADP and thrombin cause endothelium-dependent relaxations in this preparation (Figure 10; *Lüscher et al., 1988*). In contrast, in the saphenous vein, aggregating platelets cause contractions only, which are facilitated in the presence of the endothelium (*Yang et al., 1991b*). Thrombin and ADP also do not cause endothelium-dependent relaxations in the saphenous vein (*Lüscher et al., 1988*).

Stimulation of the production of endothelium-derived nitric oxide and the concomitantly

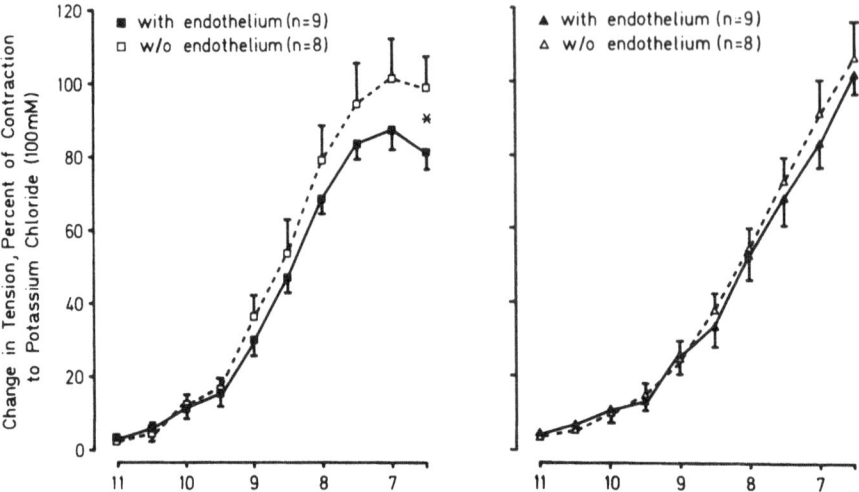

FIGURE 8: Graphs showing effects of endothelium-derived relaxing factor released by acetylcholine in human internal mammary arteries (left panel) and by bradykinin in saphenous veins (right panel) on endothelin-1-induced contractions. Note that, in arteries, endothelium-derived relaxing factor potently inhibits contractions to either endothelin-1 (□) or norepinephrine (■). In contrast, in veins, endothelin-1 (△) significantly inhibits endothelium-derived relaxing factor-induced relaxations ($p<0.01$). (*By permission of the American Heart Association*).

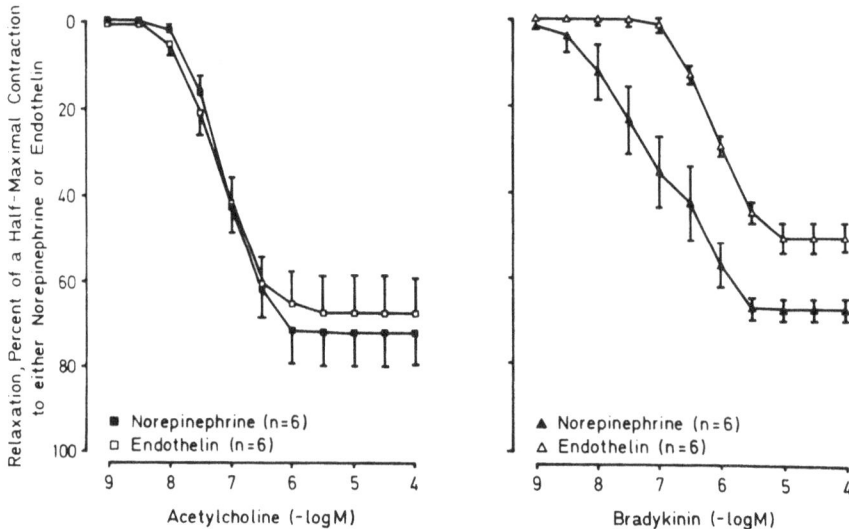

FIGURE 9: Endothelium-dependent relaxations in human internal mammary arteries (left panel) and saphenous veins (right panel) contracted with either norepinephrine (closed symbols) or endothelin-1 (open symbols). Note that in the artery, acetylcholine is similarly effective in preparations contracted with norepinephrine or endothelin-1. In contrast, endothelium-dependent relaxations to bradykinin are less pronounced in the vein as compared to those evoked by acetylcholine in the artery and they are further inhibited by endothelin-1 as compared to norepinephrine (*from Yang et al., 1990 in press, by permission*).

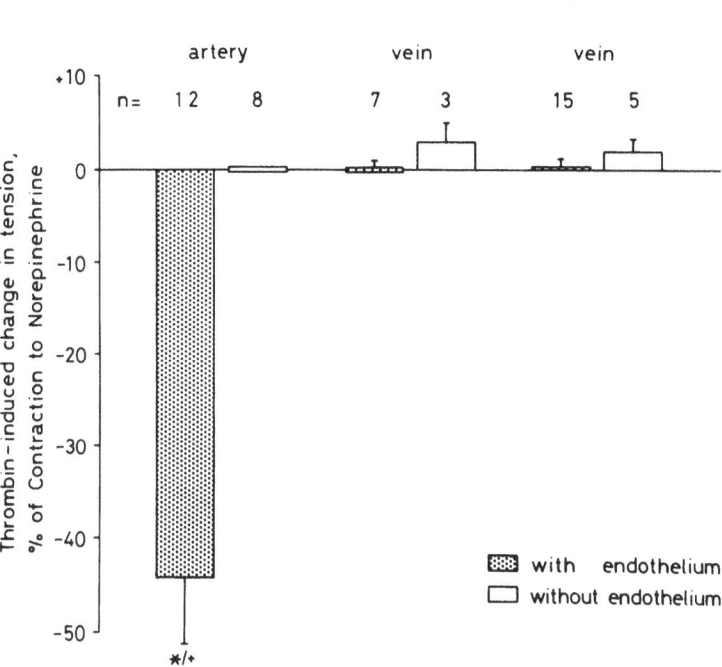

FIGURE 10: Endothelium-dependent relaxations to thrombin in the human internal mammary artery, vein and saphenous vein (*from Luscher et al., 1988, by permission*).

released prostacyclin by platelet-derived substances and coagulation products would increase local blood flow and inhibit platelet adhesion and aggregation at sites where platelets are activated (Figure 2). This would limit or prevent thrombus formation and disaggregate and flush away a nascent platelet clot in the mammary artery, while in the saphenous vein blood flow would be endangered by aggregating platelets and contraction of the graft by platelet-derived mediators such as thromboxane A_2 and serotonin. Therapeutic inhibition of platelet function (by aspirin and other drugs) does indeed enhance the patency rate of venous grafts which lack this protective mechanism (see below and *Chesebro, et al., 1982; Chesebro et al., 1984*).

(b) Effects on vascular smooth muscle: In quiescent mammary arteries with endothelium, platelets exert only small contractions which are enhanced after removal of the endothelium (*Yang et al., 1991b*). The platelet-induced contractions are markedly more pronounced in the saphenous vein and are mediated by thromboxane A_2 and in part by serotonin. As the saphenous vein is more sensitive to exogenous serotonin (*Figure 11; Yang et al. 1989*) and the thromboxane analogue U46619 than the mammary artery, the different responses can be explained by differences in endothelium function and an augmented responsiveness of venous as compared to arterial vascular smooth muscle to platelet-derived products.

CHRONIC CHANGES OF ENDOTHELIUM FUNCTION IN GRAFTS

The difference between arterial and venous endothelial cells probably persists after implantation of the grafts. Indeed, in the dog, grafting of the femoral vein into the arterial circulation does not enhance endothelium-dependent relaxations to acetylcholine, ADP and thrombin (*Miller et al., 1987*). In contrast, at sites where myointimal thickening develops- mostly at the distal part of the venous graft- endothelium-dependent relaxations tend to be reduced (*Miller et al., 1987*). Arterialization of venous endothelium also does not enhance the luminal production of prostacyclin by the venous wall (*Bush et al., 1986*) The reduced production of endothelium-derived relaxing factor and prostacyclin, particularly in areas of myointimal thickening, helps to explain why superimposed thrombi at these sites contribute to late closure of the grafts (*Fuster and Chesebro, 1986*).

CARDIOVASCULAR RISK FACTORS AND ENDOTHELIUM FUNCTION

Cardiovascular risk factors such as hypercholesteremia, diabetes and hypertension are associated with an increased risk of myocardial infarction and bypass graft occlusion in patients who underwent coronary bypass surgery and they also impair endothelium-dependent relaxations in isolated blood vessels (*Lüscher and Vanhoutte, 1990*).

(a) Hypertension: Hypertension impairs endothelium-dependent relaxations in most blood vessels obtained from rats with spontaneous and secondary forms of hypertension (*Lüscher and Vanhoutte, 1990; Lüscher et al., 1987*). Similarly, a reduced vasodilator response to acetylcholine occurs in the forearm circulation of patients with essential hypertension (*Linder et al., 1990*). Hypertension can reduce the release of endothelium-derived relaxing factor(s), stimulate the production of contracting factors (most likely prostaglandin H_2; *Kato et al., 1990; Lüscher and Vanhoutte, 1986*) and interfere with the ability of vascular smooth muscle to relax in response to nitric oxide (due to medial hypertrophy; *Lüscher et al., 1987*).

(b) Hyperlipidemia: Low density lipoproteins (LDL), but not high density lipoproteins (HDL), reduce endothelium-dependent relaxations in the rabbit aorta and porcine coronary artery (*Kugiyama et al., 1990; Tomita et al., 1990*). The oxidized form of the lipoprotein appears to be the active component. In porcine endothelial cells in culture, oxidized LDL inhibits the release of endothelium-derived relaxing factor(s) under basal conditions and after stimulation with bradykinin (*Boulanger et al., 1989*). Native LDL is (only active at very high concentrations). LDL also reduces prostacyclin formation by endothelial cells in culture (*Nordoy et al., 1978*).

(c) Diabetes: Controversial results on the effects of diabetes on endothelium-dependent relaxations have been published (*Lüscher and Vanhoutte, 1990*). In human corpora cavernosa tissues, however, the relaxations induced by acetylcholine are reduced in diabetic men with impotence (*De Tejada et al., 1989*). As the response to sodium nitroprusside is maintained, this suggests that the release of endothelium-derived relaxing factor(s) is reduced in human

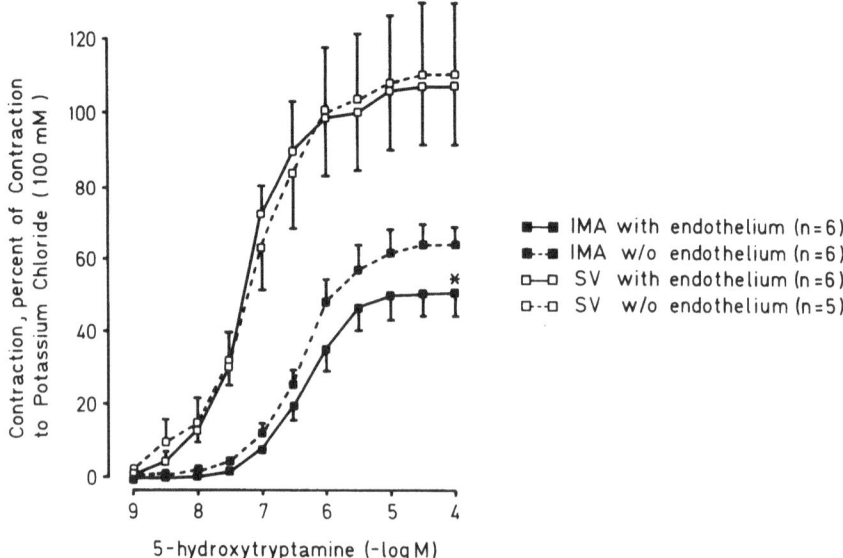

FIGURE 11: Contractions induced by serotonin (5-hydroxytryptamine) in the human internal mammary artery (IMA) and saphenous vein (SV) with or without endothelium. Note the enhanced sensitivity and potency of the monoamine in the veins as compared to the arteries. Endothelium removal leaves the response to the unaffected in the saphenous vein, but augments the maximal response to the monoamine in the artery ($*=p<0.05$) from Yang, 1989, by permission).

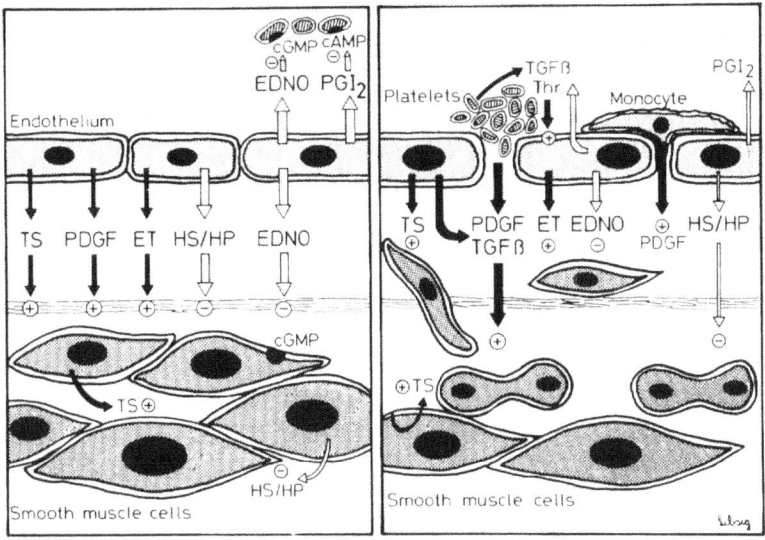

FIGURE 12: Putative role of the endothelium in the regulation of vascular growth: Under normal conditions (left panel) endothelial cells can release growth inhibitors such as heparin and heparin sulfates (HP/HS) and nitric oxide (NO) as well as growth promoters such as platelet-derived growth factor and endothelin-1 (which may facilitate the effects of growth promoter(s). Most likely, the effects of growth inhibitors dominate as the normal blood vessel wall remains quiescent. In the case of tend to adhere and aggregate at the endothelial layer and they may release potent growth factor such as PDGF and transforming growth factor βTGFβ. Dysfunctional endothelial cells will release less nitric oxide and possibly also small amounts of other growth inhibitors. This may contribute to proliferative responses occurring after endothelial injury and dysfunction.

diabetes. In the rabbit, diabetes causes the occurrence of endothelium-dependent contractions in response to acetylcholine due to the formation of thromboxane A_2 in diabetic endothelial cells (*Tesfamariam et al., 1989*).

Thus, factors associated with vascular occlusive events interfere with the release and/or action of endothelium-derived relaxing factors and (at least under certain conditions) promote the formation of endothelium-derived contracting factors. As blood vessels with less potent endothelium-dependent responses, such as the saphenous vein, have a higher occlusion rate than the mammary artery, this indicates that the endothelium plays in important protective role against bypass graft occlusion.

ATHEROSCLEROSIS

(a) Pathogenesis: Although the exact cause of atherosclerosis is still unknown, most modern concepts focus on functional changes of the vascular endothelium and its interaction with monocytes and platelets as a primary step in the disease process (*Ross, 1986*). Growth factors synthesized and released from endothelial cells, macrophages and vascular smooth muscle cells play an important role in the formation of the atherosclerotic lesion. Although the endothelial layer remains intact, a marked polymorphism with giant endothelial cells and, as a consequence, a decreased cell number per area occurs in diseased human blood vessels (*Chazov et al., 1986*).

Mechanical removal of the endothelium, on the other hand, is a well-established experimental model to induce the formation of atherosclerotic lesions (*Baumgartner and Studer, 1963*). Although, this probably does not occur *in vivo*, endothelial denudation is a frequent event during surgical preparation of coronary bypass grafts (*Lüscher et al., 1988*).

Endothelial denudation is associated with rapid platelet adhesion and release of platelet-derived substances such as platelet-derived growth factor (PDGF), a potent mitogen (Figure 12; *Fuster and Chesebro, 1986; Ross, 1986*). In vascular smooth muscle cells in culture, increases in cyclic GMP by various compounds which liberate nitric oxide inhibit the thymidine uptake and proliferation of the cells (*Garg and Hassid, 1989*). This suggests that endothelium-derived nitric oxide may act as a growth inhibitor in the blood vessel wall.

(b) Atherosclerosis in Bypass grafts: The mammary artery is remarkably resistant to the development of atherosclerosis both as a native vessel and as a bypass graft (*Singh, 1983; Barbour and Roberts, 1985*). Indeed, in patients undergoing coronary bypass surgery, the incidence of obstructive atherosclerotic changes is about 5% in this blood vessel (*Reul, 1986*). This is particularly remarkable since patients with coronary artery disease often demonstrate atherosclerotic vascular disease in other vascular beds such as the cerebral and limb circulation (*Feussuer and Matchar, 1988*).

In contrast, venous grafts undergo structural changes when used as a bypass graft. Implantation of veins into the arterial site of the circulation is associated with platelet disposition, intimal hyperplasia, plaque formation and thrombotic events (Figure 2; *Grondin et al., 1984; Fuster and Chesebro, 1986; Campeau et al., 1984; Spencer, 1986*). Atherosclerotic changes of venous grafts are more pronounced in patients with cardiovascular risk factors such as hyperlipidemia, smoking and hypertension (*Campeau et al., 1984; Spencer, 1986*).

CLINICAL IMPLICATIONS

(a) Surgical Techniques: Surgical handling of blood vessels commonly is associated with larger or smaller defects of the endothelial layer (*Lüscher et al., 1988*). With newer techniques to prepare the internal mammary artery (leaving the blood vessel in the pedicle of the thoracic fascia), endothelial coverage of arterial grafts prepared by experienced cardiac surgeons is excellent (Figure 13a; *Lüscher et al., 1988*). In contrast, endothelial damage tends to be more pronounced in saphenous vein grafts (Figure 13b, *Lüscher et al., 1988*). This may be related to more extensive surgical preparation of venous grafts (which are dissected free and completely removed from their original anatomical location) and/or to a smaller viability of the venous endothelium. Although the endothelial layer recovers within a few days, the regenerate state - as judged from the porcine coronary artery - is associated with reduced endothelium-dependent relaxations to platelets and serotonin (*Shimokawa et al., 1989*). The mechanism appears to involve loss of pertussis toxin-sensitive G-protein (*Shimokawa et al., 1989*).

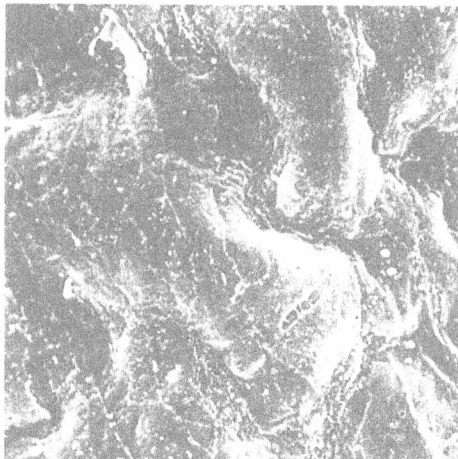 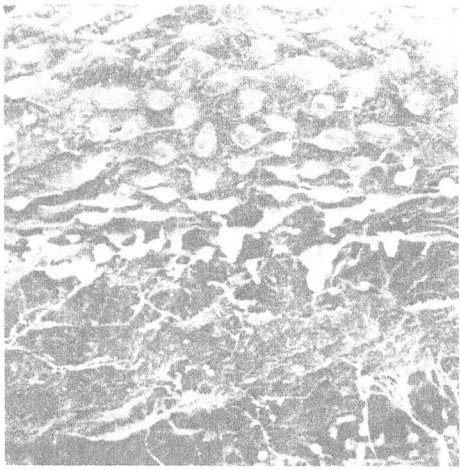

FIGURE 13: Scanning electronmicroscopy of the endothelial layer of an internal mammary artery (left) and the saphenous vein (right). Note the complete endothelial coverage in the mammary artery, but the area of endothelial denudation after surgical preparation of the saphenous vein (*from Luscher, et al., 1988, by permission*).

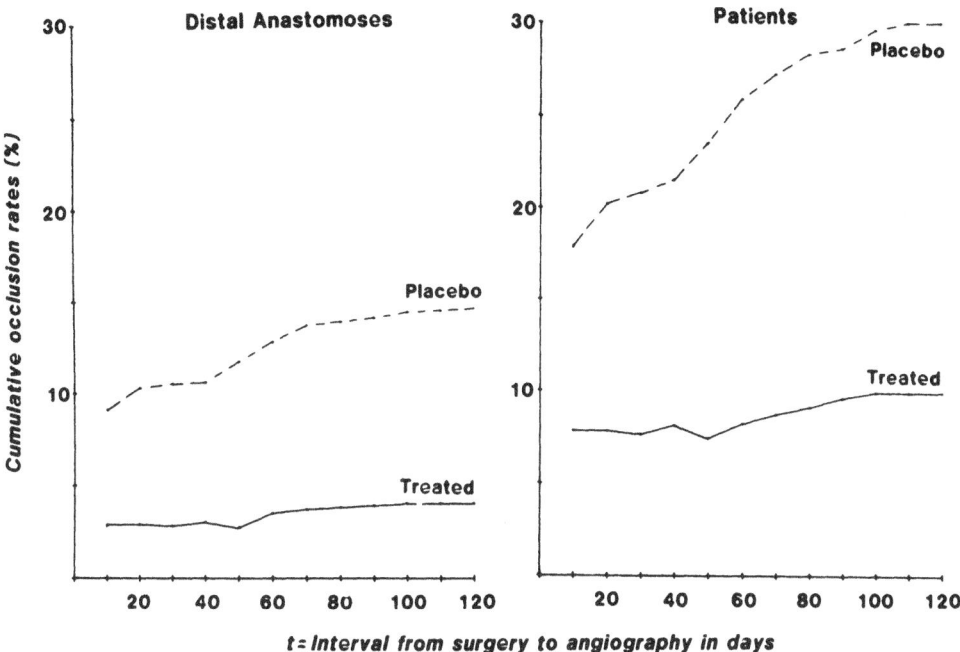

FIGURE 14: Patency of venous coronary bypass grafts in patients receiving placebo (dashed lines) or aspirin 3 x 325 mg/day plus dipyridamole (3 x 75 mg/day). Note the remarkably higher patency rate of coronary bypass grafts (whether expressed as percent of occluded anastomosis, left panel, or percent of patients with one or more occluded anastomosis, right panel) in patients receiving active drug (from Chesebro et al., 1982, by permission).

(b) Choice of arterial and venous grafts: Both the clinical data and the biological properties of the internal mammary artery suggest that arterial grafts are superior to venous grafts and should be recommended as treatment of choice in patients selected for coronary bypass surgery.

(c) Prevention of graft failure and occlusion: The postoperative care of patients receiving coronary bypass grafts focusses on platelet inhibition and the treatment of cardiovascular risk factors such as hypertension, hyperlipidemia and diabetes (*Fuster and Chesebro, 1986; Lüscher, 1988; Yang and Lüscher, 1989*). Platelet inhibitors such as aspirin have been shown to reduce the incidence of venous bypass graft occlusion (Figure 14; *Chesebro et al., 1982; Chesebro et al., 1984*). Even low dosages of aspirin (100 mg/day) are as effective as oral anticoagulation in prevention occlusion of bypass vessels (*Pfisterer et al., 1989*). Prolonged treatment for up to one year after implantation improves venous graft patency as compared to patients receiving therapy for three months only (*Pfisterer et al., 1989*). Thus, most likely, prolonged antiplatelet therapy should be performed at least in patients with venous coronary bypass grafts.

SUMMARY

The internal mammary artery and saphenous vein are used as grafts in patients undergoing coronary bypass surgery. Different biological properties of arterial and venous grafts might contribute to the better graft function, lower patient mortality and higher patency rate of the mammary artery as compared to the saphenous vein. We investigated the role of endothelium-derived relaxing factor (EDRF), which is a vasodilator and inhibitor of platelet function, in arterial and venous grafts. In mammary arteries, acetylcholine, histamine and platelet-derived products such as adenosine diphosphate and thrombin evoke endothelium-dependent relaxations which can be inhibited by the inhibitor of nitric oxide formation from L-arginine L-N^G-monomethyl arginine (L-NMMA) as well as by hemoglobin or methylene blue, delineating EDRF as the mediator. Since cyclooxygenase inhibitors do not affect the response, prostacyclin does not contribute to it. In the saphenous vein, endothelium-dependent relaxations to all agonists are much weaker, but augmented by inhibitors of cyclooxygenase. As L-NMMA prevents the relaxations and unmasks endothelium-dependent contractions to acetylcholine, the venous endothelium appears to release both EDRF and cyclooxygenase-dependent endothelium-derived contracting factor(s). Endothelin-1 has a similar potency in both the mammary artery and saphenous vein as compared to the artery. The greater release of EDRF in the internal mammary artery, particularly in response to platelet-derived products could contribute to the higher patency rate of arterial as compared to venous grafts and give new insight into mechanisms involved in graft function and vascular occlusion in man.

ACKNOWLEDGEMENT

The authors wish to thank Sabine Bohnert and Bernadette Libsig, Department of Research, University Hospitals Basel for their invaluable help during the preparation of the manuscript and its figures and Draguljeb Popovic and Silvia Distel, Operating Room Laboratory, Department of Cardiovascular Surgery, University Hospital Zurich for technical assistance. Own research reported in this manuscript was supported by grants of the Swiss National Research Foundation (Grant Nr. 32-25468.88 and SCORE-grant Nr. 3231-025150), the Swiss Cardiology Foundation, and an educational grant os Helmut Horten Foundation and Janssen Pharmaceutica, Beerse, Belgium and Baar, Switzerland.

REFERENCES

Azuma, H., Ishikawa, M., Sekizaki, S. Endothelium-dependent inhibition of platelet aggregation. *Br. J. Pharmacol.* 88:411-415, 1986.

Barbour, D.J., Roberts, W.C. Additional evidence for the relative resistance to atherosclerosis of the internal mammary artery compared to saphenous vein when used to increase myocardial blood supply. *Am. J. Cardiol.* 56:488, 1985.

Baumgartner, H.R. and Studer, A. Gezielte Überdehnung der Aorta abdominalis am normo- und hypercholesterämischen Kaninchen. Pathol Microbiol (Basel) 26:129-148, 1963.

Boulanger, C., Bühler, F.R. and Lüscher, T.F. Low density lipoproteins impair the release of endothelium-derived relaxing factor from cultured porcine endothelial cells. (Abstract) *Eur. Heart J.* 10:1883, 1989.

Bush, Jr., H.L., Jakubowski, J.A., Curl, G.R., Deykin, D. and Nabseth, D.C. the natural history of endothelial structure and function in arterialized vein grafts. *J. Vasc. Surg.* 3:204-215, 1986.

Busse, R., Lückhoff, A. and Bassenge, E. Endothelium-derived relaxant factor inhibits platelet activation. *Naunyn-Schmiedeberg's Arch. Pharmacol.* 336:566-571, 1987.

Campeau, L., Enjalbert, M., Lespérance, J., Bourassa, M.G., Kwiterovich, P. Jr, Vacholder, S., Sniderman, A. The relation of risk factors to the development of atherosclerosis in saphenous-vein bypass grafts and the progression of disease in the native circulation. *N. Eng. J. Med.* 311:1329-1332, 1984.

Chazov, E.I., Repin, V.S., Orekhov, A.N., Antonov, A.S., Preobrazhensky, S.N., Soboleva, E.L., Smirnov, V.N. Atherosclerosis: What has been learned studying human arteries. In: *Atherosclerosis Reviews 14*, A.M. Goto and R. Paoletti, eds, pp. 7-60, 1986.

Chesebro J.H., Clements, I.P., Fuster, V., Elveback, L.R., Smith, H.C., Bardsley, W.T., Frye, R.L., Homes, Jr. D.R., Vliestra, R.E., Plutz, J.R., Wallace, R.B., Puga, F.J., Orszulak, T.A., Piehler, J.M., Schaff, H.V., Danielson, G.K. A platelet-inhibitor trial in coronary bypass operation. *N. Engl. J. Med.* 307: 73-78, 1982.

Chesebro, J.H., Fuster, V., Elveback, L.R., Clements, I.P., Smith, H.C., Holmes Jr, D.R., Bardsley, W.T., Pluth, J.R., Wallace, R.B., Puga, F.J., Orszulak, T.A., Piehler, J.M., Danielson, G.K., Schafff, H.V. and Frye, R.L. Effect of dipyridamole and aspirin on late vein-graft patency after coronary bypass operation. *N. Engl. J. Med.* 10:209, 1984.

D'Souza, V.J., Velasquez, G., Kahl, F.R., Hackshaw, B.T. and Amplatz, K. Spasm of aortocoronary venous graft. *Radiol.* 151:83-84, 1984.

De Tejada, I.S., Goldstein, I., Azadzoi, K., Krane, R.J. and Cohen, R.A. Impaired neurogenic and endothelium-mediated relaxation of penile smooth muscle from diabetic men with impotence. *N. Engl. J. Med.* 320:1025-1030, 1989.

DiCorletto, P.E. and Fox, P.L. Growth factor production by endothelial cells. In: *Endothelial cells. Volume II*, U.S. Ryan, ed, CRC Press, Boca Raton, FL, U.S.A. 1987.

Effler, D.B., Groves, L.K., Sones, F.M. and Shirey, E.K. Increased myocardial perfusion by internal mammary artery implant. Vineberg's operation. *Ann. Surg.* 158:526-536, 1963.

Feussner, J.R. and Matchar, D.B. When and how to study the carotid arteries. *Ann. Intern. Med.* :805-818, 1988.

Förstermann, U., Mügge, A., Alheid, U., Haverich, A. and Frölich, J.C. Selective attenuation of endothelium-mediated vasodilation in atherosclerotic human coronary arteries. *Circ. Res.* 62:185-190, 1988.

Furchgott, R.F. and Zawadzki, J.V. The obligatory role of endothelial cells in the relaxation of arterial smooth muscle by acetylcholine. *Nature* 299:373-376, 1980.

Fuster, V. and Chesebro, J.G. Role of platelets and platelet inhibitors in aortocoronary artery vein-graft disease. *Circulation* 73:227-232, 1986.

Garg, U.C. and Hassid, A. Nitric oxide-generating vasodilators and 8-bromo-cyclic guanosine monophosphate inhibit mitogenesis and proliferation of cultured rat vascular smooth muscle cells. *J. Clin. Invest.* 83:1774-1777, 1989.

Green, G.E., Stertzer, S.H., Gordon, R.B. and Tice, D.A. Anastomosis of internal mammary artery to distal left anterior descending coronary artery. *Circulation* 41 (Suppl. 2):79-89, 1970.

Grondin, C.M., Campeau, L., Lesperence, J., Enjalbert, M. and Bourassa, M.G. Comparison of late changes in internal mammary artery and saphenous vein grafts in two consecutive series of patients 10 years after operation. *Circulation* 70 (Suppl. I):208-212, 1984.

Houston, D.S., Shepherd, J.T. and Vanhoutte, P.M. Aggregating human platelets cause direct contraction and endothelium-dependent relaxation in isolated canine coronary arteries: role of serotonin, thromboxane A_2, and adenine nucleotides. *J. Clin. Invest.* 78:539-544, 1986.

Ignarro, L.J., Byrns, R.E., Buga, G.M. and Wood, K.S. Endothelium-derived relaxing factor from pulmonary artery and vein possesses pharmacological and chemical properties identical to those of nitric oxide radical. *Circ. Res.* 61:866-879, 1987.

Kalsner, S. Cholinergic mechanisms in human coronary artery preparation implications of species differences. *J. Physiol.* 358:509-516, 1985.

Kato, T, Iwama, Y., Okumura, K., Hashimoto, H., Ito, T. and Satake, T. Prostaglandin H_2 may be the endothelium-derived contracting factor released by acetylcholine in the aorta of the rat. *Hypertension* 15:475-481, 1990.

Kugiyama, K., Kerns, S.A., Morrisett, J.D., Roberts, R. and Henry, P.D. Impairment of endothelium-dependent arterial relaxation by lysolecithin in modified low-density lipoprotein. *Nature* 344:60-162, 1990.

Linder, L., Kioski, W., Bühler, F.R. and Lüscher, T.F. Indirect evidence for the release of endothelium-derived relaxing factor in the human forearm circulation: Blunted response in essential hypertension. *Circulation* 81:1762-1767, 1990.

Loop, F.D., Lytle, B.W., Cosgrove, D.M., Stewart, R.W., Goormostic, M., Williams, G.W. and Bolding, L.A.R., Gill, C.C., Taylor, P.C., Sheldon, W.C. and Proudfit, W.L. Influence of the internal mammary artery graft on 10-year survival and other cardiac events. *N. Engl. J. Med.* 314:1-6, 1986.

Lüscher, T.F., Richard, V. and Yang, Z. Interaction between endothelium-derived nitric oxide and SIN-1 in human and porcine blood vessels. *J. Cardiovasc. Pharmacol.* 14(Suppl. 11):76-80, 1989.

Lüscher, T.F. and Vanhoutte, P.M. Endothelium-dependent contractions to acetylcholine in the aorta of the spontaneously hypertensive rat. *Hypertension* 8:344-348, 1986.

Lüscher, T.F. and Vanhoutte, P.M. *The Endothelium: Modulator of Cardiovascular Function?* CRC Press, Boca Raton, FL, U.S.A., 1990.

Lüscher, T.F., Raij, L. and Vanhoutte, P.M. Endothelium-dependent responses in normotensive and hypertensive Dahl rats. *Hypertension* 9:157-163, 1987.

Lüscher, T.F., Yang, Z., Tschudi, M., von Segesser, L., Stulz, P., Boulanger, C., Siebenmann, R., Turina, M. and Bühler, F.R. Interaction between endothelin-1 and endothelium-derived relaxing factor in human arteries and veins. *Circ. Res.* 66:1088-1094, 1990.

Lüscher, T.F., Cooke, J.P., Houston, D.S., Neves, R. and Vanhoutte, P.M. Endothelium-dependent relaxations in human arteries. *Mayo Clin. Proc.* 62:601-606, 1987.

Lüscher, T.F. and Vanhoutte, P.M. Endothelium-dependent responses in human blood vessels. *Trends Pharm. Sc.* 9:181-184, 1988.

Lüscher, T.F., Diederich, D., Siebenmann, R., Lehmann, K., Stulz, P., von Segesser, L., Yang, Z., Turina, M., Grädel, E., Weber, E. and Bühler, F.R. Difference between endothelium-dependent relaxations in arterial and in venous coronary bypass grafts. *N. Engl. J. Med.* 319: 462-467, 1988.

Lüscher, T.F., Siebenmann, R., Stulz, P., von Segesser, L., Schneider, K., Diederich, D., Lehmann, K., Bauer, E., Bertschmann, W., Grädel, E., Turina, M. and Bühler, F.R. Bedeutung des Gefässendothels in der Herzchirurgie. *Helv. Chir. Acta.* 55:529-534, 1988.

Lytle, B.W., Loop, F.D., Cosgrove, D.M., Ratliff, N.B., Easley, K. and Taylor, P.C. Long-term (5 to 12 years) serial studies of internal mammary artery and saphenous vein coronary bypass grafts. *J. Thorac. Cardiovasc. Surg.* 89:248-258, 1985.

Martin, W., Villani, G.M., Jothianandan, D. and Furchgott, R.F. Selective blockade of endothelium-dependent and glyceryl trinitrate-induced relaxation by hemoglobin and by methylene blue in the rabbit aorta. *J. Pharmacol. Exp. Ther.* 232:708-716, 1985.

Miller, V.M., Reigel, M.M., Hollier, L.H. and Vanhoutte, P.M. Endothelium-dependent responses in autogenous femoral veins grafted into the arterial circulation of the dog. *J. Clin. Invest.* 80:1350-1357, 1987.

Miller, V.M., Komori, K., Burnett, J.C. and Vanhoutte, P.M. Differential sensitivity to endothelin in canine arteries and veins. *Am. J. Physiol.*, 257:H1127-1131, 1989.

Miller, V.M. and Vanhoutte, P.M. Is nitric oxide the only endothelium-derived relaxing factor in canine femoral veins: *Am. J. Physiol.* 257:H1910-H1916, 1989.

Miller, D.W. Jr, Ivey, T.D., Bailey, W.W., Johnson, D.D., Hessel, E.A. The practice of coronary artery bypass surgery in 1980. *J. Thorac. Cardiovasc. Surg.* 81:423-427, 1981.

Myers, P.R., Minor, R.L. Jr, Guerra, R. Jr, Bates, J.N. and Harrison, D.G. Vasorelaxant properties of the endothelium-derived relaxing factor more closely resemble S-nitrosocysteine than nitric oxide. *Nature* 345:161-163, 1990.

Nordoy, A., Svensson, B., Wiebe, D. and Hoak, J.C. Lipoprotein and the inhibitory effect of human endothelial cells on platelet function. *Circ. Res.* 43:527-534, 1978.

Okies, J.E., Scott, P.U., Bigelow, J.C., Krause, A.H. and Salomon, N.W. The left internal mammary artery:the graft of choice. *Circulation* 70 (suppl I):213-221, 1984.

Palmer, R.M.J., Ferrige, A.G. and Moncada, S. Nitric oxide release accounts for the biological activity of endothelium-derived relaxing factor. *Nature* 327:524-526, 1987.

Palmer, R.M.J., Rees, D.D., Aston, D.S. and Moncada, S. L-arginine is the physiological precursor of the formation of nitric oxide in endothelium-dependent relaxation. *Biochem. Biophys. Res. Commun.* 153:1251-1256, 1988.

Pfisterer, M, Burkart, F., Jockers, G., Meyer, B., Regenass, S., Burckhardt, D., Schmitt, H.E., Müller-Brand, J., Skarvan, K., Stulz, P., Hasse, J., Grädel, E. Trial of low-dose aspirin plus dipyridamole versus anticoagulants for prevention of aortocoronary vein graft occlusion. *Lancet i*:1-7 1989.

Radomski, M.W., Palmer, R.M.J. and Moncada, S. The anti-aggregating properties of vascular endothelium: interactions between prostacyclin and nitric oxide. *Br. J. Pharmacol.* 92:639-646, 1987a.

Radomski, M.W., Palmer, R.M.J. and Moncada, S. Endogenous nitric oxide inhibits human platelet adhesion to vascular endothelium. *Lancet II*:1057-1068, 1987b.

Rapoport, P.M. and Murad, F. Agonist-induced endothelium-dependent relaxation in rat thoracic aorta may be mediated through cGMP. *Circ. Res.* 52:352-357, 1983.

Reul, G.J. The problem of the internal mammary artery. *Texas Heart Inst. J.* 13:171-172, 1986.

Ross, R. The pathogenesis of atherosclerosis - an update. *N. Engl. J. Med.* 314:488-500, 1986.

Sarabu, M.R., McClung, J.A., Fass, A. and Reed, G.E. Early postoperative spasm in left internal mammary artery bypass grafts. *Ann. Thorac. Surg.* 44:199-200, 1987.

Shimokawa, H., Flavahan, N.A. and Vanhoutte, P.M. Natural course of the impairment of endothelium-dependent relaxations after balloon endothelium-removal in porcine coronary arteries. *Circ. Res.* 65:740-753, 1989.

Singh, R.N. Atherosclerosis and internal mammary arteries. *Cardiovasc. Intervent. Radiol.* 6:72-77, 1983.

Spencer, F.C. The internal mammary artery: the ideal coronary bypass graft? *N. Engl. J. Med.* 314:50-51, 1986.

Tector, A.J., Schmahl, T.M., Canino, V.R., Kallies, J.R. and Sanfilippo, D. The role of the sequential internal mammary artery graft in coronary surgery. *Circulation* 70 (Suppl. I): 222-225, 1984.

Tector, A.J., Schmahl, T.M., Janson, B., Kallies, J.R. and Johnson, G. The internal mammary artery graft. Its longevity after coronary bypass. *J. Am. Med. Assoc.* 246:2181-2183, 1981.

Tesfamariam, B., Jakubowski, J.A. and Cohen, R.A. Contraction of diabetic rabbit aorta due to endothelium-derived PGH_2/TXA_2. *Am. J. Physiol.* 257: H1327-H1333, 1989.

Tomita, T., Ezaki, M., Miwa, M., Nakamura, K. and Inoue, Y. Rapid and reversible inhibition by low density lipoprotein of the endothelium-dependent relaxation to hemostatic substances in porcine coronary arteries. *Circ. Res.* 66:18-27, 1990.

Verkkala, K., Järvinen, A., Keto, P., Virtanen, K., Lehtola, A. and Pellinen, T. Right gastroepiploic artery as a coronary bypass graft. *Ann. Thorac. Surg.* 47:716-719, 1989.

Yang, Z., Diederich, D., Schneider, K., Siebenmann, R., Stulz, P., von Segesser, L., Turina, M., Bühler, F.R. and Lüscher, T.F. Endothelium-derived relaxing factor and protection against contractions induced by histamine and serotonin in the human internal mammary artery and in the saphenous vein. *Circulation* 80:1041-1048, 1989.

Yang, Z. and Lüscher, T.F. Endothelium-dependent regulatory mechanisms in human coronary bypass grafts: Possible clinical implications. *Z. Kardiol.* 78:80-84, 1989.

Yang, Z., Siebenmann, R., Studer, M., Egloff, L. and Lüscher, T.F. Similar endothelium-dependent relaxation, but enhanced contractility of the right gastro epiploic artery as compared to internal mammary artery. *Circulation,* submitted.

Yang, Z., von Segesser, L., Bauer, E., Stulz, P., Turina, M. and Lüscher, T.F. Different activation of endothelial L-arginine and cyclooxygenase pathway in human internal mammary artery and saphenous vein. *Circ. Res.* 68:52-60, 1991a.

Yang, Z., von Segesser, L., Bauer, E., Stulz, P., Turina, M. and Lüscher, T.F. Different interaction of platelets with arterial and venous coronary bypass grafts: role of nitric oxide and antiplatelet drugs. (abstract) *J. Am. Coll. Cardio.* 1991b (in press).

ALTERED RENOVASCULAR ENDOTHELIAL FUNCTIONS DURING NEPHROTOXICITY

N. Perico, C. Zoja and Giuseppe Remuzzi

Mario Negri Institute for Pharmacological Research
24100 Bergamo, Italy

INTRODUCTION

Until recently the vascular endothelium was considered as little more than a semipermeable non-thrombogenic membrane which lined blood vessels. However, in the past decade the endothelium has been regarded in its own role as an important metabolic, endocrine and immunological organ. Major advances have included the steady improvement in endothelial cell culture techniques (*Jaffe, et al., 1973; Gimbrone Jr., 1976*) and the discovery that endothelial monolayer is also capable of intrinsic modulation of vascular tone by elaborating several potent vasoactive substances, including vasorelaxants and vasoconstrictors (*Brenner et al., 1989; Vane et al., 1990*). In view of the extensive endothelium lining renal blood vessels and glomerular capillaries and of the increasing evidence of the role of intact endothelium in vascular homeostasis, it is not surprising that factors suspected of being important in the pathogenesis of many glomerular diseases have been examined for their capacity to cause endothelial injury, either functional or structural. Among these, bacterial endotoxins, when infused intravenously to normal rabbits, cause acute renal failure with concomitant morphologic changes of glomerular capillaries characterized by focal irregularities of endothelial fenestrae, areas of endothelial swelling, leukocyte and platelet infiltration and occlusive fibrin thrombi (*Bertani et al., 1989*), a pattern closely resembling human hemolytic uremic syndrome. That in this case endothelial injury is the initial event of the acute renal function deterioration is supported by the findings that lipopolysaccharides derived from Escherichia Coli, Salmonella minnesota and Salmonella typhosa directly damage endothelial cells in culture (*Harlan et al., 1983; McGrath and Stewart, 1969*).

In addition to bacterial endotoxins, several drugs may cause renal endothelial damage. Among these, mitomycin, an alkylating agent (*Crooke and Bradner, 1976*), has received recent attention in that it causes severe cortical infarction when infused into the rat kidney (*Cattell, 1986*). In this study, the appearance of glomerular endothelial damage 6 hours after mitomycin infusion, followed by widespread platelet thrombi in the capillary lumens, suggests a direct effect of this agent on endothelium. In this regard, exposure of human umbilical vein endothelial cells in culture to mitomycin has been shown to determine a time- and dose-dependent cell injury, which results in cell detachment from the culture substrate and subsequent lysis (*Murasawa, 1983*).

Besides mitomycin, cyclosporine (CyA) has been recently recognized to have a toxic effect on renal vascular endothelium. CyA, an active endecapeptide of fungal origin, is a potent immunosuppressive agent in widespread use for prolonging the survival of various allogeneic organ transplants (*Cohen et al., 1984; Kahan, 1984; Kahan, 1985; Bennett and Norman, 1986*). Its immunosuppressive action is predominantly mediated by the inhibition of T-helper cell production of interleukin 2, a T-cell growth factor essential for B cell and cytotoxic T-cell proliferation (*Cohen et al., 1984; Kahan, 1984*). While it has become the immunosuppressant of choice in clinical transplantation in the eighties, CyA therapy is associated with renal and liver toxicity, and with the development of hypertension (*Canadian Multicentre Transplant Study Group, 1983; European Multicentre Trial Group., 1983; Shulman et al., 1981*). Of all the known complications, nephrotoxicity is the most frequent and clinically important, and may ultimately limit the use of CyA (*Canadian Multicentre Transplant Study Group, 1983; European Multicentre Trial Group., 1983; Shulman et al., 1981; Myers et al., 1984*). Renal dysfunction can occur at any time in the course of treatment and is characterized by a wide spectrum of renal involvement from early, often reversible renal damage, to later irreversible chronic renal failure (*Myers, 1986; Remuzzi and Bertani, 1989*).

This review will focus on the renal vascular effects of CyA, as an example of nephrotoxicity, in which altered endothelial function and/or structure is regarded as the initial event.

FUNCTIONAL ABNORMALITIES OF RENAL VASCULAR ENDOTHELIUM INDUCED BY CyA

The administration of CyA is complicated in most transplant (*Canadian Multicentre Transplant Study Group, 1983*) and nontransplant patients (*Tegzess, 1988*) by episodes of acute renal insufficiency, usually rapidly reversible after the daily dose of CyA is reduced (*Flechner, et al., 1983; von Willebrand and Hayry, 1983*). The mechanism leading to acute renal failure during CyA therapy remains ill-defined. Initially, a direct toxic effect of CyA to renal tubular epithelium was considered (*von Willebrand and Hayry, 1983*). However, the absence of tubular cell necrosis in most patients with CyA-induced reversible acute renal failure raised doubts on whether CyA is, in fact, a tubular toxin. Recent evidence supports the possibility that the acute nephrotoxic effects of CyA are mediated by its effect on renal vessels. Thus renal plasma flow and glomerular filtration rate (GFR) fall after CyA is given to humans due to an acute renal vasoconstriction (*Kahan, 1985; Powell-Jackson et al., 1983*). In experimental animals renal perfusion is reduced after CyA with an increase in renal vascular resistance (*Murray, 1985*). These changes are associated with a parallel decline in GFR that occurs after intravenous (*Sullivan et al., 1984*) or intraperitoneal (*Murray et al., 1985*) administration before signs of tubular damage develops (*Dieperink et al., 1983*). In isolated perfused rat kidney preparation CyA but not vehicle caused a dose-dependent fall in renal perfusate flow or increase in perfusion pressure when kidney were perfused at constant pressure (*Perico et al., 1990c*) or flow (*Rossi et al., 1989*). English and coworkers (*English et al., 1987*) have recently provided in rats anatomical correlates to these functional observations, showing by scanning electron microscopy a focal narrowing in afferent arteriolar diameter which progressed with time of CyA administration and paralleled the decrease in inulin clearance.

Altogether these observations challenge the view of tubular toxicity as primary cause of CyA-induced acute renal failure and raise the possibility that CyA primarily affects renal vessels.

The precise mechanism by which CyA mediates renal vasoconstriction is still not understood clearly. Activation of the renin-angiotensin system (*Barros et al., 1985*) as well as renal sympathetic nervous system (*Moss et al., 1985*) has been implicated, but no definitive evidence of this is available. The most favored hypothesis suggests that CyA alters the balance between the various products of arachidonic acid metabolism with vasoactive properties. Since arachidonate metabolites generated via the cyclooxygenase enzyme have been recognized to play an important role in regulating glomerular function, one would expect that these compounds may contribute to the acute renal vasoconstriction induced by CyA administration. Although effects on vasodilatory prostaglandins are controversial (*Murray et al., 1985; Neild et al., 1983b; Lau et al., 1989*), CyA consistently augments the generation of thromboxane (Tx) A_2 (*Kawaguchi et al., 1985*), a potent vasoconstrictor compound which has been shown to increase glomerular afferent and efferent arteriolar resistances and reduce GFR, when infused into normal anesthetized rats (*Baylis et al., 1987*).

In this context, recent studies showed that at variance with vehicle, incubation of confluent monolayer of aortic endothelial cells with CyA caused a dose- and time-dependent increase in TxA_2 production measured as its stable breakdown product TxB_2 (*Zoja et al., 1986*), which was inhibited when incubations were performed in the presence of aspirin. That CyA may actually stimulate TxA_2 production is also supported by *in vivo* studies showing increased urinary excretion of immunoreactive TxB_2 in rats given short-term CyA (*Kawaguchi et al., 1985; Perico et al., 1986; Smeesters et al., 1988; Coffman et al., 1987*). The renal hemodynamic significance of the increase in TxA_2 generation during CyA administration has been recently addressed using a specific TxA_2 receptor antagonist, GR32191, that blocks the binding of TxA_2

and endoperoxide PGH_2 to the receptor (*Pasini et al., 1990*). This pharmacological manipulation partially prevented CyA-induced acute decline in RPF and GFR in normal rats (*Pasini et al., 1990*), suggesting that TxA_2 is one of the mediator(s) of such phenomenon. Evidence supporting the role for TxA_2 as a major hemodynamic mediator of renal toxicity comes also from the studies of Elzinga et al (*Elzinga et al., 1987*), who showed that substitution of fish oil rich in eicosapentaenoic acid, a known inhibitor of the renal eicosanoid synthesis (*Kelley et al., 1985*) for the conventional olive oil CyA vehicle, increased GFR and reduced renal TxA_2 content in rats with experimentally induced CyA toxicity. However, the fact that after pharmacological manipulations of TxA_2 synthesis or biological activity renal function ameliorated but did not reach control values suggests that in addition to TxA_2 other factor(s) are involved in the acute renal vasoconstriction caused by CyA.

Of interest in this regard, is the recent observation that endothelial cells synthesize a novel powerful vasoconstrictor, endothelin (*Yanagisawa et al., 1988*). Evidence is now available that in rats endothelin infusion elicits a long-lasting constriction of renal vessels associated with a reduction in RPF and GFR (*King et al., 1989*). Moreover, in isolated perfused rat kidneys a fall in GFR and renal perfusate flow has been reported over a wide range of endothelin doses (*Firth et al., 1988; Perico et al., 1990b*). In addition, endothelin has recently been found to have a pivotal role in the renal vasoconstriction characteristic of post-ischemic renal failure (*Kon et al., 1989*).

Given the notion that endothelin release is stimulated by endothelial stress or injury and since the effects of endothelin on glomerular hemodynamics (*Badr et al., 1989*) are comparable with those reported for CyA (*Barros et al., 1985*), it is tempting to speculate that endothelin participates in the acute renal vasoconstriction induced by CyA. This issue has been recently addressed in isolated, perfused rat kidney preparation in which pre-exposure to a specific anti-endothelin antibody, but not a non-immunized rabbit serum, markedly prevented CyA-induced reduction in GFR and renal perfusate flow and increase in renal vascular resistance (*Perico et al., 1990c*). Furthermore, evidence obtained in a more physiological setting showed that, an anti-endothelin antibody prevented the decline in GFR induced by giving CyA to normal rats (*Perico et al., 1990c*). These findings are supported by recent micropuncture studies of Kon and coworkers (*Kon et al., 1990a*) who also documented a protective effect of an anti-endothelin antibody on CyA-induced acute glomerular vasoconstriction. The demonstration that in all animals serum endothelin levels, 20 min after acute CyA infusion, were remarkably higher than in rats given vehicle (*Kon et al., 1990a*), supports the notion that intense microvascular constriction following CyA may reflect a perturbed endothelin synthesis and/or metabolism. This hypothesis is consistent with data showing that human aortic and glomerular endothelial cells in culture exposed to CyA released endothelin in a dose-dependent manner (*Bunchman and Brookshire et al., 1990*). Of note, the recent finding that, as compared to coronary, bronchial, and femoral, renal vascular bed is the most sensitive to the vasoconstrictory effect of endothelin (*Pernow et al., 1988*), provides an explanation for the consistent clinical observation that CyA has a peculiar toxic effect on the kidney. An additional reason for the selective toxic effect of CyA on the kidney is the recent demonstration that in CyA-treated rats a selective upregulation of high affinity endothelin receptors occurs in the kidney, but not in hepatic tissue (*Awazu et al., 1990*).

STRUCTURAL ABNORMALITIES OF RENAL VASCULAR ENDOTHELIUM INDUCED BY CyA

CyA-induced arteriolopathy

The main clinical problem related to CyA therapy is the form of renal injury that develops after long-term exposure to the drug (*Myers et al., 1984; Myers et al., 1988b*) that may possess the potential to progress irrevocably to end-stage renal disease (*Myers, 1986; Remuzzi and Bertani, 1989*). Besides tubular atrophy and interstitial fibrosis, the pattern of pathological changes that are associated with chronic CyA nephrotoxicity includes vascular damage. These lesions have been described in kidney (*Mihatsch et al., 1983*), heart (*Myers et al., 1988a*), liver (*Dische et al., 1988*), and bone marrow (*Nizze et al., 1988*) transplant recipients as well as in patients with autoimmune diseases (*Von Graffenried and Harrison, 1985*), clearly indicating that CyA therapy leads to chronic vascular-interstitial kidney damage irrespective of the underlying disease. The vascular lesions predominate in the peripheral vascular tree and are

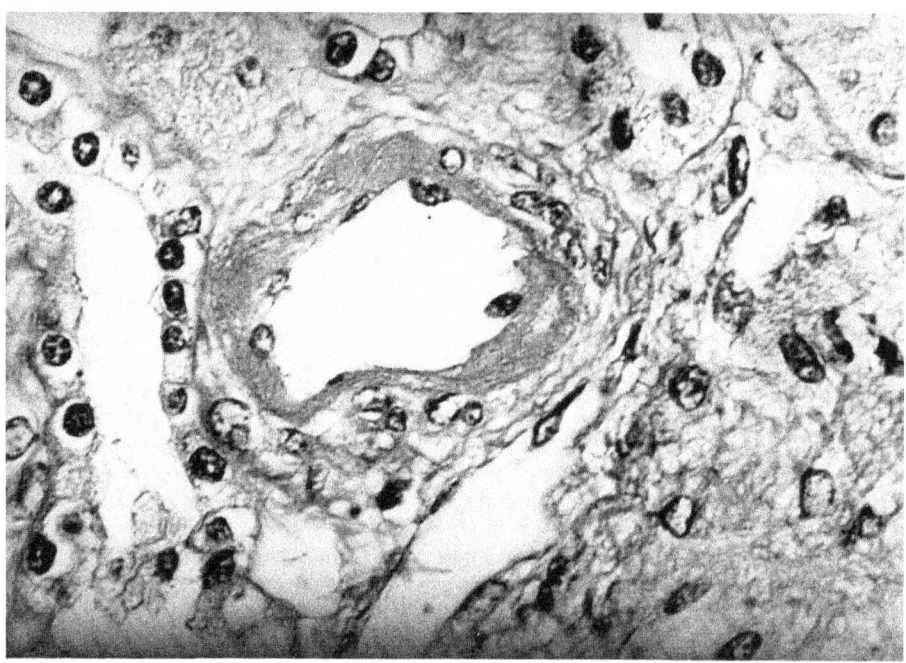

FIGURE 1: Typical CyA-associated arteriolopathy with protein deposits and narrowing of the vascular lumen.

almost exclusively located in the arterioles (afferent vessels) and arteries with up to two layers of smooth muscle cells.

Vascular lesions are preceded by changes which can only be detected by electron microscopy, mainly endothelial and/or myocyte damage with vacuolization, inclusion bodies, and single cell necrosis. Later on microthrombi consisting of fibrin and platelets in arterioles are seen by light microscopy. Arteriolopathy develops immediately after and results in insudation of macromolecular plasma protein into the damaged vessel wall, leading to nodular and circular protein deposits often at sites of necrotic myocytes (Figure 1). In non-occlusive arteriolar lesions the endothelium regenerates and rests upon a thickened basement membrane. In the arteriolar tunica media the nodular protein deposits persist, and the remaining myocytes are surrounded by thickened basement membranes and embedded in a sclerotic matrix. When the arterioles are completely occluded, the glomeruli become obsolescent, whereas they collapse if the vascular lumen is merely narrowed. Segmental or global glomerulosclerosis may follow which may lead to terminal renal failure.

The precise pathogenesis of CyA-associated chronic nephropathy remains a matter for speculation. The recent availability of experimental models resembling the chronic form of CyA nephrotoxicity (*Bertani et al., 1987; Gillum et al., 1988; Dieperink et al., 1988*) offered the opportunity to investigate whether vasoactive substances previously recognized to participate in the acute form of CyA nephrotoxicity may also play a role in chronic renal vasoconstriction and ultimately contribute to the development of CyA-associated arteriolopathy. Given the notion that CyA is a potential stimulus for endothelin release (*Bunchman et al., 1990,*), we have recently evaluated whether synthesis of endothelin occurs in a chronic model of CyA nephrotoxicity (*Perico et al., 1990a*). At variance with vehicle-treated rats, 30 day CyA administration resulted in a significant elevation in the urinary excretion of endothelin, measured by radioimmunoassay, as compared to pre-treatment values. Since previous data from our laboratories have shown that exogenous labeled endothelin added to the systemic circulation of normal rats or rats with renal mass ablation was not excreted by the kidney (*Benigni et al.*, in press), the enhanced urinary excretion rate in CyA-treated animals would reflect an increased renal production of the peptide. This is consistent with the finding that, at variance with the urinary excretion rate, plasma levels of endothelin were unchanged in rats on chronic CyA therapy (*Perico et al., 1990a*). That endothelin may be released into the systemic circulation in excessive amount but rapidly metabolized can not be ruled out, however. Since in these animals we found a significant correlation between the urinary excretion of endothelin and the serum creatinine levels, a possible contribution of this vasoactive peptide to the chronic CyA nephrotoxicity has been postulated.

Given the potent vasoconstrictor activity of endothelin, a persistent renal vasoconstriction has been suspected to play a central role in the chronic CyA nephropathy. In this regard, elevated renal vascular resistance in cardiac transplant recipients treated with CyA for more than 1 year (*Myers et al., 1988a*) has been reported. In these patients, the increased renal artery-to-peritubular capillary pressure gradient between 1 and 12 months of CyA therapy (*Myers et al., 1988a*) as well as the striking elevation of transglomerular dextran transport clearly indicate the afferent arteriole as the predominant site of increased renovascular resistance in CyA-treated heart transplant recipients. Based on these findings, it has been proposed (*Myers et al., 1988a*) that administration of CyA is associated with persistent afferent arteriolar vasospasm which ultimately leads to structural alteration and eventual obliteration of a portion of these preglomerular vessels. This added impediment to glomerular perfusion may result in glomerular damage that is irreversible. When extensive, this process could lead to a loss of ultrafiltration pressure and filtration surface area to explain the end-stage renal failure which has been described in heart transplant patients given long-term CyA. That vascular endothelial damage occurs during chronic CyA administration is confirmed by the high plasma concentration of factor VIII-related antigen (factor VIIIR:Ag) found in renal allograft recipients given CyA and with clinical evidence of nephrotoxicity (*Brown et al., 1986*). Indeed factor VIIIR:Ag is one of several proteins synthesized by vascular endothelium (*Jaffe, 1977*), whose plasma levels are increased upon vascular endothelial injury. Of note in these renal allograft recipients, levels of factor VIIIR:Ag fell toward normal as the dose of CyA was reduced (*Brown et al., 1986*).

Acute Vascular Injury with HUS/TTP Stigmata Induced by CyA

In addition to cause chronic arteriolopathy as a result of a persistent renal vasoconstriction, CyA may induce acute vascular injury by a direct toxic effect on endothelial cells. Indeed, since 1981 an increasing number of case reports have documented the finding that treatment with CyA may be followed by severe renal impairment with pathological changes characterized by glomerular thrombosis and severe arteriolar damage. This lesions resembling hemolytic-uremic syndrome (HUS) has been mainly reported in patients receiving CyA for bone marrow transplantation (*Powles et al., 1980*). Thrombocytopenia, hemolytic anemia, and deterioration of renal function have been described in a patient treated with CyA for preventing graft-versus-host disease after allogenic bone marrow transplantation (*Powles et al., 1980*). Similar observations were made by Shulman et al (*Shulman et al., 1981*) who noted severe renal insufficiency and diffuse thrombi in the glomerular capillary tufts and arteries in 3 of 16 allogeneic bone marrow recipients treated with CyA. De novo occurrence of HUS has been observed also in recipients of solid organ allografts treated with CyA. Disseminated intravascular coagulation and renal function deterioration during CyA treatment and recovery coincident with CyA withdrawal have been described in a patient who underwent orthotopic liver transplant for primary hepatocellular carcinoma (*Bonser et al., 1984*). Only limited number of cases with well-identified de novo or recurrent CyA-induced HUS have been reported in renal allograft recipients treated with CyA (*Van Buren et al., 1985; Giroux et al., 1987*) suggesting that this syndrome occurs infrequently in these patients.

At morphological examination, the most striking changes are the presence of an extensive thrombotic process in the renal microcirculation (*Giroux et al., 1987*). Indeed, in many glomeruli, several capillary lumens are occluded by thrombi extending from the afferent arterioles, which contain platelet aggregates. Direct immunofluorescent studies give negative results for immunoglobulins and complement products, but conspicuous fibrin deposits may be present in numerous glomerular capillaries. The glomerular basement membrane is often denudated and enlarged subendothelial spaces are filled with flocculent material and cellular debris. In some cases the thrombotic process is not restricted to capillary lumens of the glomerular tuft, but pre-glomerular arterioles are occluded by fibrinous thrombi containing platelet aggregates.

The mechanism by which CyA can induce this fulminant obliterating arteriolopathy remains ill-defined. The fact that the more dramatic expression of CyA-associated arteriolopathy resembles HUS, a disease in which vascular endothelial damage is considered the first event in the pathogenetic cascade (*Heptinstall, 1983*), reinforces the concept that vascular endothelium is the main target in this form of CyA toxicity.

Endothelial cell injury can trigger complex biochemical changes that subsequently determine platelet adhesion to the exposed subendothelial collagen and platelet-platelet interaction with aggregation and thrombus formation. In this context, several studies have suggested that CyA inhibits the synthesis of a "prostacyclin-stimulating factor", thereby reducing the production of prostacyclin (PGI_2) by the renal vascular endothelium (*Neild et al., 1983b; Remuzzi et al., 1978*). Endothelial injury in the presence of reduced prostacyclin would then progress to local capillary thrombosis and glomerular infarction similar to that seen in the generalized Schwartzmann reaction (*Neild et al., 1983a*). However, the source of production and identity of the so-called "prostacyclin-stimulating factor" remains elusive. Experimental rabbit models have shown that CyA reduces prostacyclin generation by the vessel wall (*Neild et al., 1983a; Neild, 1983b*). This is in line with the recent finding of Voss and coworkers (*Voss et al., 1988*), who, studying the effect of CyA on PGI_2 release by cultured human umbilical vein endothelial cells, found that CyA induced a time- and concentration-dependent reduction in unstimulated and Ca_2^+-ionophore-stimulated release of PGI_2. On the other hand, bovine endothelial cells in culture generate increased amounts of TxA_2 (*Zoja et al., 1986*). Both phenomena can promote platelet activation at the site of endothelial damage.

A recent report underscores also a role of endothelin in CyA-induced microangiopathic injury (*Fogo et al., 1990*). Indeed, Fogo and coworkers (*Fogo et al., 1990*) have shown in a renal transplant recipient given two doses of rapid intravenous infusion of CyA an abrupt decrease in urine output associated with a severe vascular injury in the graft at renal biopsy examination, including glomeruli arterioles and large arteries. Of interest, the circulating level of endothelin after CyA infusions was more than double as compared to pre-transplant values, and decreased after withdrawal of CyA and removal of the graft. Moreover, the recent observation that human umbilical vein (*Bunchman and Brookshire, 1990*) and human glomerular endothelial cells in culture (*Bunchman and Brookshire, 1990*) as well as bovine pulmonary artery endothelial cells in culture (*Kon et al., 1990b,*) released endothelin upon stimulation with CyA, opens the possibility that this vasoactive peptide may contribute to the process of thrombi formation by reducing locally the blood flow through its potent vasoconstrictor action. That this can, indeed, be the case, is indicated by the fact that the main etiologic factor of HUS/TTP is bacterial endotoxin and that endotoxin-treated animals have much higher plasma endothelin than controls (*Sugiura et al., 1989*). In this context, it is of interest that cultured endothelial cells exposed to endotoxin generate significantly higher amounts of endothelin than cells challenged with the vehicle alone (*Sugiura et al., 1989*).

CONCLUSION

Endothelial cells are much more than a semipermeable barrier between blood and the vascular smooth muscle. Besides being a highly active metabolic and endocrine organ, these cells produce several potent vasoactive substances, including vasorelaxants and vasoconstrictors which are capable to modulate the underlying smooth muscle cells. Exogenous or endogenous circulating substances may cause endothelial injury, either functional or structural. This also applies to the extensive endothelium that lines the renal blood vessels.

CyA has been recently recognized to have a toxic effect on renal vascular endothelium and to induce acute vasospasm of renal microvessels. Changes in arachidonic acid metabolism, leading to increase TxA_2 formation, and the excessive generation of endothelin are probably responsible for the acute renal vasoconstriction associated with transient renal failure that follows the acute CyA administration.

Continuous CyA therapy results in chronic microvascular damage, glomerular ischemia and renal failure, as a consequence of persistent afferent arteriolar vasoconstriction and the subsequent development of irreversible arteriolopathy.

A rare form of thrombotic microangiopathy resembling HUS occurs occasionally in renal transplant patients given CyA. It has been suggested that a fulminant endothelial toxicity of CyA is the cause of such unusual manifestation of CyA nephrotoxicity. Based on the present knowledge, a strong case can be made for limiting the use of CyA to the field of some organ transplantation in which CyA nephrotoxicity is an acceptable risk as long as no better immunosuppression is available.

REFERENCES

Awazu, M., Sugiura, M., Inagami, T., Kon, V. Cyclosporine-induced glomerular dysfunction involves upregulation of glomerular receptors for endothelin. *Clin. Res.* 38:464, 1990 (abstr).

Badr, K.F.,Murray, J.J., Breyer, M.D., Takahashi, K., Inagami, T., and Harris, R.C. Mesangial cell, glomerular and renal vascular responses to endothelin in the rat kidney. *J. Clin. Invest.* 83:336-342, 1989.

Barros, E.J.G., Boim, M.A., Ajzen, H., Ramos, O.L., and Schor, N. Glomerular hemodynamics and hormonal participation on cyclosporine nephrotoxicity. *Kidney Int.* 32:19-25, 1985.

Baylis, C. Effects of administered thromboxanes on the intact, normal rat kidney. *Renal Physiol.* 10:110-121, 1987.

Benigni, A.,Perico, N., Gaspari, F., Zoja, C., Bellizzi, L., Gabanelli, M., and Remuzzi, G. Increased renal endothelin production in rats with reduced renal mass. *Am. J. Physiol.* (in press)

Bennett, W.M. and Norman, D.J. Action and toxicity of cyclosporine. *Ann. Rev. Med.* 37:215-224, 1986.

Bertani, T., Abbate, M., Zoja, C., Corna, D. and Remuzzi, G. Sequence of glomerular changes in experimental endotoxemia: a possible model of hemolytic uremic syndrome. *Nephron* 53: 330-337, 1989.

Bertani, T., Perico, N., Abbate, M., Battaglia, C., and Remuzzi, G. Renal injury induced by long-term administration of cyclosporin A to rats. *Am. J. Pathol.* 127:569-579, 1987.

Bonser, R.S., Adu, D., Franklin, I., and McMaster, P. Cyclosporin-induced hemolytic uraemic syndrome in liver allograft recipient. *Lancet* 2:1337, 1984.

Brenner, B.M., Troy, J.L. and Ballermann, B.J. Endothelium-dependent vascular responses. Mediators and Mechanisms. *J. Clin. Invest.* 84:1373-1378, 1989.

Brown, Z., Neild, G.H., Willoughby, J.J., Somia, N.V., and Cameron, S.J. Increased factor VIII as an index of vascular injury in cyclosporine nephrotoxicity. *Transplantation* 42:150-153, 1986.

Bunchman, T.E., and Brookshire, C.A. Cyclosporine stimulated synthesis of endothelin by human endothelial cells in tissue culture. *Kidney Int.* 37:365, 1990, (abstr).

Canadian Multicentre Transplant Study Group: A randomized clinical trial of cyclosporine in cadaveric renal transplantation. *N. Engl. J. Med.* 309:809-815, 1983.

Cattell, V. Endothelial cell diseases in experimental models: Mitomycin. In *Drugs and Kidney*, ed by T. Bertani, G. Remuzzi and S. Garattini, Raven Press, New York, pp.27-37, 1986.

Coffman, T.M., Carr, D.R., Yarger, W.E., and Klotman, P.E. Evidence that renal prostaglandin and thromboxane production is stimulated in chronic cyclosporine nephrotoxicity. *Transplantation* 43:282-285 1987.

Cohen, D.J., Loertscher, R., Rubin, M.F., Tilney, N.L., Carpenter, C.B. and Strom, T.B. Cyclosporine: A new immunosuppressive agent for organ transplantation. *Ann. Intern. Med.* 101:667-682, 1984.

Crooke, S.T. and Bradner, W.T. Mitomycin C. A review. *Cancer Treat. Rev.* 3:121-139, 1976.

Dieperink, H., Starklint, H., and Leyssac, P.P. Nephrotoxicity of cyclosporine - an animal model: study of the nephrotoxic effect of cyclosporine on overall and renal tubular function in conscious rats. *Transplant. Proc.* 15 (suppl 1):2736-2741, 1983.

Dieperink, H., Leyssac, P.P., Starklint, H., and Kemp, E. Long-term cyclosporin nephrotoxicity in the rat: effects on renal function and morphology. *Nephrol. Dial. Transplant.* 3:317-326, 1988.

Dische, F.E., Neuberger, J., Keating, J., Parsons, V., Calne, R.Y. and Williams, R. Kidney pathology in liver allograft recipients after long-term treatment with cyclosporin A. *Lab. Invest.* 58:395-402, 1988.

Elzinga, L., Kelley, V.E., Houghton, D.C., and Bennett, W.A. Modification of experimental nephrotoxicity with fish oil as the vehicle for cyclosporine. *Transplantation* 43:271-274, 1987.

English, J., Evan, A., Houghton, D.C., and Bennett, W.A. Cyclosporine-induced acute renal dysfunction in rat. *Transplantation* 44:135-141, 1987.

European Multicentre Trial Group: Cyclosporin in cadaveric renal transplantation: One-year follow-up of a multicentre trial. *Lancet* 2:986-989, 1983.

Firth, J.D., Ratcliffe, P.J., Raine, A.E.G., and Ledingham, J.G.G. Endothelin: an important factor in acute renal failure? *Lancet* 2:1179-1182, 1988.

Flechner, S.M., van Buren, C., Kerman, R.H., and Kahan, B.D. The nephrotoxicity of cyclosporine in renal transplant recipients. *Transplant. Proc.* 15 (suppl 1-2) :2689-2694, 1983.

Fogo, A., Hakim, R.C., Sugiura M., Inagami T., Kon V. Severe endothelial injury in a renal transplant patient receiving cyclosporine. *Transplantation* 49:1190-1192, 1990.

Gillum, D.M., Truong, L., Tasby, J., Migliore, P., and Suki, W.N. Chronic cyclosporine nephrotoxicity. A rodent model. *Transplantation* 46:285-292, 1988.

Gimbrone, M.A. Jr. Culture of vascular endothelium. In *Progress in Hemostasis and Thrombosis*, ed by T.H. Spaet, Grune and Stratton, New York, pp. 1-28, 1976.

Giroux, L., Smeesters, C., Corman, J., Paquin, F., Allaire, G., and St-Louis, G. Hemolytic uremic syndrome in renal allografted patients treated with cyclosporin. *Can. J. Physiol. Pharmacol.* 65:1125-1131, 1987.

Harlan, J.M., Harker, L.A., Reidy, M.A., Gajdusek, C.M., Schwartz, S.M., Striker, G.E. Lipopolysaccharide-mediated bovine endothelial cell injury *in vitro*. *Lab. Invest.* 48:269-274, 1983.

Heptinstall, R.H. Hemolytic uremic syndrome, thrombotic thrombocytopenic purpura, and systemic scleroderma (progressive systemic sclerosis). In *Pathology of the Kidney*, 3rd ed., Boston, Little Brown, Boston, p. 907-961, 1983.

Jaffe, E.A., Nachman, R.L., Becker, S.G. and Minick, D.R. Culture of human endothelial cells derived from umbilical veins. Identification by morphologic and immunologic criteria. *J. Clin. Invest.* 52:2745-2756, 1973.

Jaffe, E.A. Endothelial cells and the biology of factor VIII. *N. Engl. J. Med.* 296:377-383, 1977.

Kahan, B.D. Cyclosporine: A powerful addition to the immunosuppressive armamentarium. *Am. J. Kidney Dis.* 3:444-455, 1984.

Kahan, B.D. Cyclosporine: The agent and its action. *Transpl. Proc.* 17 (suppl 1): 5-18, 1985.

Kawaguchi, A., Goldman, M.H., Shapiro, R., Foegh, M.L., Ramwell, P.W., and Lower, R.R. Increase in urinary thromboxane B_2 in rats caused by cyclosporine. *Transplantation* 40:214-216, 1985.

Kelley, V.E., Ferretti, A., Izui, S., and Strom, T.B. A fish oil diet rich in eicosapentaenoic acid reduces cyclooxygenase metabolites, and suppresses lupus in MRL-lpr mice. *J. Immunol.* 134:1914-1919, 1985.

King, A.J., Brenner, B.M., and Anderson, S. Endothelin: a potent renal and systemic vasoconstrictor peptide. *Am. J. Physiol.* 256:F1051-F1058, 1989.

Kon, V., Sugiura, M., Inagami, T., Harvie, B.R., Ichikawa, I., and Hoover, R.L. Role of endothelin in cyclosporine-induced glomerular dysfunction. *Kidney Int.* 37:1487-1491, 1990a.

Kon, V., Yoshioka, T., Fogo, A., and Ichikawa, I. Glomerular actions of endothelin *in vivo*. *J. Clin. Invest.* 83:1762-1767, 1989.

Kon, V., Sugiura, M., Inagami, T., Hoover, R.L., Fogo, A., Harvie, B.R., and Ichikawa, I. Cyclosporine (Cy) causes endothelin-dependent acute renal failure. *Kidney Int.* 37:486 1990b, (abstr).

Lau, D.C.W., Wong, K-L., and Hwang, W.S. Cyclosporine toxicity on cultured rat microvascular endothelial cells. *Kidney Int.* 35:604-613, 1989.

McGrath, J.M. and Stewart, G.J. The effects of endotoxin on vascular endothelium. *J. Exp. Med.* 129:833-848, 1969.

Mihatsch, M.J., Thiel, G., Spicktin, M.P., Oberholzer, M, Brunner, F.P., Harder, F., Olivieri, V., Bremer, R., Ryffel, B., Stocklin, E., Torhorst, J., Gudat, F., Zollinger, H.U., and Loertscher, R. Morphological findings in kidney transplants after treatment with cyclosporine. *Transplant. Proc.* 15 (suppl 1):2821-2835, 1983.

Moss, N.G., Powell, S.L., and Falk, R.J. Intravenous cyclosporine activates afferent and efferent renal nerves and causes sodium retention in innervated kidneys in rats. *Proc. Natl. Acad. Sci. USA* 82:8222-8226, 1985.

Murasawa, K. The injurious effect of granulocytes and mitomycin c added to cultured human vascular endothelium. *Nippon Geka Hokan* 52:818- 827, 1983.

Murray, B.M., Paller, M.S., and Ferris, T.F. Effect of cyclosporine administration on renal hemodynamics in conscious rats. *Kidney Int.* 28:767-774, 1985.

Myers, B.D., Newton, L., Boshkos, C., Macoviak, J.A., Frist, W.H., Derby, G.C., Perlroth, M.G., and Sibley, R.K. Chronic injury of human renal microvessels with low-dose cyclosporine therapy. *Transplantation* 46:694-703, 1988a.

Myers, B.D. Cyclosporine nephrotoxicity. *Kidney Int.* 30:964-974, 1986.

Myers, B.D., Sibley, R., Newton, L., Tomlanovich, S.J., Boshkos, C., Stinson, E., Luetscher, J.A., Whitney, D.J., Krasny, D., Coplon, N.S., and Perlroth, M.G. The long-term course of cyclosporine-associated chronic nephropathy. *Kidney Int.* 33:590-600, 1988a.

Myers, B.D., Ross, J., Newton, L., Luetscher, J., and Perlroth, M. Cyclosporine-associated chronic nephropathy. *N. Engl. J. Med.* 311:699-705, 1984.

Neild, G.H., Rocchi, G., Imberti, L., Fumagalli, F., Brown, Z., Remuzzi, G., and Williams, D.G. Effect of cyclosporine on prostacyclin synthesis by vascular tissue in rabbits. *Transplant. Proc.* 15 (suppl 1):2398-2400, 1983b.

Neild, G.H., Ivory, K., and Williams, D.G. Glomerular thrombi and infarction in rabbits with serum sickness following cyclosporine therapy. *Transpl. Proc.* 15 (Suppl 1) 2782-2786, 1983a.

Nizze, H., Mihatsch, M.J., Zollinger, H.U., Brocheriou, C., Gokel, J.M., Henry, K., Sloane, J.P., and Stovin, P.G. Cyclosporine-associated nephropathy in patients with heart and bone marrow transplants. *Clin. Nephrol.* 30:248-260, 1988.

Pasini, M., Perico, N., and Remuzzi, G. Roles for thromboxane (Tx)A_2 and sulfidopeptide leukotrienes (LT) in cyclosporine (CsA)-induced acute renal failure. *Kidney Int.* 37:350, 1990, (abstr).

Perico, N., Dadan, J., and Remuzzi, G. Endothelin mediates the renal vasoconstriction induced by cyclosporine in the rat. *J.A.S.N.*, 1990c.

Perico, N., Benigni, A., Ladny, J.R., Imberti, 0., Bellizzi, L., and Remuzzi, G. Chronic cyclosporine A (CyA) administration to rats increases urinary excretion of big-endothelin and endothelin. *J.A.S.N.* 1:617, 1990a.

Perico, N., Dadan, J., Gabanelli, M., and Remuzzi, G. Cyclooxygenase products and atrial natriuretic peptide modulate the renal response to endothelin. *J. Pharmacol. Exp. Ther.* 252:1213-1220, 1990b.

Perico, N., Zoja, C., Benigni, A., Ghilardi, F., Gualandris, L., and Remuzzi, G. Effect of short-term cyclosporine administration in rats on renin-angiotensin and thromboxane A_2: Possible relevance to the reduction in glomerular filtration rate. *J. Pharmacol. Exp. Ther.* 239:229-235, 1986.

Pernow, J., Boutier, J.F., Franco-Cereceda, A., Lacroix, J.S., Matran, R., and Lundberg, J.M. Potent selective vasoconstrictor effects of endothelin in the pig kidney *in vivo*. *Acta. Physiol. Scand.* 134:573-574, 1988.

Powles, R.L., Clink, H.M., Spence, D., Morgenstern, G., Watson, J.G., Selby, P.J., Woods, M., Barrett, A., Jameson, B., Sloane, J., Lawler, S.D., Kay, H.E., Lawson, D., McElwain, T.J., and Alexander, P. Cyclosporin A to prevent graft-versus-host disease in man after allogeneic bone-marrow transplantation. *Lancet* 1:327-329, 1980.

Powell-Jackson, P.R., Young, B., Calne, R.Y., and Williams, R. Nephrotoxicity of parenterally administered cyclosporine after orthotopic liver transplantation. *Transplantation* 36:505-508, 1983.

Remuzzi, G., Misiani, R., Marchesi, D., Livio, M., Mecca, G., and de Gaetano, G. Hemolytic uremic syndrome: Deficiency of plasma factor(s) regulating prostacyclin activity? *Lancet* 2:871-872, 1978.

Remuzzi, G. and Bertani, T. Renal vascular and thrombotic effects of cyclosporine. *Am. J. Kidney Dis.* A13:261-271, 1989.

Rossi, N.F., Churchill, P.C., McDonald, F.D., and Ellis, V.R. Mechanism of cyclosporine A-induced renal vasoconstriction in the rat. *J. Pharmacol. Exp. Therap.* 250:896-901, 1989.

Shulman, H., Striker, G., Deeg, H.J., Kennedy, M., Storb, R. and Thomas, E.D. Nephrotoxicity of cyclosporin A after allogeneic marrow transplantation. Glomerular thromboses and tubular injury. *N. Eng. J. Med.* 305:1392-1395, 1981.

Smeesters, C., Chaland, P., Giroux, L., Moutquin, J.M., Etienne, P., Douglas, F., Corman, J., St-Louis, G., and Daloze, P. Prevention of acute cyclosporine A nephrotoxicity by thromboxane synthetase inhibitor. *Transplant. Proc.* 20 (suppl 3):658-664, 1988.

Sugiura, M., Inagami, T., and Kon, V. Endotoxin stimulates endothelin release *in vivo* and *in vitro* as determined by radioimmunoassay. *Biochem. Biophys. Res. Commun.* 161:1220-1227, 1989.

Sullivan, B.A., Hak, L.J., and Finn, W.F. Cyclosporine nephrotoxicity: Studies in laboratory animals. *Transplant. Proc.* 17:145-154, 1984.

Tegzess, A.M., Doorenbos, B.M., Minderhound, J.M., and Donker, A.J. Prospective serial renal function studies in patients with non renal disease treated with cyclosporine A. *Transplant Proc.* 20 (suppl 2):390-393, 1988.

VanBuren, D., Van Buren, C.T., Flechner, S.M., Maddox, A.M., Verani, R., and Kahan, B.D. De novo hemolytic uremic syndrome in renal transplant recipients immunosuppressed with cyclosporine. *Surgery* 98:54-62, 1985.

Vane, J.R., Anggard, E.E. and Botting, R.M. Regulatory functions of the vascular endothelium. *N. Engl. J. Med.* 323:27-36, 1990.

Von Graffenried, B., and Harrison, W.B. Renal function in patients with autoimmune diseases treated with cyclosporine. *Transplant. Proc.* 17:215-231, 1985.

von Willebrand, E., and Hayry, P. Cyclosporin-A deposits in renal allograft. *Lancet* 2:189-192, 1983.

Voss, B.L., Hamilton, K.K., Samara, E.N.S. and McKee, P.A. Cyclosporin suppression of endothelial prostacyclin generation. *Transplantation* 45:793-796u, 1988.

Yanagisawa, K., Kurihara, H., Kimura, S., Tomobe, Y., Kobayasni, M., Mitsui, Y., Yazaki, Y., Goto, K., and Masaki, T. A novel potent vasoconstrictor peptide produced by vascular endothelial cells. *Nature* 332:411-415, 1988.

Zoja, C., Furci, L., Ghilardi, F., Zilio, P., Benigni, A., and Remuzzi, G. Cyclosporin-induced endothelial cell injury. *Lab. Invest.* 55:455-462, 1986.

CEREBRAL ENDOTHELIAL FUNCTION: PHYSIOLOGY AND PATHOPHYSIOLOGY

Donald D. Heistad, F. Faraci, and G. Baumbach

Departments of Internal Medicine, Pharmacology, and Pathology
University of Iowa College of Medicine
Iowa City, Iowa 52242, U.S.A.

INTRODUCTION

The purpose of this chapter is to summarize some recent concepts about the role of cerebral vascular endothelium under physiological and pathophysiological conditions.

The first part of the chapter will focus on the function of cerebral endothelium as a blood-brain barrier. We will discuss mechanisms by which the blood-brain barrier accounts for unique responses of cerebral blood vessels, especially to humoral stimuli. Current concepts will be considered concerning dysfunction of the blood-brain barrier during acute hypertension.

The second part of the chapter focuses on cerebral endothelium in disease states. We will discuss alterations of the blood-brain barrier during chronic hypertension, protection of the barrier by vascular hypertrophy, and alterations of endothelium-dependent responses by chronic hypertension. Effects of atherosclerosis on endothelial function also will be examined, with discussion of implications for the pathophysiology of transient ischemic attacks (TIAs). Finally, we will provide evidence that responses to endothelium-dependent agonists may be altered by aging.

BLOOD-BRAIN BARRIER

The blood-brain barrier minimizes and regulates the passage of water-soluble molecules from the blood into the brain and cerebrospinal fluid. The barrier is characterized by tight endothelial junctions, a paucity of pinocytotic vesicles (or perhaps canaliculi) in the endothelium, and an enzymatic barrier.

In contrast to most intracranial blood vessels, endothelium of the choroid plexus, area postrema, and other circumventricular organs is fenestrated, and thus the vessels are relatively permeable to blood-borne humoral agents.

REGULATION OF BLOOD FLOW

Circulating catecholamines and vasoactive peptides have little effect on cerebral blood flow. The absence of responses is attributed to the endothelial blood-brain barrier, which minimizes the access of hormones to cerebral vascular smooth muscle. Recently, we have suggested that the view that humoral stimuli have little effect on cerebral blood vessels needs to be modified.

Several vasoactive peptides modulate resistance of large arteries to the brain. Angiotensin constricts and vasopressin dilates *(Faraci et al., 1988)* large arteries to the brain. Because responses of small vessels are in the opposite direction, however, there is little change in total cerebral vascular resistance and blood flow. Although blood flow is unchanged, constriction or dilatation of large arteries produces pronounced changes in cerebral microvascular pressure.

We have proposed that neurohumoral regulation of large cerebral arteries and microvascular pressure may have important implications *(Faraci et al., 1990)*. First, changes in microvascular pressure may affect "central baroreceptors", and thereby activate compensatory responses (Figure 1). Second, changes in microvascular pressure may affect filtration in regions of the brain in which microvessels are relatively permeable.

Hormones may have especially important effects in regions of the brain in which the blood-brain barrier is permeable. For example, vasopressin produces profound reductions in blood flow to the choroid plexus *(Faraci et al., 1988)* and a sustained reduction in production of cerebrospinal fluid *(Faraci et al., 1990)*. These effects occur at plasma levels of vasopressin that are equivalent to those observed during intracranial hypertension or hypoxia. These findings have led us to suggest that vasopressin may protect against elevations of intracranial pressure by reducing production of cerebrospinal fluid.

The area postrema is supplied by capillaries with fenestrated endothelium. One function of the area postrema, apparently, is to serve as a sensing organ for circulating angiotensin and vasopressin, and thus play an important role in autonomic control of the cardiovascular system *(Ferrario et al., 1987; Undesser et al., 1985)*. An unexpected finding was that even though the blood-brain barrier in the area postrema is permeable, regulation of vascular tone is similar in the area postrema and surrounding brain stem, in which the barrier is intact *(Faraci, F.M. et al., 1989)*. We understood this finding when it became evident that, even though endothelium of capillaries in the area postrema is fenestrated, endothelium of arterioles that supply the area postrema has tight junctions *(Faraci et al., 1989)*. Thus, blood-borne peptides are able to penetrate into interstitial fluid of the area postrema, through the capillaries, and the area postrema is able to sense blood-borne hormones, but the hormones do not compromise the sensing function of the region by altering blood flow.

The dura mater is one of the meninges that covers the brain. Vessels of the dura are relatively permeable. Thus, the brain (which is protected by the blood-brain barrier) is covered by, and in close proximity to, a structure in which the barrier is absent. Recently, we have found that dura is perfused at a relatively high level of blood flow (almost 40 ml/min per 100 gm) *(Faraci et al., 1989)*. Substance P and serotonin, which have been implicated in the pathogenesis of vascular headache, produce pronounced vasodilatation in the dura mater *(Faraci et al., 1989)*. Thus, blood-borne hormones that have little effect on cerebral blood flow may have large effects on blood flow to the dura.

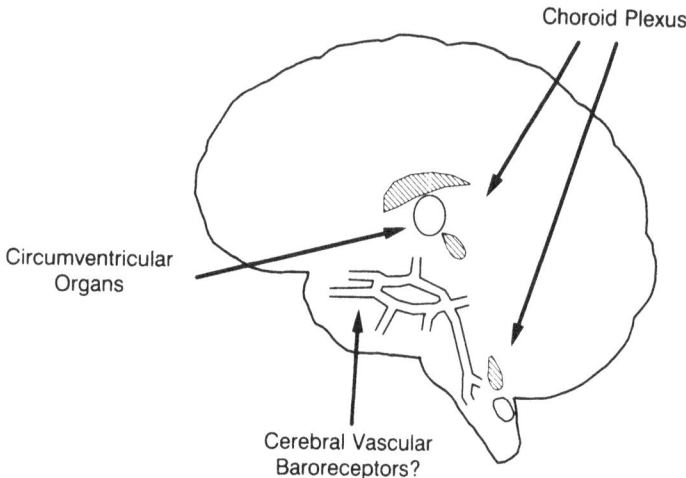

FIGURE 1: Neurohumoral stimuli appear to modulate cerebral microvascular pressure. Changes in microvascular pressure may affect filtration in the choroid plexus, filtration (and thus osmolarity) in other circumventricular organs, and putative central baroreceptors *(Faraci, F.M. et al., 1988)*.

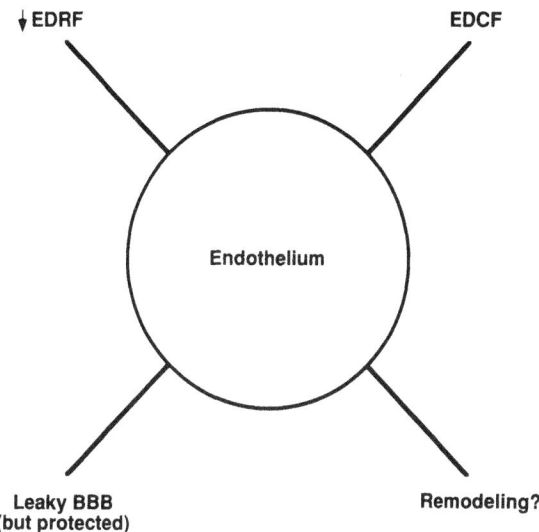

FIGURE 2: Effects of chronic hypertension on cerebral endothelium. Hypertension may impair synthesis and/or release of endothelium-derived relaxing factor (EDRF) and release an endothelium-derived contracting factor (EDCF). Chronic hypertension also appears to be associated with an intrinsically leaky blood-brain barrier (BBB), which is masked by hypertrophy of blood vessels upstream. Finally, endothelial dysfunction in chronic hypertension may contribute to "remodeling" of blood vessels.

DYSFUNCTION OF THE BLOOD-BRAIN BARRIER

Physical stimuli, drugs, infection, and neoplastic processes can increase permeability of the blood-brain barrier. Our studies have focused on dysfunction of the barrier produced by acute hypertension.

Acute hypertension produces passive dilatation of cerebral blood vessels and dysfunction of the blood-brain barrier *(Johansson et al., 1970)*. A surprising finding is that small venules, not arterioles or capillaries, are the primary site of disruption of the barrier during acute hypertension *(Mayhan and Heistad, 1985)*. Pronounced increases in cerebral venous pressure produce disruption of the barrier and may account for dysfunction of the barrier during acute hypertension *(Mayhan and Heistad, 1986)*.

The luminal surface of cerebral endothelium is coated with an anionic glycocalyx which imparts a negative charge to the endothelium. It has been suggested that dysfunction of the blood-brain barrier during acute hypertension may be associated with loss of the endothelial negative charge. Our studies, however, suggest that acute hypertension produces less disruption of the blood-brain barrier to anionic dextran than to neutral dextran *(Mayhan et al., 1989)*. These findings imply that molecular charge is preserved during hypertension, and that molecular charge is an important determinant of disruption of the blood-brain barrier during acute hypertension.

There are important regional differences in the brain in relation to susceptibility to dysfunction of the blood-brain barrier. Acute hypertension produces marked disruption of the barrier in cerebrum but little disruption in brain stem *(Baumbach and Heistad, 1985)*. Regional differences in autoregulation account for differences in susceptibility of the barrier to dysfunction.

CHRONIC HYPERTENSION

Disruption of the blood-brain barrier by a hyperosmolar solution is greater in stroke-

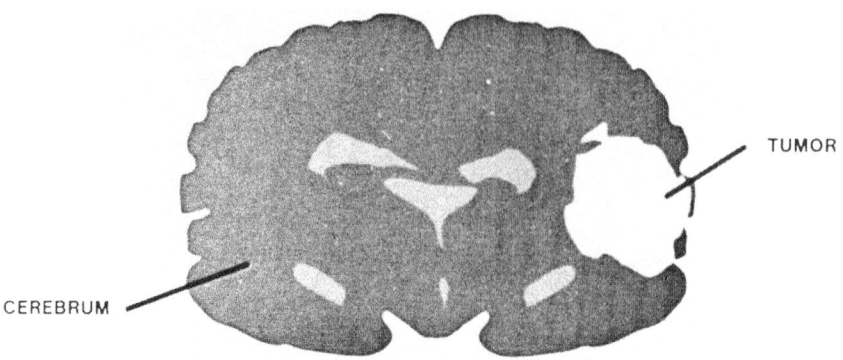

CEREBRUM	TUMOR
• Intact BBB	• Permeable BBB
• Minimal Response to Vasoactive Substances	• Highly Responsive to Vasoactive Substances
• Large Response to Hypercapnia	• Minimal Response to Hypercapnia
• Autoregulation	• No Autoregulation

FIGURE 3: Effects of brain tumors. Diagram of coronal section of the brain in a dog with a sarcoma. There are profound differences in regulation of blood flow to normal cerebrum and to brain tumors *(Panther, L.A. et al., 1985)*.

prone spontaneously hypertensive rats (SHRSP) than in normotensive rats (WKY) *(Tamaki et al., 1984)*. Furthermore, osmotic disruption of the barrier produces fatal cerebral edema in SHRSP, but not in WKY *(Tamaki et al., 1984)*. Thus, there appears to be an inherent defect of the blood-brain barrier to osmotic disruption in chronic hypertension.

In contrast to effects of osmotic disruption, acute hypertension produces less disruption of the blood-brain barrier in SHRSP than in WKY *(Mayhan et al., 1987)*. It is likely that hypertrophy of cerebral arterioles in SHRSP attenuates increases in microvascular pressure downstream *(Werber and Heistad, 1984)*, attenuates increases in venous pressure, and thereby, protects the blood-brain barrier.

Hypertensive encephalopathy, which at one time was attributed to vasospasm and ischemia, now appears to result from passive dilatation of cerebral vessels and disruption of the blood-brain barrier *(Farrar et al., 1976; Kontos et al., 1981)*. Our findings suggest that hypertensive encephalopathy occurs spontaneously in SHRSP *(Sadoshima and Heistad, 1982)*. Sympathetic nerves protect the barrier during acute hypertension, and thereby protect against hypertensive encephalopathy *(Sadoshima and Heistad, 1982)* and stroke *(Sadoshima et al., 1981)*. The protective effect of sympathetic nerves in SHRSP is related to a "trophic" effect, as the nerves promote development of vascular hypertrophy *(Baumbach et al., 1989)*.

In the aorta, endothelial dysfunction during chronic hypertension also is manifested by impairment of endothelium-dependent vasodilatation *(Konishi, 1983)* and release of an endothelium-derived contracting factor *(Luscher and Vanhoutte, 1986)*. In cerebral arterioles, dilator responses to acetylcholine are profoundly impaired in SHRSP *(Mayhan, et al., 1987)*. Impairment of relaxation in SHRSP probably is the result of release of a constricting factor from cerebral arterioles *(Mayhan, et al., 1988)*.

We have proposed recently that cerebral arterioles in SHRSP undergo a structural change, which we have described as "remodeling" *(Baumbach and Heistad, 1989)*, that may be related to endothelial dysfunction. It is generally accepted that vessels undergo hypertrophy during chronic hypertension, with encroachment on the vascular lumen *(Folkow et al., 1970)*. We have observed that external diameter, as well as internal diameter, is reduced in cerebral arterioles of SHRSP *(Baumbach and Heistad, 1989)*. Thus, the arteriole is "remodeled". Our calculations suggest that remodeling of cerebral arterioles may account for more than 75% of encroachment on the lumen in SHRSP, and that hypertrophy per se accounts for a surprisingly small portion of the encroachment.

Mechanisms that account for vascular remodeling in SHRSP are not established, but it is possible that endothelial dysfunction plays a role. Endothelial denudation impairs normal growth of the carotid artery *(Langille and O'Donnell, 1986)*. We speculate that endothelial dysfunction of cerebral vessels in SHRSP, which is manifested both by an intrinsically abnormal blood-brain barrier *(Tamaki et al., 1984)* and by abnormalities in modulation of vascular tone *(Mayhan et al., 1988)*, also may contribute to impaired structural dilatation. Thus, endothelial dysfunction may contribute to remodeling of cerebral arterioles *(Baumbach and Heistad, 1927)* (Figure 2).

ENDOTHELIUM AND VASCULAR DYSFUNCTION

Abnormalities of cerebral endothelium may be central to vascular changes in several conditions, in addition to chronic hypertension. First, many studies indicate that endothelium-dependent relaxation is impaired in atherosclerotic arteries. We have suggested that this abnormality may contribute to the pathophysiology of transient ischemic attacks with blindness (amaurosis fugax) *(Williams et al., 1989)*. The hypothesis is that alteration of responses to vasoactive substances that are released by platelets may produce vasospasm and contribute to amaurosis. Furthermore, regression of atherosclerotic lesions is accompanied by improvement in endothelium-dependent relaxation *(Harrison et al., 1987)* and reduction of susceptibility to cerebral vasospasm *(Heistad, D.D. et al., 1987)*.

Second, the blood-brain barrier often is absent in brain tumors. Thus, blood-borne vasoactive stimuli that have little effect on blood flow to normal cerebrum may produce large changes in blood flow to brain tumors (Figure 3) *(Panther et al., 1985)*. Thus, the fenestrated capillaries in brain tumors may offer a window of opportunity in treatment of brain tumors. For example, vasodilator agents may selectively increase blood flow to brain tumors and allow chemotherapeutic agents to be delivered preferentially to the tumors, with no increase in delivery to normal cerebrum.

Third, aging is characterized by impaired dilator responses of cerebral arterioles to agonists that release endothelium-derived relaxing factor *(Mayhan et al., 1990)*. In aging, in contrast to chronic hypertension *(Mayhan et al., 1988)*, impaired vasodilatation does not appear to be related to production of a cyclooxygenase constrictor substance. Finally, cerebral arterioles undergo atrophy of vascular muscle during aging *(Hajdu et al., In press)*. As described above, in relation to vascular remodeling during chronic hypertension, we speculate that endothelial dysfunction during aging may contribute to atrophy of cerebral arterioles.

ACKNOWLEDGMENTS

Original studies described in this chapter were supported by a Medical Investigator Award and Research funds from the Veterans Administration and NIH Grants HL 16066, NS 24621, HL 14230, and HL 14388 and HL 22149, and an Established Investigator Award from the American Heart Association. We thank Ms. Marjorie Keaough for typing the manuscript.

REFERENCES

Baumbach, G.L., and Heistad, D.D. Heterogeneity of Brain Blood Flow and Permeability During Acute Hypertension. *Am. J. Physiol.: Heart Circ. Physiol.*, 18:H629-H637, 1985.

Baumbach, G.L., Heistad, D.D., and Siems, J.E. Effect of Sympathetic Nerves on Composition and Distensibility of Cerebral Arterioles in Rats. *J. Physiol. (London)*, 416:123-140, 1989.

Baumbach, G.L., and Heistad, D.D. Remodeling of Cerebral Arterioles in Chronic Hypertension. *Hypertension*, 13:968-972, 1989.

Bradbury, M.W.B. The Concept of a Blood-Brain Barrier. Chichester: Wiley & Sons, 1979.

Faraci, F.M., Mayhan, W.G., Schmid, P.G., and Heistad, D.D. Effects of Arginine Vasopressin on Cerebral Microvascular Pressure. *Am. J. Physiol.: Heart Circ. Physiol.*, 255:H70-H76, 1988.

Faraci, F.M., Mayhan, W.G., Farrell, W.J., and Heistad, D.D. Humoral Regulation of Blood Flow to Choroid Plexus: Role of Arginine Vasopressin. *Circ. Res.*, 63:373-379, 1988.

Faraci, F.M., Kadel, K.A., and Heistad, D.D. Vascular Responses of Dura Mater. *Am. J. Physiol. Heart Circ. Physiol.*, 26:H157-H161, 1989.

Faraci, F.M., Choi, J., Baumbach, G.L., Mayhan, W.G., and Heistad, D.D. Microcirculation of the Area Postrema: Permeability and Vascular Responses. *Circ. Res.*, 65:417-425, 1989.

Faraci, F.M., and Heistad, D.D. Regulation of Large Cerebral Arteries and Cerebral Microvascular Pressure. *Circ. Res.*, 66:8-17, 1990.

Faraci, F.M., Mayhan, W.G., and Heistad, D.D. Effect of Vasopressin on Production of Cerebrospinal Fluid: Possible Role of Vasopressin (V1) Receptors. *Am. J. Physiol.*, 27:R94-R98, 1990.

Farrar, J.K., Jones, J.V., Graham, D.I., Strandgaard, S., MacKenzie, E.T. Evidence Against Cerebral Vasospasm During Acutely Induced Hypertension. *Brain Res.*, 104:176-180, 1976.

Ferrario, C.M., Barnes K.L., Diz, D.I., Block, C.H., Averill, D.B. Role of Area Postrema Pressor Mechanisms in the Regulation of Arterial Pressure. *Can. J. Physiol. Pharmacol.*, 65:1591-1597, 1987.

Folkow, B., Hallbeck, M., Lundgren, Y., Weiss, L. Background of Increased Flow Resistance and Vascular Reactivity in Spontaneously Hypertensive Rats. *Acta. Physiol. Scand.*, 80:93-106, 1970.

Hajdu, M.A., Heistad, D.D., Siems, J.E., and Baumbach, G.L. Effects of Aging on Mechanics and Composition of Cerebral Arterioles in Rats. *Circ. Res.* 66:1747-1754, 1990.

Harrison, D.G., Armstrong, M.L., Freiman, P.C., and Heistad, D.D. Restoration of Endothelium Dependent Relaxation by Dietary Treatment of Atherosclerosis. *J. Clin. Invest.*, 80:1808-1811, 1987.

Heistad, D.D., Breese, K., and Armstrong, M.L. Cerebral Vasoconstrictor Responses to Serotonin After Dietary Treatment of Atherosclerosis. *Stroke*, 18:1068-1073, 1987.

Johansson, B.B., Li C-L, Olsson Y., Klatzo I. The Effect of Acute Arterial Hypertension on the Blood-Brain Barrier to Protein Tracers. *Acta. Neuropathol.* 16:117-124, 1970.

Konishi, M., Su, C. Role of Endothelium in Dilator Responses of Spontaneously Hypertensive Rat Arteries. *Hypertension* 5:881-886, 1983.

Kontos, H.A., Wei, E.P., Dietrich, W.D., et al. Mechanism of Cerebral Arteriolar Abnormalities After Acute Hypertension. *Am. J. Physiol.*, 240:H511-H527, 1981.

Langille, B.L, O'Donnell F Reductions in Arterial Diameter Produced by Chronic Decreases in Blood Flow are Endothelium-Dependent. *Science* 231:405-407, 1986.

Luscher, T.F., Vanhoutte P.M. Endothelium-Dependent Responses to Platelets and Serotonin in Spontaneously Hypertensive Rats. *Hypertension* 8:II-55-II-60, 1986.

Mayhan, W.G., and Heistad, D.D. Permeability of Blood-Brain Barrier to Various Sized Molecules. *Am. J. Physiol.: Heart Circ. Physiol.*, 17:H712-H718, 1985.

Mayhan, W.G., and Heistad, D.D. Role of Veins and Cerebral Venous Pressure in Disruption of the Blood-Brain Barrier. *Circ. Res.*, 59:216-220, 1986.

Mayhan, W.G., Faraci, F.M., and Heistad, D.D. Mechanisms of Protection of the Blood-Brain Barrier During Acute Hypertension in Chronically Hypertensive Rats. *Hypertension*, 9:III-101-III-105, 1987.

Mayhan, W.G., Faraci, F.M., and Heistad, D.D. Impairment of Endothelium-Dependent Responses of Cerebral Arterioles in Chronic Hypertension. *Am. J. Physiol.: Heart Circ. Physiol.*, 22:H1435-H1440, 1987.

Mayhan, W.G., Faraci, F.M., and Heistad, D.D. Responses of Cerebral Arterioles to Adenosine 5´-Diphosphate, Serotonin, and the Thromboxane Analogue U-46619 During Chronic Hypertension. *Hypertension*, 12:556-561, 1988.

Mayhan, W.G., Faraci, F.M., Siems, J.L., and Heistad, D.D. Role of Molecular Charge in Disruption of the Blood-Brain Barrier During Acute Hypertension. *Circ. Res.*, 64:658-664, 1989.

Mayhan, W.G., Faraci, F.M., Baumbach, G.L., and Heistad, D.D. Effects of Aging on Responses of Cerebral Arterioles. *Am. J. Physiol.:Heart Circ. Physiol.* 27:H1138-H1143, 1990.

Panther, L.A., Baumbach, G.L., Bigner, D.D., Piegors, D., Groothuis, D.R., and Heistad, D.D. Vasoactive Drugs Produce Selective Changes in Flow to Experimental Brain Tumors. *Ann. Neurol.*, 18:712-715, 1985.

Sadoshima, S., Busija, D., Brody, M., and Heistad, D.D. Sympathetic Nerves Protect Against Stroke in Stroke-Prone Hypertensive Rats. *Hypertension*, 3:I-124-I-127, 1981.

Sadoshima, S., and Heistad, D.D. Sympathetic Nerves Protect the Blood-Brain Barrier in Stroke-Prone Spontaneously Hypertensive Rats. *Hypertension*, 4:904-907, 1982.

Tamaki, K., Sadoshima, S., and Heistad, D.D. Increased Susceptibility to Osmotic Disruption of the Blood-Brain Barrier in Chronic Hypertension. *Hypertension,* 6:633-638, 1984.

Undesser, K.P., Hasser, E.M., Haywood, J.R., Johnson, A.K., Bishop, V.S. Interactions of Vasopressin with the Area Postrema in Arterial Baroreflex Function in Conscious Rabbits. *Circ. Res.,* 56:410-417, 1985.

Werber, A.H., and Heistad, D.D. Effects of Chronic Hypertension and Sympathetic Nerves on the Cerebral Microvasculature of Stroke-Prone Spontaneously Hypertensive Rats. *Circ. Res.,* 55:286-294, 1984.

Williams, J.K., Baumbach, G.L., Armstrong, M.L., and Heistad, D.D. Hypothesis: Vasoconstriction Contributes to Amaurosis Fugax. *J. Cereb. Blood Flow Metab.,* 9:111-116, 1989.

MECHANISMS OF ALTERED REACTIVITY
IN THE CEREBRAL MICROCIRCULATION

Hermes A. Kontos

Department of Medicine
Medical College of Virginia
Richmond, Virginia 23298, U.S.A.

INTRODUCTION

Complete removal of the endothelium by mechanical or chemical means is the preferred method for demonstrating endothelium-dependent responses in large vessels. Attempts to apply this technique to cerebral microvessels have not been fruitful because these vessels are thin-walled and the removal of the endothelium usually resulted in damage to the vascular smooth muscle with consequent generalized depression of all responses (*Dacey and Bassett, 1987*). However, convincing evidence for the presence of endothelium-dependent relaxation in cerebral arterioles was obtained by selective damage to the endothelium using photochemical techniques (*Rosenblum, 1986; Rosenblum et al., 1987a; Rosenblum et al., 1987b*). These methods rely on the intravascular administration of a dye coupled with radiation of selected vessels with ultraviolet or laser light. The dye absorbs the light and generates heat or reactive oxygen species which damage the endothelium. By appropriate selection of the duration of radiation and its intensity, the injury can be limited to the endothelium and platelet aggregation is avoided. Under these conditions there is minimal histologically demonstrable injury to the endothelium (*Povlishock et al., 1983, Rosenblum et al., 1987b*). In these irradiated vessels, the vasodilation from topical acetylcholine, bradykinin or the calcium ionophore A23187 is either eliminated or converted to vasoconstriction. Agents which induce dilation by acting directly on the vascular smooth muscle, like nitroprusside and papaverine, are unaffected.

The dependence of the vasodilator response to acetylcholine on the endothelium was corroborated by the demonstration in bioassay experiments (*Kontos et al. 1988*) that this agent releases a transferrable unstable vasodilator agent similar to that generated by acetylcholine in large vessels or by bradykinin in cultured endothelial cells (Table 1). This agent, therefore, is believed to be the endothelium-derived relaxing factor (EDRF) generated by acetylcholine in cerebral microvessels.

IDENTITY OF EDRFs IN THE CEREBRAL MICROCIRCULATION

There is evidence for at least two different EDRFs in the cerebral microcirculation. As mentioned above, the EDRF generated by acetylcholine has characteristics similar to those of the EDRF generated by this agent in large vessels *in vitro* (*Furchgott, 1983*) or by bradykinin in cultured endothelial cells (*Palmer et al. 1987*).

Because of the strong evidence supporting the contention that the EDRF from acetylcholine in large vessels *in vitro* (*Ignarro et al. 1987*) or from bradykinin in cultured endothelial cells (*Palmer et al. 1987*) is nitric oxide, this possibility was investigated in cerebral microvessels. The EDRF from acetylcholine can be distinguished from nitric oxide by pharmacological means. First, agents which generate oxygen radicals by autoxidation in the

TABLE 1: Properties of EDRF generated by acetylcholine in cerebral microvessels of the cat

1. Destroyed by extending transit time to 1-2 minutes
2. Release prevented by blocking muscarinic receptors with atropine
3. Vasodilator action blocked by methylene blue or by hemoglobin
4. Destroyed by oxygen radical generating agents or interventions
5. Release prevented by agents which damage endothelium

extracellular space, like methylene blue and hemoglobin, inhibited the vasodilation from acetylcholine without affecting the vasodilation from nitric oxide or other nitrodilators, like nitroprusside and nitroglycerin (*Marshall et al., 1988; Marshall and Kontos, 1988*). Second, agents, which oxidize thiols in vascular smooth muscle like nitroblue tetrazolium (NBT) or hydrogen peroxide, inhibited the vasodilation from nitric oxide and other nitrodilators, but did not affect the response to nitroso-L-cysteine or to EDRF from another source or to adenosine, an agent that relaxes vascular smooth muscle via adenylate cyclase (*Marshall et al., 1988; Wei and Kontos, 1990*). These findings distinguish the EDRF from acetylcholine in cerebral microvessels from nitric oxide and suggest the possibility that it might be a nitrosothiol-like nitroso-L-cysteine.

The mechanism of the vasodilation from bradykinin in cerebral arterioles is different from that due to acetylcholine. The vasodilation from bradykinin is completely or almost completely blocked by indomethacin, a cyclooxygenase inhibitor, while the response to acetylcholine is unaffected by indomethacin (*Kontos et al., 1990*). The vasodilation from bradykinin is completely inhibited by the combination of superoxide dismutase and catalase (*Kontos et al., 1984*) and partially reduced by deferoxamine (*Kontos et al. 1990*), an agent which scavenges the catalytic iron involved in the iron-catalyzed Haber-Weiss reaction. The response to acetylcholine is unaffected by these agents. 3-Aminotriazole, an agent which inhibits the production of superoxide from cyclooxygenase, eliminates the vasodilation from bradykinin but does not affect the response to acetylcholine (*Kontos et al., 1990*). In bioassay experiments, acetylcholine induces a transferrable, unstable vasodilator agent, while bradykinin does not, suggesting that the EDRF from bradykinin is too short-lived to survive the 6-second transit time involved in the bioassay experiments (*Kontos et al., 1990*). These findings show that the EDRF from bradykinin in cerebral arterioles is an oxygen radical derived from metabolism of arachidonate via cyclooxygenase. Topical application of bradykinin on the brain of cats generates superoxide from metabolism of arachidonate via cycooxygenase (*Kontos et al., 1985*). Presumably, this takes place in the endothelium. Superoxide then emerges from the endothelium via the anion channel. Its exit from there can be blocked by anion channel blockers (*Kontos et al., 1985*). In the extracellular space, it undergoes dismutation to hydrogen peroxide which, together with superoxide in the presence of catalytic iron, generates hydroxyl radical. Since superoxide, hydrogen peroxide, and hydroxyl radical are all vasodilators in cerebral arterioles (*Wei et al., 1985a*), one or more of these contributes to the vasodilation. The contribution to the bradykinin-induced arteriolar dilation from these different products of univalent reduction of oxygen appears different in different species. In the cat, hydroxyl radical as well as its precursors contribute to the vasodilation (*Kontos et al., 1990*). In the mouse, hydroxyl radical appears to be the predominant species responsible for the vasodilation (*Rosenblum, 1987*), while in the rat hydrogen peroxide seems to be responsible for the dilation with little contribution from superoxide or from hydroxyl radical (*Yang et al., 1990*).

AGENTS AND PROCEDURES WHICH INTERFERE WITH ENDOTHELIUM-DEPENDENT RELAXATION IN CEREBRAL ARTERIOLES

A variety of agents or procedures eliminate endothelium-dependent relaxation from acetylcholine or convert it to vasoconstriction. These are shown in Table 2. A common feature for all these is that they generate oxygen radicals. Methylene blue (*Marshall et al., 1988*) and hemoglobin autoxidize spontaneously and release superoxide at a low rate. The

TABLE 2: Interventions interfering with endothelium-dependent vasodilation from acetylcholine in the cerebral microcirculation

1. Methylene Blue
2. Hemoglobin
3. Arachidonate
4. Hydrogen Peroxide
5. Bradykinin
6. Acute Hypertension
7. Fluid-Percussion Brain Injury
8. Ischemia/Reperfusion

rate of generation of superoxide by methylene blue is enhanced in the presence of reducing agents (*Marshall et al., 1988*). Bradykinin activates phospholipases and releases free arachidonate. In association with the resulting accelerated metabolism of arachidonate via cyclooxyenase, there is generation of superoxide (*Kukreja et al., 1986*). Topical application on the brain surface of arachidonate in high concentration also generates superoxide (*Kontos et al., 1985*). Hydrogen peroxide is an oxidant in itself, but it may also damage endothelium generating superoxide as a consequence of this damage (*Burke-Wolin and Gurtner, 1987*). Finally, a variety of interventions including acute hypertension (*Wei et al., 1985b*), ischemia-reperfusion (*Kontos, 1989*), and fluid-percussion brain injury (*Kontos and Wei, 1986*) have been shown to generate superoxide.

The inhibition of the vasodilation from topical acetylcholine in response to these procedures is mediated by oxygen radicals as shown by the finding that it is prevented by pretreatment with oxygen radical scavengers. Also, in some cases, the normal vasodilator response to acetylcholine is restored by subsequent treatment with oxygen radical scavengers following its reversal by radicals. An interesting feature of the inhibition of the vasodilation from acetylcholine by acute hypertension, fluid-percussion brain injury, and ischemia-reperfusion, is that the effect is transient and the normal vasodilator response returns spontaneously after a period of 2 - 4 hours (*Ellison et al., 1989*; unpublished observations). This appears to be coincident with the cessation of superoxide production.

MECHANISM OF INHIBITION OF ENDOTHELIUM-DEPENDENT RESPONSES BY OXYGEN RADICALS

The products of univalent reduction of oxygen, superoxide, hydrogen peroxide, and hydroxyl radical, are capable of inhibiting the endothelium-dependent relaxation from acetylcholine by several mechanisms. These involve action on the endothelium, effects on EDRF after its release in the extracellular space, and effects on vascular smooth muscle. Frequently, multiple mechanisms co-exist.

An important distinction is between damage to the endothelium with inability to secrete EDRF and an attack on EDRF itself after it is released into the extracellular space. This distinction can be made by testing the effect of specific oxygen radical scavengers, such as SOD plus catalase. Since these agents are macromolecules and do not easily penetrate into the interior cells, a rapid reversal of the abnormal responses following the application of the scavengers implies an action in the extracellular space. On the other hand, if the effect is on the ability of the endothelium to produce EDRF, these agents do not alter the abnormal responses. Using this approach, we have found that methylene blue and hemoglobin inhibit the dilation from acetylcholine by attacking EDRF after its release in the extracellular space (*Marshall et al., 1988; Marshall and Kontos, 1988*). The agent immediately responsible appears to be hydroxyl radical because either SOD or catalase or deferoxamine restores the response to normal. On the other hand, arachidonate inhibits the vasodilation from acetylcholine or converts it to vasoconstriction by generating oxygen radicals which damage the endothelium and inhibit the generation of EDRF (*Kontos et al., 1989*). The mediation of this effect by oxygen radicals is shown by the fact that the abnormality is inhibited by pretreatment with SOD and catalase or by deferoxamine, implicating the hydroxyl radical as the mediator (*Kontos et al., 1984; Kontos et al., 1989*). Once the abnormality has occurred, however, the application of SOD plus catalase or deferoxamine does not restore the response to normal (*Kontos et al., 1989*). Bioassay experiments show that, indeed, the vascular smooth muscle remains responsive to EDRF from another source, but no EDRF production can be identified from the tissue that has been exposed to arachidonate (*Kontos et al., 1989*).

Hydrogen peroxide can act by either inhibition of production of EDRF or by destroying EDRF in the extracellular space depending on the dose (*Wei and Kontos, 1990*). In low dose, it causes injury to the endothelium and generates oxygen radicals. The normal response to acetylcholine can be restored after application of hydrogen peroxide by scavenging these radicals with SOD and catalase, while the effect on the endothelium can be prevented by pretreatment with deferoxamine. Bioassay experiments confirm the elimination of EDRF and its restoration following application of SOD and catalase. If the dose of hydrogen peroxide is increased, however, then the elimination of the vasodilator response to acetylcholine and EDRF production are not restored to normal by SOD and catalase.

Since all the products of univalent reduction of oxygen dilate cerebral arterioles (*Wei et al., 1985a*), the blood vessels display dilation under baseline conditions in the presence of oxygen radicals. This dilation may affect endothelium-dependent responses in a non-specific way since dilated vessels might be expected to dilate less in response to any given stimulus. Also, oxygen radicals may damage vascular smooth muscle and affect its responsiveness to many agents. Under these conditions, it is necessary to carry out more complex experiments to identify the mechanisms involved. Bioassay experiments are particularly helpful in this respect. For example, following acute hypertension or ischemia-reperfusion, endothelium-dependent cerebral arteriolar vasodilator responses to acetylcholine are converted to vasoconstriction as a result of the generation of oxygen radicals. Pretreatment with SOD and catalase inhibits this effect, but the responses, although they remain vasodilatory, are depressed (*Wei et al., 1985b; Kontos, 1989*). Under these conditions, bioassay experiments show that the amount of EDRF generated is increased (unpublished observations). Therefore, the depression of the responses must be attributed to non-specific factors including the baseline arteriolar dilation.

ROLE OF ENDOTHELIUM IN THE REGULATION OF THE CEREBRAL MICROCIRCULATION

Attempts to identify a role of endothelium-dependent responses in the regulation of the cerebral microcirculation have been disappointing. It is clear that the elimination of endothelium-dependent responses affects seriously the response to specific agonists which act via the endothelium. The important responses of the cerebral microcirculation, however, do not seem to be affected. This includes the response to arterial hypoxia, responses to hypocapnia and hypercapnia and autoregulatory vasodilation in response to arterial hypotension. The relaxation of large cerebral vessels induced by hypercapnia *in vitro* was shown to be endothelium-independent (*Toda et al., 1989*).

Also, moment to moment regulation of the cerebral microcirculation via the endothelium could not be verified. Basal secretion of EDRF could not be demonstrated in bioassay experiments (*Kontos et al., 1988*), a clear difference from what has been found in *in vitro* studies with large vessels or endothelial cells in culture. Second, procedures which eliminated endothelium-dependent relaxation, like methylene blue, did not alter baseline vascular caliber (*Marshall et al., 1988*). Such a change would be expected if there was a tonic influence from basal EDRF secretion.

ACKNOWLEDGEMENTS

Supported by grants HL21851, MS 19316 and NS 25630 from the National Institutes of Health.

REFERENCES

Burke-Wolin, T., and Gurtner, G.H. Peroxides elicit superoxide anion release from bovine pulmonary arterial endothelial cells. *Faseb J.* 3:A1308, 1989.

Dacey, G., Jr., and Bassett, J.E. Cholinergic vasodilation of intracerebral arterioles in rats. *Am. J. Physiol.* 253:H1253-H1260, 1987.

Ellison, M.D., Erb, D.E., Kontos, H.A., and Povlishock, J.T. Recovery of impaired endothelium-dependent relaxation after fluid-percussion brain injury in cats. *Stroke* 20:911-917, 1989.

Furchgott, R.F. Role of endothelium in responses of vascular smooth muscle. *Circ. Res.* 53:557-573, 1983.

Ignarro, L.J., Byrns, R., Buga G.M., Woods, K.S., and Chaudhuri, G. Pharmacologic evidence that endothelium-derived relaxing factor is nitric oxide-elicited vascular smooth muscle relaxation. *J. Pharmacol. Exp. Ther.* 244:181-189, 1987.

Kontos, H.A., Wei, E.P., Povlishock, J.T., and Christman, C.W. Oxygen radicals mediate the cerebral arteriolar dilation from arachidonate and bradykinin in cats. *Circ. Res.* 55:295-303, 1984.

Kontos, H.A., Wei, E.P., Ellis, E.F., Jenkins, L.W., Povlishock, J.T., Rowe, G.T., and Hess, M.L. Appearance of superoxide anion radical in cerebral extracellular space during increased prostaglandin synthesis in cats. *Circ. Res.* 57:142-151, 1985.

Kontos, H.A., and Wei, E.P. Superoxide production in experimental brain injury. *J. Neurosurg.* 64:803-807, 1986.

Kontos, H.A., Wei, E.P., Marshall, J.J. In vivo bioassay of endothelium-derived relaxing factor. *Am. J. Physiol.* 255:H1259-H1262, 1988.

Kontos, H.A. Oxygen radicals in cerebral ischemia, In *Cerebrovascular Diseases*, ed. by M.D. Ginsberg and W.D. Dietrich, Raven Press, New York, pp. 365-371, 1989.

Kontos, H.A., Wei, E.P., Povlishock, J.T., Kukreja, R.C., and Hess, M.L. Inhibition by arachidonate of cerebral arteriolar dilation from acetylcholine. *Am. J. Physiol.* 256:H665--H671, 1989.

Kontos, H.A., Wei, E.P., Kukreja, R.C., Ellis, E.F., and Hess, M.L. Differences in endothelium-dependent cerebral dilation by bradykinin and acetylcholine. *Am. J. Physiol.* 258:H1261-H1266, 1990.

Kukreja, R.J., Kontos, H.A., Hess, M.L., and Ellis, E.F. PGH synthase and lipoxygenase generate superoxide in the presence of NADH or NADPH. *Circ. Res.* 59:612.619, 1986.

Marshall, J.J., and Kontos, H.A. Independent mechanisms of blockade of endothelium-dependent and nitroprusside-induced dilation by hemoglobin (Abstract). *Faseb J.* 2:A710, 1988.

Marshall, J.J., Wei, E.P., and Kontos, H.A. Independent blockade of cerebral vasodilation from acetylcholine and nitric oxide. *Am. J. Physiol.* 255:H847-H854, 1988.

Misra, H.P., and Fridovich, I. The generation of superoxide radical during the autooxidation of hemoglobin. *J. Biol. Chem.* 247:6960, 1972.

Palmer, R.M.J., Ferrige, A.G., Moncada, S. Nitric oxide release accounts for the biological activity of endothelium-derived relaxing factor. *Nature* 327:524-526, 1987.

Povlishock, J.T., Rosenblum, W.I., Sholley W.M., and Wei, E.P. An ultrastructural analysis of endothelial change paralleling platelet aggregation in a light/dye model of microvascular insult., *Am. J. Pathol.* 110:148-160, 1983.

Rosenblum, W.I. Endothelial dependent relaxation demonstrated *in vivo* in cerebral arterioles. Stroke 17:494-497, 1986.

Rosenblum, W.I. Hydroxyl radical mediates the endothelium-dependent relaxation produced by bradykinin in mouse cerebral arterioles. *Circ. Res.* 61:601-603, 1987.

Rosenblum, W.I., Nelson, G.H., and Povlishock, J.T. Laser-induced endothelial damage inhibits endothelium-dependent relaxation in the cerebral microcirculation of the mouse. *Circ. Res.* 60:169-176, 1987a.

Rosenblum, W.I., Povlishock, J.T., Wei, E.P., Kontos, H.A., and Nelson, G.H. Ultrastructural studies of pial vascular endothelium following damage resulting in loss of endothelium-dependent relaxation. *Stroke* 18:927-931, 1987.

Toda, N., Hatano, Y., and Mori, K. Mechanisms underlying response to hypercapnia and bicarbonate of isolated dog cerebral arteries. *Am. J. Physiol.* 257:H141-H146, 1989.

Wei, E.P., Christman, C.W., Kontos, H.A., and Povlishock, J.T. Effects of oxygen radicals on cerebral arterioles. *Am. J. Physiol.* 248:H157-H162, 1985.

Wei, E.P., Kontos, H.A., Christman, C.W., DeWitt, D.S., and Povlishock, J.T. Superoxide generation and reversal of acetylcholine-induced cerebral arteriolar dilation after acute hypertension. *Circ. Res.* 57:781-787, 1985.

Wei, E.P., Ellison, M.D., Kontos, H.A., and Povlishock, J.T. O_2 radicals in arachidonate-induced increased blood-brain barrier permeability to proteins. *Am. J. Physiol.* 251:H693--H699, 1986.
Wei, E.P., and Kontos, H.A. H_2O_2 and endothelium-dependent cerebral arteriolar dilation: Implications for the identity of EDRF generated by acetylcholine. *Hypertension* (in press).
Yang, S.T., Mayhan, W.G., and Heistad, D.D. Mechanisms of impaired endothelium-dependent responses of cerebral arterioles during chronic hypertension. *Faseb J.* 4:A556, 1990.

VI. MECHANISMS OF ENDOTHELIAL CELL DYSFUNCTION

THE ROLE OF APOLIPOPROTEIN E AND APOLIPOPROTEIN B IN ATHEROSCLEROSIS

Thomas L. Innerarity

Gladstone Foundation Laboratories
Cardiovascular Research Institute
Department of Pathology
University of California San Francisco
San Francisco, California 94140-0608, U.S.A.

INTRODUCTION

Two apolipoproteins (apo-), apo-B100 and apo-E, are extremely important in lipoprotein metabolism and in the development of accelerated atherosclerosis. Their significance stems from their role in binding to the low density lipoprotein (LDL) receptor and mediating the catabolism of lipoproteins via the LDL receptor pathway. Much has been learned about accelerated atherosclerosis from studies of human genetic diseases involving these apolipoproteins and their receptors. The genetic disorders to be discussed are caused by defects in the LDL receptor or in its ligands, *i.e. apo-B100 and apo-E*. These genetic abnormalities disrupt the LDL receptor-mediated clearance of apo-B100- and apo-E-containing lipoproteins, causing these lipoproteins to accumulate in the plasma. These same apo-B100- and apo-E-containing lipoproteins are induced to elevated levels in the plasma by diets high in cholesterol and saturated fat.

Plasma lipoproteins are a metabolically diverse group of macromolecules that transport lipids (including cholesterol and triglycerides) and have one or more protein constituents called apolipoproteins (or apoproteins) (Figure 1). Lipoproteins are spherical particles, as visualized by electron microscopy, that consist of a core and a shell. The core consists of neutral lipids, such as triglycerides and cholesteryl esters, whereas the shell is made up of unesterified cholesterol, phospholipids, and apolipoproteins. Apolipoproteins transport plasma lipids and regulate the metabolism of lipoproteins by acting as co-factors of various enzymes involved in lipoprotein metabolism or as ligands for lipoprotein receptors (*Innerarity, in press; Mahley, 1990*).

The focus in this discussion will be on the pathway for the receptor-mediated catabolism of apo-B100- and apo-E-containing lipoproteins and the evidence that lipoproteins in this pathway are involved in atherogenesis.

Chylomicrons, the largest of the lipoprotein particles, are synthesized by the intestine to transport dietary triglycerides and cholesterol (Figure 2A). In the capillary beds, lipoprotein lipase hydrolyses the triglycerides, converting the chylomicrons into chylomicron remnants. Normally, the chylomicron remnants are rapidly cleared from the plasma by the liver. The apo-E on the surface of these particles mediates their interaction with liver receptors (Figure 2A) (*Innerarity, in press; Mahley, 1990*).

The pathway for the catabolism of endogenously derived lipoproteins is very similar (Figure 2B). In the capillary bed, very low density lipoproteins (VLDL) synthesized by the liver are converted by lipoprotein lipase into smaller, more cholesterol-rich lipoproteins, *i.e.* VLDL remnants and intermediate density lipoproteins (IDL). A fraction of these lipoproteins are cleared from the plasma by binding (via apo-E) to hepatic LDL receptors and, the remainder are converted to LDL. The LDL formed in the plasma are also cleared from the circulation by the LDL receptor pathway, but apo-B100 serves as the ligand (*Innerarity, in press; Mahley, 1990*).

Two structurally related forms of apo-B, apo-B100 and apo-B48, are produced by the same gene. Synthesized in the liver, apo-B100 is a huge glycoprotein consisting of 4536 amino acids and having a molecular weight of 550,000, of which about 10% is carbohydrate. In addition to its role as a ligand for the LDL receptor, apo-B100 is necessary for the synthesis and secretion of VLDL by the liver. In humans, apo-B48 is synthesized in the intestine and consists of 2152 amino acids. It plays a role analogous to that of apo-B100 in the liver in that it is necessary for the assembly and secretion of chylomicrons by the intestine (*Innerarity, 1990; Young, 1990; Innerarity, et al., 1990*).

The other ligand for the LDL receptor is apo-E, a single polypeptide chain containing 299 amino acids with a molecular weight of 34,000. The predicted secondary structure of apo-E is shown in Figure 3. The amino- and carboxy-terminal portions of the protein have been predicted to be highly ordered and are connected by a hinge region of predicted random structure (*Mahley and Rall, Jr., 1989; Mahley, 1988*). The presence of two highly ordered domains has been verified experimentally by limited proteolysis and guanidine denaturation of apo-E (*Wetterau et al., 1988*).

This review will cover three specific genetic abnormalities that disrupt the receptor-mediated pathways of lipoprotein catabolism, causing the accumulation of lipoproteins associated with premature atherosclerosis. Type III hyperlipoproteinemia is characterized by the accumulation of chylomicron and VLDL remnants in the plasma, as a result of abnormalities in apo-E. These remnant lipoproteins, collectively referred to as ß-VLDL, are

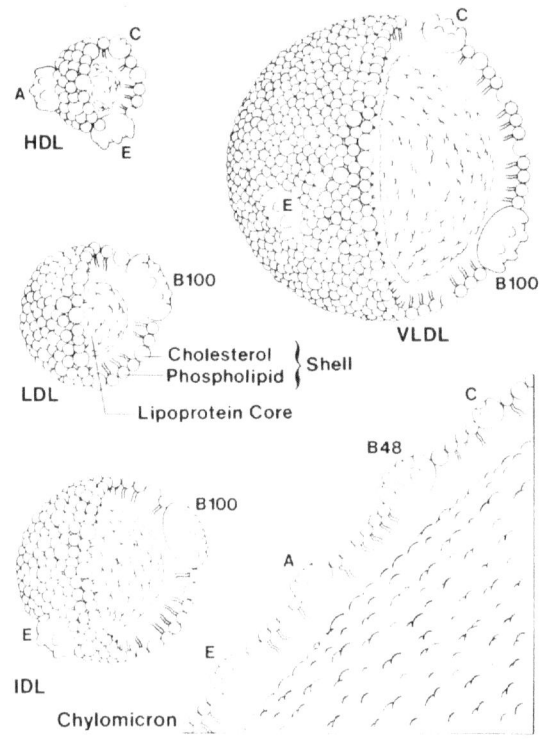

FIGURE 1: Schematic structures of plasma lipoproteins. The drawing is roughly indicative of the differences in size among the pipoprotein classes. HDL, high density lipoprotein; LDL, low density lipoprotein; IDL, intermediate density lipoprotein; VLDL very low density lipoprotein. A, B48, B100, C, and E are apolipoproteins, which represent the protein constituents of the various particles (*modified from "Scientific American Medicine", Section 9, 1989, Scientific American, Inc.*).

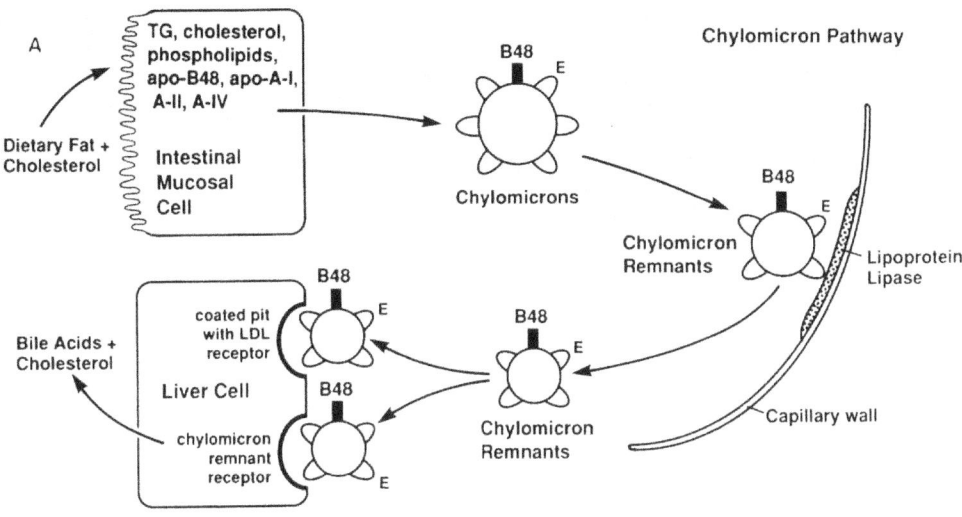

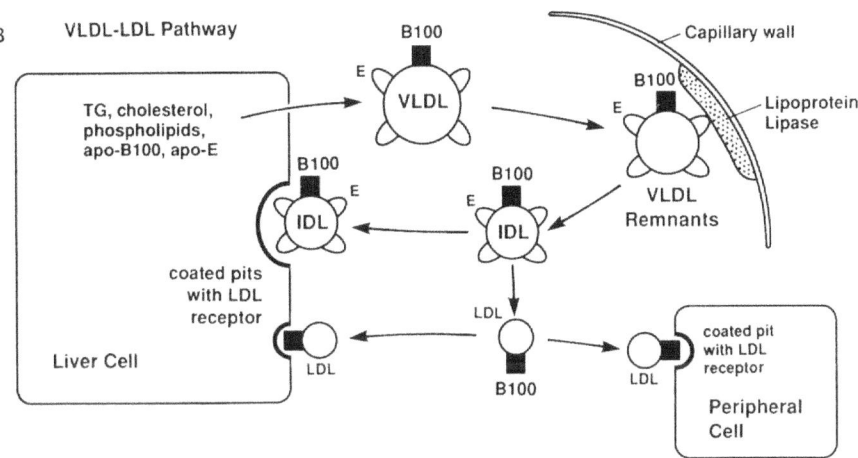

FIGURE 2: General scheme of lipoprotein metabolism that focuses on the central role of receptor-mediated catabolism. Only apo-B and apo-E are shown, because these apolipoproteins serve as ligands for these receptors. (A) Synthesis and catabolism of intestinally derived lipoproteins. (B) Synthesis and catabolism of hepatically derived lipoproteins. TG, triglycerides (*modified with permission, R.J. Havel Med. Clin. North Am. 1982*).

avidly taken up by macrophages at the sites of lesions in the arterial wall, where they contribute to the formation of lipid-filled cells known as foam cells. Familial hypercholesterolemia is characterized by the accumulation of very high levels of LDL cholesterol and is the result of abnormalities in the LDL receptor. The most recently discovered disorder, known as familial defective apo-B100, is characterized by a moderate to severe accumulation of LDL in the plasma and is due to an abnormality in the apo-B100 molecule.

TYPE III HYPERLIPOPROTEINEMIA

About one in every thousand adults is affected with type III hyperlipoproteinemia (*Mahley and Rall, Jr., 1989*). These subjects have both hypertriglyceridemia and hypercholesterolemia, as demonstrated by the accumulation of chylomicron and VLDL remnants (ß-VLDL) in the plasma. High levels of ß-VLDL predispose these individuals to the premature development of atherosclerosis, particularly in the peripheral arteries, and causes xanthomas of the palmar creases. Because these subjects usually have normal or low plasma levels of LDL, their accelerated atherosclerosis cannot be attributed to elevated LDL levels. The underlying genetic abnormality in this disease is a mutation of apo-E that causes it to bind defectively to lipoprotein receptors, thus disrupting the normal catabolism of VLDL and chylomicron remnants (*Mahley and Rall, Jr., 1989*).

The vast majority of type III hyperlipoproteinemic subjects are homozygous for a particular form of mutant apo-E, apo-E2(158 Arg Cys), one of three common genetically determined forms of apo-E. ApoE2 binds defectively to LDL receptors, whereas the other two forms (apo-E3 and apo-E4) bind normally. In addition to the point mutation of a neutral amino acid in place of a basic amino acid at residue 158 (as is found in apo-E2), the sequencing of apo-E from a number of type III hyperlipoproteinemic individuals has demonstrated other mutations (*Rall, Jr., S.C., et al., 1990*). These occur at apo-E residues 136, 142, 145, and 146 (Figure 4). All of these mutations involve a neutral amino acid in place of a basic amino acid in the center of the molecule. These data emphasize the importance of certain key basic amino acid residues (arginine and lysine) in binding and helped to localize the receptor-binding region of apo-E to the middle of the 299-amino acid apo-E molecule (*Rall, Jr., S.C., et al., 1990*).

From this and from other lines of evidence, we know that the midportion of apo-E (in the vicinity of amino acids 140 to 160) interacts directly with the LDL receptor (*Mahley, R.W., 1988*). As shown in Figures 3 and 4, this region of the molecule is thought to form an alpha-helix. These positively charged arginine and lysine residues very likely interact with a cluster of negatively charged glutamic and aspartic acid residues on the LDL receptor (*Brown and Goldstein, 1986*).

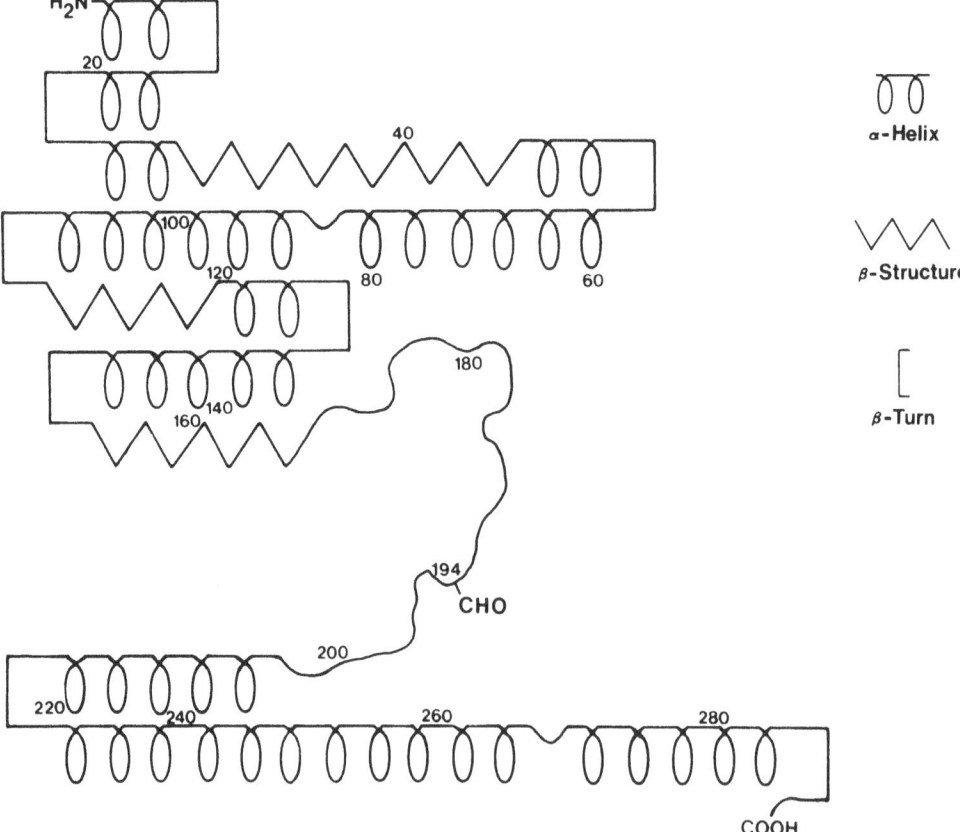

FIGURE 3: Schematic representation of the predicted secondary structure of apo-E (*modified with permission, J.R. Wetterau et al., 1988*).

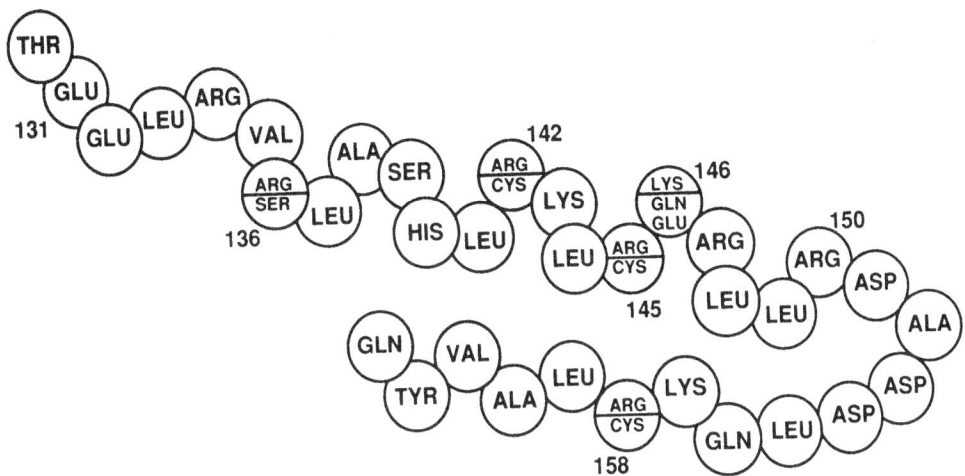

FIGURE 4: Schematic representation of the predicted secondary structure of the receptor-binding region of apo-E, indicating the locations of the naturally occurring amino acid substitutions in this region of the molecule that disrupt receptor binding. (In defective apo-E, the bottom amino acid substitutes for the top one) (modified with permission, from R.W. Mahley and T.L. Innerarity, 1983, Lipoprotein receptors and cholesterol homeostasis. Biochim. Biophys. Acta. 737; 197-222.)

In type III hyperlipoproteinemia, the receptor binding-defective apo-E impairs the uptake of chylomicron remnants by hepatic lipoprotein receptors (*Mahley and Rall, Jr., 1989*). Normally, apo-E mediates the clearance of chylomicron remnants via the LDL receptor and a poorly defined receptor known as the apo-E, or chylomicron remnant, receptor. At present, the mechanism responsible for chylomicron remnant clearance is not completely understood, but it is clear that apo-E is the ligand responsible for uptake of these lipoproteins by the liver (*Innerarity, in press; Mahley, 1990*). The remnant lipoproteins that accumulate (the β-VLDL) are of two types. The chylomicron remnants consist of partially lipolyzed chylomicrons that are enriched in cholesterol and possess apo-B48 and apo-E as their major apolipoprotein components. The other type, VLDL remnants, consists of partially lipolyzed VLDL (*i.e.* cholesterol-enriched VLDL and IDL) that contain primarily apo-B100 and apo-E. Very low density lipoprotein (VLDL) remnants and IDL accumulate in the plasma because their defective apo-E impairs their interaction with the hepatic LDL receptor (*Mahley and Rall, Jr., 1989*). In addition, the presence of the abnormal apo-E also appears to impair the lipolytic conversion of VLDL to LDL. Therefore, VLDL remnant accumulation appears to be secondary to impaired uptake of these remnants and to impeded processing to LDL (*Mahley, and Rall, Jr., 1989*).

Most subjects with abnormal apo-E do not develop gross hyperlipoproteinemia, and in those that do, a second genetic or an environmental factor is necessary to precipitate marked hyperlipoproteinemia. However, the mutation within apo-E is the primary defect required for the development of type III hyperlipoproteinemia (*Mahley and Rall, Jr., 1989*).

The most striking clinical feature of type III hyperlipoproteinemia is the occurrence of xanthomas. Roughly half of untreated patients have *palmar xanthomas,* yellowish lipid deposits in the palmar creases of the hand. Type III hyperlipoproteinemic subjects may also have other types of xanthomas. Regardless of the appearance or location of the xanthomas, they all contain macrophages filled with cholesteryl ester (foam cells). In addition, premature atherosclerosis is common among these subjects. Interestingly, there is a predisposition for lesions of the peripheral arteries in this disorder: peripheral vascular disease involving the lower limbs is almost as common as coronary artery disease. In both the development of xanthomas and the development of lipid-laden atheromas, there is reason to believe that β-VLDL play a unique role (*Mahley and Rall, Jr., 1989*).

In vitro studies have shown that mouse peritoneal macrophages incubated with β-VLDL accumulate massive amounts of cholesteryl ester, making them resemble foam cells seen in atherosclerotic lesions (*Goldstein, et al., 1980*). Shown in Figure 5 is a cultured mouse peritoneal macrophage that had been incubated with canine β-VLDL and stained with oil red

0. The cholesteryl ester content of the macrophages incubated *in vitro* with ß-VLDL typically increases 100- to 200-fold. ß-VLDL (including both chylomicron remnants and VLDL remnants) possess the unique ability to cause massive cholesteryl ester accumulation in peritoneal macrophages and blood monocyte-derived macrophages. Other equally cholesterol-rich lipoproteins, such as normal LDL or cholesterol-induced LDL, lack this ability (*Goldstein et al., 1980*).

The accumulation of the remnant lipoproteins in the plasma is undoubtedly responsible for the accelerated atherosclerosis seen in these subjects. The mechanism making these lipoproteins atherogenic may involve an alternative pathway for the uptake of these cholesterol-enriched lipoproteins, *i.e.*, the uptake by macrophages in the atherosclerotic lesion that become foam cells.

Further support for remnant lipoproteins being atherogenic comes from the observation that one of the changes in the lipoproteins induced by diets high in saturated fat and cholesterol is the accumulation of chylomicron and VLDL remnants in the plasma. This is true to some extent in virtually all animals studied (*Mahley, 1982*). These diet-induced remnants are very similar to those described in type III hyperlipoproteinemia: they are of intestinal and hepatic origin, and they possess the ability to cause massive cholesteryl ester accumulation in macrophages. The diet-induced remnants, of course, do not have the abnormal apo-E, but they accumulate in the plasma via other mechanisms (*Mahley, 1982*).

There are several possible mechanisms for the appearance of remnant lipoproteins induced by high-fat, high-cholesterol diets. In addition to the induction of chylomicron remnants by increased intestinal synthesis and secretion as a result of a high dietary load of fat and cholesterol, the accumulation may be augmented to some degree by reduced clearance. Likewise, these atherogenic diets enhance VLDL production by the liver (and hence, VLDL remnants). In addition, VLDL remnant accumulation appears to result from

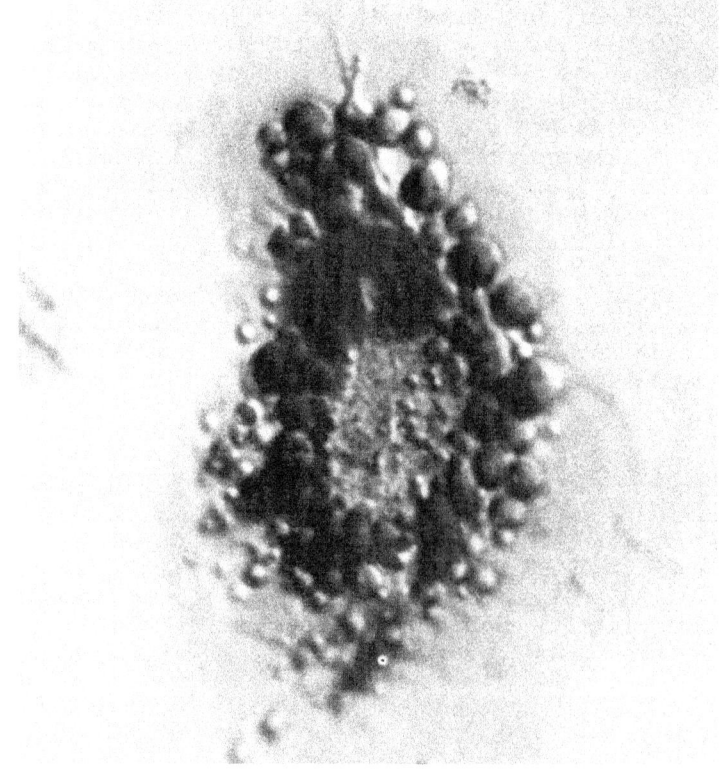

FIGURE 5: A cultured mouse peritoneal macrophage that was incubated with canine ß-VLDL and stained with oil red 0 (*reproduced with permission, R.W. Mahley and S.C. Rall, Jr., 1989*).

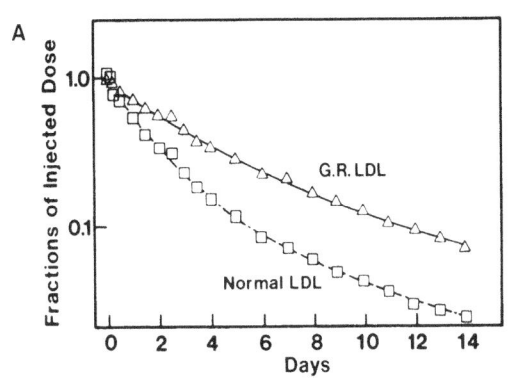
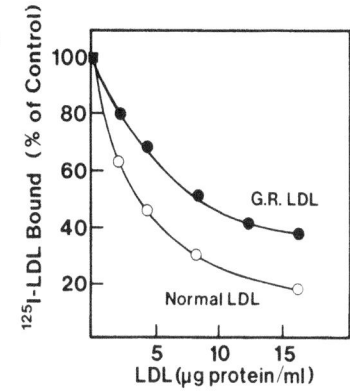

FIGURE 6: *In vivo* (A) and *in vitro* (B) evidence for defective receptor-binding LDL.(A) Plasma decay curve of G.R. LDL and normal LDL intravenously injected into subject G.R. (*reproduced from G.L. Vega and S.M. Grundy, 1986*) (B) Ability of G.R. LDL and normal LDL to compete with normal ^{125}I-labeled LDL for binding to LDL receptors on normal human fibroblasts (*reproduced with permission, T.L. Innerarity et al., 1987*).

a decrease in hepatic LDL receptor expression, that occurs in cholesterol-fed animals (*Angelin et al., 1983*). Any decrease in receptor expression, especially in the liver, would also result in increased plasma levels of LDL and IDL. Moreover, the infusion of chylomicron remnants down-regulated the expression of LDL receptors in canine livers within 2 to 4 h after intravenous administration was begun (*Zilversmit, 1979*). This is a very rapid response to circulating lipoprotein levels. Of the different classes of lipoproteins, remnant lipoproteins are the most effective in down-regulating hepatic LDL receptors.

Thus, there are several lines of evidence supporting the concept that remnant lipoproteins from subjects with type III hyperlipoproteinemia (abnormal apo-E) and diet-induced hyperlipoproteinemia (normal apo-E) are atherogenic. It is entirely possible that the transient appearance of a large load of remnant lipoproteins induced by diets high in fat and cholesterol regularly over many years, as occurs in the Western diets, may be the direct link between diet and accelerated coronary artery disease in Western societies (*Zilversmit, 1979*).

ROLE OF LOW DENSITY LIPOPROTEINS IN ATHEROSCLEROSIS

Two genetic disorders demonstrate the role of LDL in the development of accelerated atherosclerosis. These are familial hypercholesterolemia, which, Brown and Goldstein (*Brown and Goldstein, 1983b*) demonstrated, is caused by mutations in the LDL receptor, and familial defective apo-B100, which is caused by a mutation in one of the ligands of the LDL receptor, apo-B100.

Familial hypercholesterolemia

Familial hypercholesterolemia (FH) is characterized clinically by the severe elevation of plasma LDL and by the presence of xanthomas and coronary heart disease. The primary defect in FH is the presence of one or another mutation in the gene coding for the LDL receptor that gives rise to a defect in binding of LDL to the receptor. The LDL receptor pathway, more specifically the hepatic LDL receptor pathway, is the principal route for the clearance of LDL from the plasma. An alternate, poorly defined, non-receptor-mediated pathway also catabolizes LDL, but with a much lower efficiency. Normally, about three-fourths of the plasma LDL is cleared by the receptor-mediated pathway and about one-fourth by the non-receptor-mediated pathway (*Brown and Goldstein, 1983*). Mutations in the LDL receptor that prevent it from functioning normally and disrupt the efficient receptor-mediated clearance of LDL result in the accumulation of LDL in the plasma; they can then be cleared from the circulation only by the non-receptor-mediated pathway. Many different types of studies have shown that in normal animals and humans, the concentration of LDL in the plas-

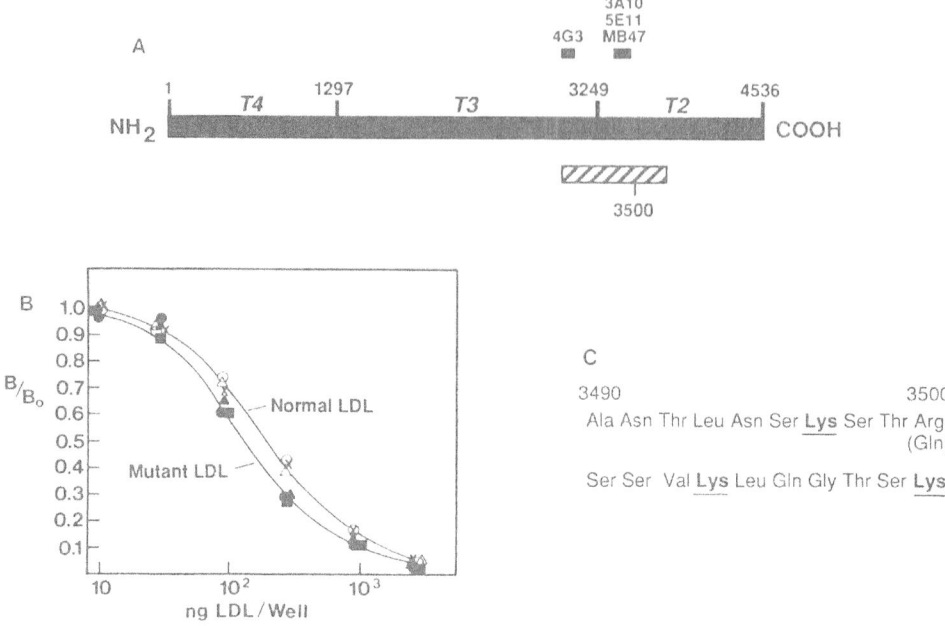

FIGURE 7: The mutation responsible for familial defective apo-B100 is the substitution of a glutamine for an arginine at apo-B100 residue 3500, which occurs in the epitope for monoclonal antibody MB47. (A) Linear representation of apo-B100 showing the location of four monoclonal antibodies that inhibit LDL receptor binding (4G3, 3A10, 5E11, MB47) and thereby define the region of the receptor-binding domain of apo-B100 (hatched box). The "3500" represents the location of the amino acid substitution of glutamine for arginine that causes FDB. (B) Ability of G.R.'s LDL and affected family members' LDL (lower curve) to compete with normal LDL (upper curve) for binding to apo-B100 monoclonal antibody MB47. B/B_o = ratio of the amount of antibody found in the presence of competitor LDL to the amount bound the absence of competitor LDL (*modified with permission, K.H. Weisgraber et al., 1988*). (C) The sequence of the probable epitope of monoclonal antibody MB47. The (Gln) denotes the glutamine-for-arginine substitution that occurs at residue 3500 in FDB heterozygotes. The underlined lysines are highlighted to reflect the known role of basic residues of apo-B100 in binding to the LDL receptor.

ma is inversely related to the number of hepatic LDL receptors (*Brown and Goldstein, 1986; Brown and Goldstein, 1983*).

Individuals who are homozygous for FH (about one in every million persons) have few if any functional LDL receptors. Their plasma LDL accumulate to extremely high levels (650 to 1000 mg/dl). Clinically, these subjects usually have cutaneous xanthomas within the first 5 years of life and frequently die of coronary heart disease in their teenage years. For example, it has been reported that one homozygote experienced an acute myocardial infarction at the age of 18 months and that another died from the same cause at 3 years of age. Rarely do FH homozygotes survive past the age of 30 (*Goldstein and Brown, 1989*).

Individuals who are heterozygotes for FH (about one in every 500 persons) possess half the number of normal LDL receptors. These individuals typically have plasma cholesterol levels of 350-450 mg/dl. The excess cholesterol in the plasma of FH heterozygotes and homozygotes is found entirely in the LDL fraction. Although there are minor compositional differences in the LDL from FH subjects compared with the LDL from normal individuals, it is more likely that the increased risk of atherosclerosis in FH subjects is due to elevated levels of LDL rather than to minor alterations in the LDL. Familial hypercholesterolemia heterozygotes usually develop coronary heart disease much later than FH homozygotes. Nevertheless, the risk of atherosclerosis is much greater for FH heterozygotes than for normal individuals. For example, male FH heterozygotes have a 50% probability of having a myocardial infarction by the age of 50. Although most FH heterozygotes develop tendon xanthomas, this usually occurs much later in life than is observed in FH homozygotes (*Goldstein and Brown, 1989*).

Familial defective apolipoprotein-B100

Familial defective apo-B100 (FDB) is a genetic disorder associated with hypercholesterolemia and is caused by a mutation in apo-B100 that disrupts the binding of LDL to the LDL receptor (*Innerarity, et al., 1990*). The first evidence of defective receptor-binding LDL came from results of *in vivo* turnover studies that determined the rate at which intravenously injected radiolabeled LDL were cleared from the plasma. Clearance rates of autologous and normal LDL were determined in subjects who had moderate primary hypercholesterolemia. The LDL from one of these subjects, G.R., were striking in that their clearance from the plasma was only about 50% of normal (*Vega and Grundy, 1986*) (Figure 6A). Since about three-fourths of normal LDL are cleared from the plasma by the hepatic LDL receptor pathway, these results suggested that G.R.'s LDL bound defectively to the LDL receptor (*Brown and Goldstein, 1983b*).

To test this possibility directly, we examined the ability of G.R.'s LDL to bind to LDL receptors on cultured human fibroblasts in a competitive binding assay. As shown in Figure 6B, the LDL from subject G.R. were not as effective as were normal LDL in competing with ^{125}I-labeled LDL for receptor binding. The LDL from subject G.R. had only about 32% of normal receptor-binding activity (*Innerarity, et al., 1987*).

Based on a number of different types of experiments, the LDL from G.R. were found to be normal except for the defect in receptor binding. Thus, it seemed likely that the defect was located in or near the receptor-binding domain of apo-B100. One approach to defining the receptor-binding domain of apo-B100 has been derived from the use of monoclonal antibodies. The mapping of the epitopes of apo-B monoclonal antibodies that inhibit the binding of LDL to the LDL receptor has shown that the region illustrated in Figure 7A is important in receptor binding (*Milne, et al., 1989*). It was postulated that if the binding defect did result from a mutation in the apo-B100 receptor-binding domain, then the use of these same inhibitory monoclonal antibodies against this region might be useful in detecting the mutation. Four monoclonal antibodies were used to probe for the mutation (Figure 7B). Three of these reacted equally well with normal and defective LDL. However, in the case of MB47, there was a reproducible difference between the binding of MB47 to the LDL from affected family members (three solid symbols) and non-affected family members and normal LDL (two open symbols). The abnormal LDL were more effective competitors than were normal LDL (Figure 7B) (*Weisgraber et al., 1988*). These results suggested that the mutation may occur in the region of the MB47 epitope, which would result in a higher binding affinity of MB47 to defective apo-B100. The MB47 epitope is shown in Figure 7C (*Pease et al., 1990*). Genomic clones that included the MB47 epitope were sequenced, and it was found that G.R. and affected family members had one unique mutation at codon 3500. G.R. is heterozygous for the glutamine-for-arginine substitution at this site, with the normal allele coding for arginine and the mutant allele for glutamine. This was the only unique amino acid change in the receptor-binding region of apo-B100, and this mutation has been found in every subject whose LDL have a binding defect but in no individuals with normal binding (*Soria, et al., 1989*).

Three investigations have examined the impact of this mutation on plasma cholesterol levels and the prevalence of the disorder in their separate clinical populations (Table 1). In one study, 1100 individuals from four different geographic regions were screened for this disorder by one or more of the following procedures: 1) the binding of monoclonal antibody MB47 to LDL in a competitive RIA using plasma, 2) the detection of a single base change at codon 3500 using allele-specific probes on polymerase chain reaction (PCR)-amplified genomic DNA, and 3) an *in vitro* competitive receptor-binding assay on LDL isolated from plasma. Eleven subjects with the 3500 mutation were identified, and pedigree analysis of these eleven probands uncovered an additional 30 FDB heterozygotes, making a total of 41. The average plasma cholesterol level of these 41 FDB heterozygotes was 269 mg/dl. As compared with the 50th percentile plasma cholesterol levels of the 48,000 volunteers who participated in the Lipid Research Clinics' population studies, the FDB heterozygotes had an average plasma cholesterol level 81 mg/dl higher than age- and sex-matched controls (*Innerarity, et al., 1990*).

In a second study, Tybjaerg-Hansen et al. screened for the 3500 mutation using allele-specific probes on PCR-amplified genomic DNA from 374 unrelated English subjects with hypercholesterolemia and another 371 unrelated English and Danish subjects with a clinical diagnosis of atherosclerosis. Ten FDB heterozygotes were identified from this study, and seven of the 10 had evidence of coronary heart disease. The mean plasma cholesterol level

TABLE 1: Frequency of Familial Defective Apolipoprotein B-100 and Its Impact on Cholesterol Levels

	Subjects Screened	Probands	Total Affected	Average Plasma Cholesterol Level (mg/dl)	Plasma Cholesterol Level Above Age-and-Sex-Matched Controls (mg/dl)	Estimated Frequency in the General Population
Innerarity, et al., 1990	1100	11	41	269	81	~1/500
Tybjaerg-Hansen, et al., 1990	374	10	10	369	163	~1/600
Schuster, et al., 1990	300	08	18	328	134	~1/700

for the 10 FDB heterozygotes was 369 mg/dl, which is remarkably similar to the plasma cholesterol levels typically found for FH heterozygotes (*Tybjaerg-Hansen, A., et al., 1990*).

In a third study, Schuster *et al.* screened a group of 300 hypercholesterolemic individuals from Munich, Germany, that included a subgroup of 57 subjects with FH. Although no FDB mutations were found in the FH subgroup, eight FDB probands were identified among the other hypercholesterolemic individuals. Pedigree analysis identified an additional 10 family members with the 3500 mutation. The FDB subjects from this study had an average plasma cholesterol level of 328 mg/dl, which was 134 mg/dl higher than age- and sex-matched controls (*Schuster et al., 1990*).

These three studies demonstrate that the 3500 mutation has a profound effect on plasma cholesterol level that is due exclusively to increased LDL cholesterol. In fact, with the exception of mutations in the LDL receptor, no other known genetic mutation causes such a large increase in the plasma LDL cholesterol level. The clinical features of FDB appear to be dependent upon the degree of hypercholesterolemia. Familial defective apo-B100 heterozygotes with moderate hypercholesterolemia did not display any clinical features typical of FH heterozygotes, such as xanthomas and corneal arcus. However, individuals with severely elevated plasma cholesterol levels had clinical manifestations strikingly similar to those of FH heterozygotes. For example, many of the FDB heterozygotes in the Tybjaerg-Hansen *et al.* and Schuster *et al.* studies had xanthomas, atherosclerotic plaques in the carotid arteries, and clinically relevant coronary artery disease.

Because the populations in the three studies were selected and/or were hypercholesterolemic, it was not possible to obtain a precise determination of the frequency of this disorder. However, all three studies estimated its prevalence in the general population: one in 500, one in 600, and one in 700 for the studies by Innerarity *et al.* (*Innerarity et al., 1990*), Tybjaerg-Hansen *et al.* (*Tybjaerg-Hansen, et al., 1990*), and Schuster *et al.* (*Schuster, et al., 1990*), respectively.

This disorder has been identified in the United States, Canada, England, Germany, Denmark, Austria, and Italy. Not only does this genetic mutation appear to be widespread, but if our estimates of the frequency are correct, then this disorder is as common as FH and is one of the most prevalent single-gene mutations known that cause a clinical abnormality. With regard to the frequency of the phenotype, however, there is a significant difference between FH and FDB. Familial hypercholesterolemia is caused by many different mutations of the LDL receptor gene. To date, investigators have established the molecular basis for over 40 different FH mutations (*Brown and Goldstein, 1983a*). In contrast, only one mutation has been discovered that causes FDB. How many other mutations exist that alter LDL metabolism and account for elevations in the total and LDL cholesterol levels in the plasma? This remains to be determined, but if other apo-B100 mutations are found that cause FDB, then mutations of apo-B100 will be identified as an even more significant cause of hypercholesterolemia in the general population than has been determined to date.

DIETARY INDUCTION OF ELEVATED LOW DENSITY LIPOPROTEIN LEVELS

The accumulation of LDL can be induced in animals and in humans by diets high in saturated fat and cholesterol. These diets decrease the expression of LDL receptors in the liver, which could be the mechanism whereby plasma LDL levels are increased as a result of fat and cholesterol feeding. Again, there is a parallel between a genetic defect and diet-induced hyperlipoproteinemia; in FH, LDL levels increase because of defective or absent LDL receptors, and after fat and cholesterol feeding, LDL levels increase because of LDL receptor down-regulation.

ATHEROGENIC POTENTIAL OF LOW DENSITY LIPOPROTEINS

At the present time, we do not understand the precise mechanism by which LDL cause atherosclerosis. However, apo-B100 and LDL-like particles are found in atherosclerotic lesions and do contribute to the deposition of cholesterol there. The complexity of the problem results from the fact that native LDL (normal LDL isolated from plasma) do not cause massive cholesterol loading of macrophages or smooth muscle cells, the two cells that accumulate cholesterol in the atherosclerotic lesion (*Brown and Goldstein, 1983a*).

However, there is an interesting and intriguing observation that may have physiological significance in the development of atherosclerosis: chemically modified LDL are taken up by macrophages through a unique receptor(s), and these LDL cause marked cholesteryl ester accumulation (*Brown and Goldstein, 1983*). Chemical modification of LDL by procedures that derivatize lysine residues will cause them to be taken up by a receptor specific for chemically modified LDL. For example, it has been postulated by Fogelman and associates that malondialdehyde modification of LDL could occur *in vivo* in the region of lipid peroxidation within the arterial wall (*Fogelman, et al., 1980*). More recently, Haberland *et al.* demonstrated that malondialdehyde-modified proteins, and presumably malondialdehyde-modified LDL, could be detected within the aortic atherosclerotic lesions of WHHL rabbits (*Haberland et al., 1988*). Watanabe heritable hyperlipidemic (WHHL) rabbits are animals that possess defective LDL receptors similar to those in humans with FH, and the rabbits also develop accelerated atherosclerosis (*Goldstein, et al., 1983*).

Another important type of chemical modification of LDL is oxidation. Oxidation can be cell-mediated (via incubation of LDL with endothelial cells, smooth muscle cells, or macrophages (all of which are arterial components) or non-cell-mediated (via copper and iron ions) (*Steinberg et al., 1989*). Oxidized LDL have been found in atherosclerotic plaques and in the blood of experimental animals at risk for the development of atherosclerosis (*Morton et al., 1986*). Moreover, treatment of WHHL rabbits with the potent antioxidants probucol or butylated hydroxytoluene retards the development of lesions (*Kita, et al., 1987; Carew, et al., 1987*).

Oxidized LDL can also have a number of other physiological actions that may be a part of the atherogenic process. When incubated with endothelial cells in culture, oxidized LDL induce the synthesis of a chemotactic factor for monocytes. In addition, oxidized endothelial cells incubated with LDL synthesize a binding protein or receptor for monocytes (*Berliner, et al., 1990*). These *in vitro* results mimic the early events in the formation of fatty streaks. For example, in experimental animals it has been shown that a short time after cholesterol feeding there is an increased entry of monocytes into the vessel wall. Thus, two biochemical events defined *in vitro* (the attraction of the monocytes to the endothelium and the binding of the monocytes to the endothelial surface) may be important initial events in atherogenesis. Recently, Rajavashisth *et al.* have shown that oxidized LDL, when incubated with endothelial cells *in vitro*, induce high levels of mRNAs of granulocyte and macrophage colony-stimulating factors. Since these growth factors affect the proliferation, migration, and metabolism of macrophages, granulocytes, and endothelial cells, cells that are involved in the development of atherosclerosis, this may be another possible mechanism by which oxidized LDL are atherogenic (*Rajavashisth et al., 1990*). Oxidized LDL are also toxic to endothelial and other cells in culture (*Morel et al., 1984*). It is also of interest that ß-VLDL (which are associated with the development of type III hyperlipoproteinemia), when incubated with cultured endothelial cells, increase the production of a chemotactic factor and the adhesion of monocytes to the endothelium (*Territo et al., 1989*).

REFERENCES

Angelin, B., Raviola, C.A., Innerarity, T.L., and Mahley, R.W. Regulation of hepatic lipoprotein receptors in the dog. Rapid regulation of apolipoprotein B,E receptors, but not of apolipoprotein E receptors, by intestinal lipoproteins and bile acids. *J. Clin. Invest.* 71:816-831, 1983.

Berliner, J.A., Territo, M.C., Sevanian, A., Ramin, S., Kim, J.A., Bamshad, B., Esterson, M., and Fogelman, A.M. Minimally modified low density lipoprotein stimulates monocyte endothelial interactions. *J. Clin. Invest.* 85:1260-1266, 1990.

Brown, M.S. and Goldstein, J.L. A receptor-mediated pathway for cholesterol homeostasis. *Science* 232:34-47, 1986.

Brown, M.S. and Goldstein, J.L. Lipoprotein metabolism in the macrophage: implications for cholesterol deposition in atherosclerosis. *Annu. Rev. Biochem.* 52:223-261, 1983a.

Brown, M. S. and Goldstein, J. L. Lipoprotein receptors in the liver. Control signals for plasma cholesterol traffic. *J. Clin. Invest.* 72:743-747, 1983b.

Carew, T.E., Schwenke, D.C., and Steinberg, D. Antiatherogenic effect of probucol unrelated to its hypocholesterolemic effect: evidence that antioxidants *in vivo* can selectively inhibit low density lipoprotein degradation in macrophage-rich fatty streaks and slow the progression of atherosclerosis in the WHHL rabbit. *Proc. Natl. Acad. Sci. U.S.A.* 84:7725-7729, 1987.

Fogelman, A.M., Shechter, I., Seager, J., Hokom, M., Child, J.S., and Edwards, P.A. Malondialdehyde alteration of low density lipoproteins leads to cholesteryl ester accumulation in human monocyte-macrophages. *Proc. Natl. Acad. Sci. U.S.A.* 77:2214-2218, 1980.

Goldstein, J.L., Kita, T., and Brown, M.S. Defective lipoprotein receptors and atherosclerosis. Lessons from an animal counterpart of familial hypercholesterolemia. *N. Engl. J. Med.* 309:288-296, 1983.

Goldstein, J.L. and Brown, M.S. Familial hypercholesterolemia in *The Metabolic Basis of Inherited Disease, 6th Edition*, C.R. Scriver, A.L. Beaudet, W.S. Sly, and D. Valle, Eds., McGraw-Hill, New York pp. 1215-1250, 1989.

Goldstein, J.L., Ho, Y.K., Brown, M.S., Innerarity, T.L., and Mahley, R.W. Cholesterol ester accumulation in macrophages resulting from receptor-mediated uptake and degradation of hypercholesterolemic canine ß-very low density lipoproteins. *J. Biol. Chem.* 255:1839-1848, 1980.

Haberland, M.E., Fong, D., and Cheng, L. Malondialdehyde-altered protein occurs in atheroma of Watanabe heritable hyperlipidemic rabbits. *Science* 241:215-218, 1988.

Innerarity, T.L. Plasma lipoproteins, in *Encyclopedia of Human Biology*, R. Dulbecco, Ed., Academic Press, Inc., San Diego (in press).

Innerarity, T.L., Weisgraber, K.H., Arnold, K.S., Mahley, R.W., Krauss, R.M., Vega, G.L., and Grundy, S.M. Familial defective apolipoprotein B-100: low density lipoproteins with abnormal receptor binding. *Proc. Natl. Acad. Sci. U.S.A.* 84:6919-6923, 1987.

Innerarity, T.L. Familial hypobetalipoproteinemia and familial defective apolipoprotein B_{100}. Genetic disorders associated with apolipoprotein B. *Curr. Opinion Lipidol.* 1:104-109, 1990.

Innerarity, T.L., Mahley, R.W., Weisgraber, K.H., Bersot, T.P., Krauss, R.M., Vega, G.L., Grundy, S.M., Friedl, W., Davignon, J., and McCarthy, B.J. Familial defective apolipoprotein B100: A mutation of apolipoprotein B that causes hypercholesterolemia, *J. Lipid Res.*, 31:1337-1349, 1990.

Kita, T., Nagano, Y., Yokode, M., Ishii, K., Kume, N., Ooshima, A., Yoshida, H., and Kawai, C. Probucol prevents the progression of atherosclerosis in Watanabe heritable hyperlipidemic rabbit, an animal model for familial hypercholesterolemia. *Proc. Natl. Acad. Sci. U.S.A.* 84:5928-5931, 1987.

Mahley, R.W. and Rall, Jr., S.C. Type III hyperlipoproteinemia (dysbetalipoproteinemia): the role of apolipoprotein E in normal and abnormal lipoprotein metabolism, in *The Metabolic Basis of Inherited Disease, 6th Edition*, C.R. Scriver, A.L. Beaudet, W. S. Sly, and D. Valle, Eds. McGraw-Hill, New York, pp.1195-1213, 1989.

Mahley, R.W. Biochemistry and physiology of lipid and lipoprotein metabolism, in Principles and Practice of Endocrinology and Metabolism, K.L. Becker and C.R. Kahn, Eds., J.B. Lippincott, Philadelphia, pp. 1219-1229, 1990.

Mahley, R.W. Atherogenic hyperlipoproteinemia. The cellular and molecular biology of plasma lipoproteins altered by dietary fat and cholesterol. *Med. Clin. North Am.* 66:375-402, 1982.

Mahley, R.W. Apolipoprotein E: cholesterol transport protein with expanding role in cell biology. *Science* 240:622-630, 1988.

Milne, R., Théolis, Jr., R., Maurice, R., Pease, R.J., Weech, P.K., Rassart, E., Fruchart, J.C., Scott, J., and Marcel, Y.L. The use of monoclonal antibodies to localize the low density lipoprotein receptor-binding domain of apolipoprotein B. *J. Biol. Chem.* 264:19754-19760, 1989.

Morel, D.W., DiCorleto, P.E., and Chisolm, G.M. Endothelial and smooth muscle cells alter low density lipoprotein *in vitro* by free radical oxidation. *Arteriosclerosis* 4:357-364, 1984.

Morton, R.E., West, G.A., and Hoff, H.F. A low density lipoprotein-sized particle isolated from human atherosclerotic lesions is internalized by macrophages via non-scavenger-receptor mechanism. *J. Lipid Res.* 27:1124-1134, 1986.

Pease, R.J., Milne, R.W., Jessup, W.K., Law, A., Provost, P., Fruchart, J.C., Dean, R.T., Marcel, Y.L., and Scott, J. Use of bacterial expression cloning to localize the epitopes for a series of monoclonal antibodies against apolipoprotein B100. *J. Biol. Chem.* 265:553-568, 1990.

Rajavashisth, T.B., Andalibi, A., Territo, M.C., Berliner, J.A., Navab, M., Fogelman, A.M., and Lusis, A.J. Induction of endothelial cell expression of granulocyte and macrophage colony-stimulating factors by modified low-density lipoproteins. *Nature* 344:254-257, 1990.

Rall, Jr., S.C., Innerarity, T.L., Weisgraber, K.H., Wardell, M.R., and Mahley, R.W. The type of mutation in apolipoprotein E determines whether type III hyperlipoproteinemia is expressed as a dominant or recessive trait, in *Proceedings of the VII Atherosclerosis and Cardiovascular Disease Conference*, G.C. Descovich, Ed., Kluwer Academic Publishers, Dordrecht, The Netherlands, pp. 81-88.

Schuster, H., Rauh, G., Kormann, B., Hepp, T., Humphries, S., Keller, C., Wolfram, G., and Zöllner, N. Familial defective apolipoprotein B-100. Comparison with familial hypercholesterolemia in 18 cases detected in Munich. *Arteriosclerosis*, 10:577-581, 1990.

Soria, L.F., Ludwig, E.H., Clarke, H.R.G., Vega, G.L., Grundy, S.M., and McCarthy, B.J. Association between a specific apolipoprotein B mutation and familial defective apolipoprotein B-100. *Proc. Natl. Acad. Sci. U.S.A.* 86:587-591, 1989.

Steinberg, D., Parthasarathy, S., Carew, T.E., Khoo, J.C., and Witztum, J.L. Beyond cholesterol. Modifications of low-density lipoprotein that increase its atherogenicity. *N. Engl. J. Med.* 320:915-924, 1989.

Territo, M.C., Berliner, J.A., Almada, L., Ramirez, R., and Fogelman, A.M. ß-very low density lipoprotein pretreatment of endothelial monolayers increases monocyte adhesion. *Arteriosclerosis* 9:824-828, 1989.

Tybjaerg-Hansen, A., Gallagher, J., Vincent, J., Houlston, R., Talmud, P., Dunning, A.M., Seed, M., Hamsten, A., Humphries, S.E., and Myant, N.B. Familial defective apolipoprotein B-100: detection in the United Kingdom and Scandinavia, and clinical characteristics of ten cases. *Atherosclerosis* 80:235-242, 1990.

Vega, G.L. and Grundy, S.M. *In vivo* evidence for reduced binding of low density lipoproteins to receptors as a cause of primary moderate hypercholesterolemia. *J. Clin. Invest.* 78:1410-1414, 1986.

Weisgraber, K.H., Innerarity, T.L., Newhouse, Y.M., Young, S.G., Arnold, K.S., Krauss, R.M., Vega, G.L., Grundy, S.M., and Mahley, R.W. Familial defective apolipoprotein B-100: enhanced binding of monoclonal antibody MB47 to abnormal low density lipoproteins. *Proc. Natl. Acad. Sci. U.S.A.* 85:9758-9762, 1988.

Wetterau, J.R., Aggerbeck, L.P., Rall, Jr., S.C., and Weisgraber, K.H. Human apolipoprotein E3 in aqueous solution. I. Evidence for two structural domains. *J. Biol. Chem.* 263:6240-6248, 1988.

Young, S.G. Recent progress in understanding apolipoprotein B. *Circulation*, 82:1574-1594, 1990.

Zilversmit, D.B. Atherogenesis: a postprandial phenomenon. *Circulation* 60:473-485, 1979.

ENDOTHELIAL CELL-MATRIX INTERACTIONS IN HEALTH AND DISEASE

Elisabetta Dejana, A. Zanetti, C. Dominguez-Jimenez and G. Conforti

Laboratory of Vascular Biology
Istituto di Ricerche
Farmacologiche Mario Negri
Milano, Italy

INTRODUCTION

Human endothelial cells (EC) adhere, spread and organize their cytoskeleton on a variety of molecules of the extracellular matrix such as fibronectin, vitronectin, laminin and collagen (*Form et al., 1983; Ingber and Folkman, 1989*). EC express receptor molecules which on the outer side of the membrane, recognize and bind different components of the extracellular matrix and, on the cytoplasmic side, transmit intracellular signals and link a chain of proteins of the membrane-microfilament interaction complex involved in the mechanism of adhesion and cytoskeletal organization (*Burridge, 1986*). Most of the EC receptors for extracellular matrix components belong to a recently discovered superfamily of adhesive membrane proteins denominated "integrins" (*Ruoslahti and Pierschbacher, 1987; Hynes, 1986*). These receptors have several structural and functional homologies so that it is believed that they differentiated from a common ancestral gene. They are all heterodimers of two non-covalently linked subunits. The larger subunit has been termed "α chain" and the smaller subunit "β chain". The name "integrins" arose because they are integral membrane proteins, i.e. each subunit has a transmembrane segment, a small C-terminal cytoplasmic domain and a large N-terminal extracellular domain (*Ruoslahti and Pierschbacher, 1987; Hynes, 1986*). Some of these receptors specifically recognize in the ligand proteins a sequence of only three aminoacids (arginine-glycine-aspartic acid, RGD). Many proteins contain this sequence but not all of them are recognized by an integrin receptor; however, this sequence is the cell recognition site of a large number of extracellular matrix and plasma proteins including fibrinogen, vitronectin, fibronectin, thrombospondin and von Willebrand factor (*Ruoslahti and Pierschbacher, 1987*). Despite the similarities in the cell binding sequences in ligand proteins, the cell can recognize them individually through specific and separate receptors. This suggests that other aminoacids surrounding the RGD sequence, and the RGD steric conformation, confer specificity to the interaction with one integrin receptor or another (*Ruoslahti and Pierschbacher, 1987*).

The number of the integrin family members is expanding. A subclassification of the family has been attempted based on the observation that some members of the integrin group have the same β chain but different α chains. This has resulted in the definition of three subfamilies: the β_1 or VLA (very late antigens) (*Hemler et al.,1987*); the β_2 or leu-cam (leukocyte adhesion molecules) (*Sanchez-Madrid et al., 1983*) ; the β_3 or cytoadhesins (*Ginsberg et al., 1988*). However, at least five additional and novel β chains have been recently described: β_4 (*Kajiji et al., 1989*), β_5 (*Cheresh et al., 1989; Freed et al., 1989; Ramaswamy and Hemler, 1990*), β_6 (*Sheppard et al., in press*), β_p (*Holzmann et al., 1989*) and the β chain of the melanoma laminin receptor (*Kramer et al., 1989*).

ENDOTHELIAL INTEGRINS

EC possess at least four receptors belonging to the β_1 subfamily. The VLA-2 or $\alpha_2 \beta_1$ integrin is identical to the platelet GpIa-IIa complex, which is the receptor for collagen in platelets and in other cells of hematopoietic and non hematopoietic origin (for review see *Dejana and Lauri, 1990*). In EC it behaves differently, acting as a major receptor for laminin and also binding (though rather less efficiently) to collagen and fibronectin (*Languino et al., 1989*). VLA-3, or $\alpha_3\beta_1$, which is expressed in lower amounts, appear to be a multifunctional receptor: it recognizes fibronectin, collagen and laminin (*Wagner and Carter, 1987*). VLA-5, or $\alpha_5\beta_1$, has been described as the fibronectin receptor in many types of cells including EC (*Conforti et al., 1989*). VLA-6, or $\alpha_6 \beta_1$, is present in a very low amount on EC. This molecule is the laminin receptor in platelets and possibly plays a similar role in EC (for review see *Dejana and Lauri, 1990*).

EC express only one integrin belonging to the β_3 or cytoadhesin subfamily which was isolated and defined as vitronectin receptor. This molecule has the same β chain than the platelet complex GpIIb-IIIa but a distinct α chain (*Ginsberg et al., 1988*). When it was isolated and reconstituted into artificial phospholipid membranes, it had a very low specificity and recognized (beside vitronectin), von Willebrand factor, fibrinogen and fibronectin (*Conforti et al., 1990*). It behaves, however, in a different way in different cell types. In EC (*Cheresh, 1987*) it maintains its multifunctional characteristics (*Ignotz et al., 1989*) while in MG63 osteosarcoma cells or in vascular smooth muscle cells it only recognizes vitronectin (*Dejana et al., 1986; 1989*).

Finally, a novel integrin has been identified (*Freed et al., 1989*), which has the same α chain as the vitronectin receptor but a distinct β chain. The function of this molecule is still unknown but it has the property to become heavily phosphorylated when the cells are activated with phorbol esters.

Most of our data come from studies performed on cultured human umbilical vein or bovine thoracic aorta EC. However, when a comparison has been attempted among EC from different parts of the vasculature the integrin composition appeared to be essentially similar (*Albelda et al., 1989*). Also, EC integrins were not significantly modified by the time the cells spent in culture or by passaging the cells (*Albelda et al., 1989*). Interestingly, some evidence indicates that microvascular EC express VLA-1 while large vessels' EC are negative (G. Tarone personal communication). The biological meaning of this observation remains to be clarified.

ACTIVITY MODULATION

From what has been reported above, it appears that the same receptor (e.g. VLA-2 and the vitronectin receptor) might behave differently in EC compared with other types of cells (for example, platelets or tumor cells). This suggests that the diversity of the integrin system could be further augmented by a cell specific type of regulation. Possible mechanisms of such regulation could include alternative mRNA splicing, post-translational modification of the receptor, or the association of the receptor with some modifying components (for example gangliosides or glycosaminoglycans...). However, no evidence is so far available documenting differences in the structure or in the synthetic pathways of endothelial integrins compared with other cells. Alternative splicings for integrin subunits have been described but these processes were present equally in all the cell types studied, including EC (*van Kuppevelt et al., 1989*). A detailed study of synthesis of the vitronectin receptor shows that EC follow a pathway similar to that described for GpIIb-IIIa in megakariocytes (*Polack et al., 1989*).

Modulation of integrin synthesis and expression in EC is still a relatively unexplored area of research. Tarone et al. (in press) reported that the combination of tumor necrosis factor and interferon induces a 50-70% decrease in vitronectin receptor number while no change was detected in the β_1 subgroup of integrin molecules. This effect is particularly interesting considering that these cytokines are able to induce a dramatic modification of EC shape and matrix composition (*Stolpen et al., 1986*).

Cell adhesion to matrix or plasma proteins seems to be more efficiently regulated by modulation of the activity of integrin receptors than by changes in their number. This phenomenon has been widely studied for platelet GpIIb-IIIa (*Ginsberg et al., 1988*) and leukocyte leu-cam integrins (*Vedder and Harlan, 1988*). Activation of these cells by aggregating agents or chemotactic stimuli can change integrin receptor conformation and in this way modify their binding capacity. Ion concentration and the phospholipid composition of the membrane also modify integrin receptor affinity for ligand proteins in purified systems (*Conforti et al., 1990; Gailit and Ruoslahti, 1988*) and can change integrin receptor conformation.

Finally, as with other types of receptors, phosphorylation could be another mechanism of regulation of integrin activity. Tyrosine phosphorylation of fibronectin receptor in oncogene transformed cells has been described, and this was associated with an altered interaction with the cytoskeleton (*Hirst et al., 1986*). However, it is still unknown if this mechanism of regulation is present in normal cells too.

INTEGRIN MEDIATED CELL ACTIVATION AND CYTOSKELETAL ASSEMBLY

Cell recognition of extracellular matrix proteins is a complex phenomenon which involves a variety of adhesive receptors and specific cell responses. It has been recognized for many years that cell interaction with the matrix is not simply a phenomenon of cell attachment but is followed by specific responses which lead to cell differentiation, migration and growth. This implies the transmission of intracellular messages through specific receptor activation. Little is known about the biochemical signalling pathways that integrin receptors activate inside the cells. The platelet integrin GpIIb-IIIa regulates Ca^{++} and Na^+/H^+ exchange (*Brass, 1985; Banga et al., 1986*) and tyrosine specific protein phosphorylation (*Farrell and Martin, 1989*). However, we still do not have any evidence that these pathways are activated after integrin receptor occupancy in EC or other cell types.

EC express a variety of integrin molecules. An attractive possibility is that each receptor could induce specific cellular responses and/or that the engagement of more than one receptor is required for full cell activation. The same matrix protein can be bound by different integrins. For example, fibronectin is linked by VLA-2, VLA-3, VLA-5 and the vitronectin receptor, whereas laminin is recognized by VLA-2, VLA-3 and VLA-6. The receptors appear to bind to different domains of the molecules (i.e. RGD or other sequences), suggesting a multiple type of interaction between the ligand proteins and separate binding sites on the cell membrane. It has been recently reported that laminin can induce EC differentiation into tubular structures and that this phenomenon requires cell attachment both to an RGD sequence via an integrin receptor and to a Tyr-Ile-Gly-Ser-Arg (YIGSR) sequence through a non integrin binding site (*Grant et al., 1989*). In addition to chemical signalling, integrin molecules may convey regulatory information through interactions with cytoskeletal proteins. During attachment, the basal cell surface forms several types of contacts (known as focal contacts, or adhesion plaques) which represent the area of closest interaction between the substratum, the cell membrane and the membrane insertion sites of actin microfilament bundles (*Burridge, 1986*). In EC (*Dejana et al., 1988*) and other cells (*Singer et al., 1988*) the fibronectin receptor (VLA-5, $\alpha_5\beta_1$) and the vitronectin receptor are clustered in focal contacts during cell adhesion. The organization of these receptors is strictly dependent on the specific molecules in the substratum. When EC are plated on vitronectin, only the vitronectin receptor is clustered, while when the cells are plated on fibronectin, only the fibronectin receptor is organized in focal contacts. When EC are seeded on fibrinogen and von Willebrand factor, both receptors are organized in adhesion plaques although only the vitronectin receptor is able to recognize these substrata. This paradox can be explained by the release of endogenous matrix proteins (mainly fibronectin) by EC, which cause binding and clustering of the fibronectin receptor (*Albelda et al., 1989; Dejana et al., 1989; 1990*).

The mechanism of integrin receptor clustering and its consequences are still unknown. An attractive possibility is that integrin receptor clustering (as with other types of receptors) involves signal transmission across the membrane, which induces assembly of the cytoskeletal proteins vinculin and talin, which in turn mediate actin microfilament organization. It has been suggested fibronectin receptors could be interconnected with actin fibers by directly binding to talin (*Horwitz et al., 1986*).

The biological meaning of integrin clustering and cytoskeletal organization is still a matter of debate. It is probably essential for the maintenance of cell adhesion, cell shape and motility but not for the first phases of cell attachment. When cytoskeletal protein assembly is inhibited by increasing cAMP in EC (*Lampugnani et al., in press*), cell adhesion to different substrata was not modified. In addition, when recombinant fibronectin receptor lacking the cytoplasmic domain (and therefore the possibility to bind cytoskeletal proteins) was inserted into cells, there was no clustering and cytoskeletal organization, although the cells could still bind to fibronectin (*Salowska et al., 1989*).

ACKNOWLEDGEMENTS

This work was supported by the Italian National Research Council (P.F. Biotechnologie e Biostrumentazione e P.F. Tecnologie Biomediche e Sanitarie) and by the Associazione Italiana per la Ricerca sul Cancro (AIRC) and by NATO Research Grant CT-0429/87.

REFERENCES

Albelda, S.M., Daise, M., Levine, E.M. and Buck, C.A. Identification and characterization of cell-substratum adhesion receptors on cultured human endothelial cells. *J. Clin. Invest.* 83:1992-2002, 1989.

Banga, H.S., Simons, E., Brass, L.F. and Rittenhouse, S.E. Activation of phospholipases A and C in human platelets exposed to epinephrine: role of glycoproteins IIb-IIIa and dual role of epinephrine. *Proc. Natl. Acad. Sci. U.S.A.* 83:9197-9201, 1986.

Brass, L.F. Ca^{++} transport across the platelet plasma membrane. A role for membrane glycoproteins IIb-IIIa. *J. Biol. Chem.* 260:2231-2238, 1985.

Burridge, K. Substrate adhesion in normal and transformed fibroblasts: Organization and regulation of cytoskeletal, membrane and extracellular matrix components at focal contacts. *Cancer Rev.* 4:18-78, 1986.

Cheresh, D.A. Human endothelial cells synthesize and express an Arg-Gly-Asp-directed adhesion receptor involved in attachment to fibrinogen and von Willebrand factor. *Proc. Natl. Acad. Sci. U.S.A.* 84:6471-6475, 1987.

Cheresh, D.A., Smith, J.W., Cooper, H.M. and Quaranta, V. A novel vitronectin receptor integrin ($\alpha_v\beta_x$) is responsible for distinct adhesive properties of carcinoma cells. *Cell* 57:59-69, 1989.

Conforti, G., Zanetti, A., Colella, S., Abbadini, M., Marchisio, P.C., Pytela, R., Giancotti F., Tarone, G., Languino, L.R. and Dejana, E. Interaction of fibronectin with cultured human endothelial cells. Characterization of the specific receptor. *Blood* 73:1576-1585, 1983.

Conforti, G., Zanetti, A., Pasquali-Ronchetti, I., Quaglino, D., Neyroz, P. and Dejana, E. Modulation of vitronectin receptor binding by membrane lipid composition. *J. Biol. Chem.* 265:4011-4019, 1990.

Dejana, E., Colella, S., Languino, L.R., Balconi, G., Corbascio, G.C. and Marchisio, P.C. Fibrinogen induces adhesion, spreading and microfilament organization of human endothelial cells *in vitro*. *J. Cell. Biol.* 104:1403-1411, 1986.

Dejana, E., Colella, S., Conforti, G., Abbadini, M., Gaboli, M. and Marchisio, P.C. Fibronectin and vitronectin regulate the organization of their respective Arg-Gly-Asp adhesion receptors in cultured human endothelial cells. *J. Cell. Biol.* 107:1215-1223, 1988.

Dejana, E., Lampugnani, M.G., Giorgi, M., Gaboli, M., Federici, A.B., Ruggeri, Z.M. and Marchisio P.C. Von Willebrand factor promotes endothelial cell adhesion via an Arg-Gly-Asp-dependent mechanism. *J. Cell. Biol.* 109:367-375, 1989.

Dejana, E., Lampugnani, M.G., Giorgi, M., Gaboli, M. and Marchisio, P.C. Fibrinogen induces endothelial cells adhesion and spreading via the release of endogenous matrix proteins and the recruitment of more than one integrin receptor. *Blood* 75:1509-1517, 1990.

Dejana, E. and Lauri, D. Biochemical and functional characteristics of integrins: a new family of adhesive receptors present in hematopoietic cells. *Hematologica* 75:1-6, 1990.

Farrell, J.E. and Martin, G.S. Tyrosine specific protein phosphorylation is regulated by glycoprotein IIb-IIIa in platelets. *Proc. Natl. Acad. Sci. U.S.A.* 86:2234-2238, 1989.

Form, D.M., Pratt, B.M., and Madri, J.A. Endothelial cell proliferation during angiogenesis. *In vitro* modulation by basement membrane components. *Lab. Invest.* 55:521-25, 1983.

Freed, E., Gailit, J., van der Geer, P., Ruoslahti, E. and Hunter,T. A novel integrin β subunit is associated with vitronectin receptor α subunit (α_v) in a human osteosarcoma cell line and is a substrate for protein kinase C. *EMBO J.* 8:2955-2965, 1989.

Gailit, J. and Ruoslahti, E. Regulation of the fibronectin receptor affinity by divalent cations. *J. Biol. Chem.* 263:2065-2067, 1988.

Ginsberg, M.H., Loftus, J.C. and Plow, E.F. Cytoadhesins, integrins and platelets. *Thromb. Haemost.* 59:1-20, 1988.

Grant, D.S., Tashiro, K.I., Segul-Real, B., Yamada, Y., Martin, G.R. and Kleinman, H.K. Two different laminin domains mediate the differentiation of human endothelial cells into capillary-like structures *in vitro*. *Cell* 58:933-943, 1989.

Hemler, M.E., Huang, C. and Schwarz, L. The VLA protein family: characterization of five different surface heterodimers each with a common 130,000 Mr subunit. *J. Biol. Chem.* 262:3300-3309, 1987.

Hirst, R., Horwitz, A., Buck, C. and Rohrschneider, L. Phosphorylation of the fibronectin receptor complex in cells transformed by oncogenes that encode tyrosine kinases. *Proc. Natl. Acad. Sci. U.S.A.* 83:6470-6474, 1986.

Holzmann, B., McIntyre, B.W. and Weissman, I.L. Identification of a murine Peyer's Patch-specific lymphocyte homing receptor as an integrine molecule with an α chain homologous to human VLA-4. *Cell* 56:37-46, 1989.

Horwitz, A., Duggan, K., Buck, C., Beckerle, M.C. and Burridge, K. Interaction of plasma membrane fibronectin receptor with talin a transmembrane linkage. *Nature (Lond)* 320: 531-534, 1986.

Hynes, R.O. Integrins: a family of cell surface receptors. *Cell* 48:549-554, 1986.

Ignotz, R.A., Heino, J. and Messague, J. Regulation of cell adhesion receptors by transforming growth factor β. Regulation of vitronectin receptor and LFA-1. *J. Biol. Chem.* 264:389-392, 1989.

Ingber, D.E. and Folkman, J. How does extracellular matrix control capillary morphogenesis? *Cell* 58:803-805, 1989.

Kajiji, S., Tamura, R.N. and Quaranta, V. A novel integrin ($\alpha_E\beta_4$) from human epithelial cells suggests a fourth family of integrin adhesion receptors. *EMBO J.* 8:673-680, 1989.

Kramer, R.H., Mc Donald, K.A. and Vu, M.P. Human melanoma cells express a novel integrin receptor for laminin. *J. Biol. Chem.* 264:15642-15649, 1989.

Lampugnani, M.G., Giorgi, M., Gaboli, M., Dejana, E. and Marchisio P.C. Endothelial cell motility, integrin receptor clustering and microfilament organization are inhibited by agents that increase intracellular cAMP. *Lab. Invest.* 63:521-531, 1990.

Languino, L.R., Gehlsen, K.R., Wayner, E., Carter, W.G., Engwall, E. and Ruoslahti, E. Endothelial cells use $\alpha_2\beta_1$ integrin as a laminin receptor. *J. Cell. Biol.* 109:2455-2462, 1989.

Polack, B., Duperray, A., Troesch, A., Berthier, R. and Marguerie, G. Biogenesis of the vitronectin receptor in human endothelial cell: evidence that the vitronectin receptor and GpIIb-IIIa are synthesized by a common mechanism. *Blood* 73:1519-1524, 1989.

Ramaswamy, H. and Hemler, M.E. Cloning, primary structure and properties of a novel human integrin β subunit. *EMBO J.* 9:1561-1568, 1990.

Ruoslahti, E. and Pierschbacher, M.D. New perspectives in cell adhesion: RGD and integrins. *Science* 238:491-497, 1987.

Solowska, J., Guan, J.L., Marcantonio, E.E., Trevithick, J.E., Buck, C.A. and Hynes, R.O. Expression of normal and mutant avian integrin subunits in rodent cells. *J. Cell. Biol.* 109:853-861, 1989.

Sanchez-Madrid, F., Nagy, J.A., Robbins, E., Simon, P. and Springer, T.A. A human leukocyte differentiation antigen family with distinct α-subunits and a common β subunit. The lymphocyte function-associated antigen (LFA-1), the C3bi complement receptor (OKM1/Mac-1), and the Gp 150,95 molecule. *J. Exp. Med.* 158:1785-803, 1983.

Singer. I., Scott, S., Kawka, D.W., Kazazis, D.M., Gailit, J. and Ruoslahti, E. Cell surface distribution of fibronectin and vitronectin receptors depends on substrate composition and extracellular matrix accumulation. *J. Cell. Biol.* 106:2171-2182, 1988.

Sheppard, D., Rozzo, C., Starr, L., Quaranta, V., Erle, D.J. and Pytela, R. Complete amino acid sequence of a novel integrin β subunit (β_6) identified in epithelial cells using the polymerase chain reaction. *J. Biol. Chem.* (in press)

Stolpen, A.H., Guinan, E.C., Fiers, W. and Pober, J.S. Recombinant tumor necrosis factor and immune interferon act singly and in combination to reorganize human vascular endothelial cell monolayers. *Am. J. Pathol.* 123:16-24, 1986.

Tarone, G., Stefanuto, G., Mascarello, P. and Defilippi, P. Expression of receptors for extracellular matrix proteins in human endothelial cells. *J. Lipid Mediators*. (in press) van Kuppevelt ,T., Languino, L.R., Gailit, J.O. and Ruoslahti, E. An alternative cytoplasmic domain of the integrin β_3 subunit. *Proc. Natl. Acad. Sci. U.S.A.* 86:5415-5418, 1989.

Vedder, N.B. and Harlan, J.M. Increased surface expression of CD11b/CD18 (Mac-1) is not required for stimulated neutrophil adherence to cultured endothelium. *J. Clin. Invest.* 81: 676-682, 1988.

Wayner, E.A. and Carter, W.G. Identification of multiple cell adhesion receptors for collagen and fibronectin in human fibrosarcoma cells possessing unique α and common β subunits. *J. Cell. Biol.* 105:1873-1884, 1987.

BIOSYNTHESIS AND ASSEMBLY OF VON WILLEBRAND FACTOR BY VASCULAR ENDOTHELIAL CELLS: RELEVANCE TO PATHOPHYSIOLOGY

Jan A. van Mourik

Central Laboratory of the Netherlands Red Cross
Blood Transfusion Service Amsterdam
The Netherlands

INTRODUCTION

The von Willebrand factor (vWF) plays an essential role in the events that lead to normal arrest of bleeding. This plasma protein mediates platelet-vessel wall interactions at sites of vascular injury (*Tschopp et al., 1974; Sakariassen et al. 1979*), and it functions as a stabilizing carrier protein of factor VIII (*Weiss et al., 1977; Brinkhous et al., 1985*), an essential cofactor of the intrinsic pathway of blood coagulation (*van Dieijen et al. 1981; Mertens et al., 1985*). In addition to its well-defined role in platelet adhesion and regulation of the coagulation system, vWF may help to anchor endothelial cells to the extracellular matrix (*Cheresh, 1987*). It thus seems clear that the function of vWF is of broad physiological significance, and insights into its mode of action, biosynthesis and processes that control its production are of importance for understanding the process of normal hemostasis and the pathophysiology of thrombosis and atherosclerosis.

The cloning of the vWF gene and the expression of wild-type and mutant cDNAs in different heterologous cells has provided the basis of significant advancements in the understanding of the molecular events associated with the synthesis, processing, assembly and secretion by vascular endothelial cells, the major production site of vWF.

BIOSYNTHESIS, STORAGE AND SECRETION OF VON WILLEBRAND FACTOR

Immunohistochemical studies have revealed that endothelial cells of probably all human tissues contain vWF (*Bloom et al., 1973; Hoyer et.al., 1973*). Studies with cultured cells showed that most normal endothelial cells synthesize and secrete vWF, including endothelial cells isolated from large and smaller veins, capillaries, aorta and arteries (*Jaffe et al., 1974; Folkman et al., 1979*). Besides megakaryocytes (*Nachman et al., 1977; Sporn et al.1985*), the endothelial cell is the only cell type that synthesizes vWF.

vWF distinguishes itself from many other endothelial proteins in that it can be secreted by the cell by more than one pathway. vWF released into the medium is either directly linked to protein synthesis (the constitutive pathway) or is induced by endothelial cell agonists that may trigger release of previously synthesized vWF from storage vesicles (regulated pathway). The constitutive secretory nature of the endothelium is reflected by the observation that soon after synthesis vWF accumulates extracellularly in the absence of a stimulus (*Jaffe et al., 1974*). No external stimulus or trigger is required for this type of secretion. On the other hand, if endothelial cells are exposed to stimuli such as thrombin, the calcium ionophore A23187 or phorbol esters, vWF rapidly accumulates outside the cell (*Levine et al., 1982; Loesberg et al., 1983; Sporn et al., 1986*). These observations indicate that vWF, released when the endothelial cell receives an appropriate signal, originates from a storage pool. Previously

(*Wagner et. al., 1982*), it has been shown that vWF is located intracellularly in the endothelial cell-specific Weibel-Palade bodies (*Weibel and Palade, 1964*). These rod-shaped vesicles, which most likely originate from the Golgi apparatus (*Sengel and Stoebner, 1970*), contain vWF in a highly condensed form (Figures 1 and 2) and disappear from endothelial cells upon stimulation (*Reinders et al., 1984*). In addition, the secretory vesicles containing vWF that sediment at high density upon density gradient centrifugation (*Reinders et al., 1984*), are morphologically similar to Weibel Palade bodies examined *in situ* by electron microscopy (*Ewenstein et al., 1987; Reinders et al., 1988*). Weibel-Palade bodies contain neither any marker enzymes of other cell organelles nor other adhesive proteins such as fibronectin and thrombospondin (*Reinders et al, 1984; 1985*). Taking these findings together, it seems justifiable to conclude that the Weibel-Palade body serves as the secretory vesicle involved in regulated secretion of vWF, whereas other vesicles, which most likely also originate from the Golgi apparatus, are involved in a passive flow of vWF to the cell surface. As has been shown with other cell types, it is to be expected that the transport vesicles of both the constitutive and the regulated pathway fuse with the plasma membrane to release their contents by exocytosis (*McNiff and Gil, 1983; Kelly, 1985*). Unlike the vesicles of the constitutive pathway, the secretion of the Weibel-Palade bodies is prevented from fusing until the level of cytoplasmic calcium is raised (*Loesberg et al., 1983; de Groot et al., 1984*).

It is likely that the storage pool of vWF serves an important role in supporting platelet adhesion to a damaged vessel wall. This pool is biologically active (*Reinders et al., 1988*), whereas constitutively released vWF, because of its limited degree of polymerization (see below), is probably less effective in supporting platelet adhesion. Upon activation of endothelial cells (for instance by thrombin formed locally as a result of vascular damage), local accumulation of stored vWF could, therefore, be of importance in attaining effective hemostasis. Pertinent to this point could be the finding that vWF secreted by the regulated pathway is predominantly released to the luminal site of the endothelium (*van Buul-Wortelboer et al., 1989*), thus providing a subset of biologically active polymers that is readily available in the vicinity of the injured vessel.

STRUCTURE OF VON WILLEBRAND FACTOR

Molecular cloning of the full-length vWF cDNA has revealed that the vWF mRNA is translated as a pre-pro-polypeptide, composed of a signal peptide (22 amino acid residues),

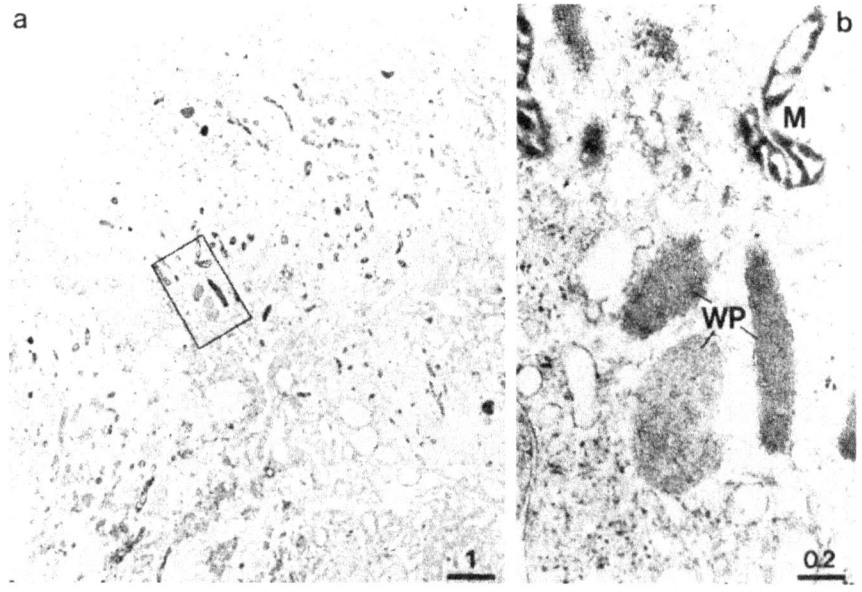

FIGURE 1: Electron micrographs of cultured endothelial cells. b, magnification of a, showing a few Weibel-Palade bodies (WP) and a mitochondrion (M). (*From Reinders et al. 1988*).

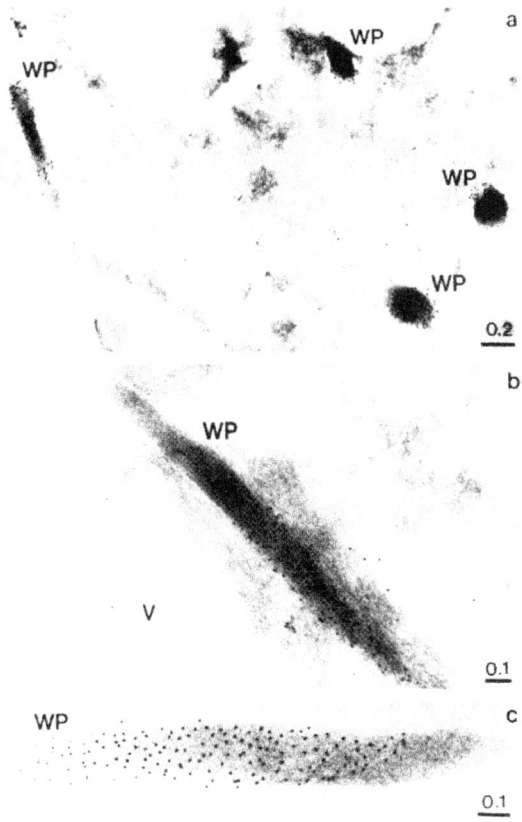

FIGURE 2: Immuno-electron microscopy of endothelial cells, stained for vWF. Gold- labeled protein A (dots) indicate the presence of vWF. Cellular structures staining for vWF are predominantly Weibel-Palade bodies (WP) and occasionally unidentified vesicles (V in b). In b and c longitudinal sectioned Weibel-Palade bodies are shown that stain for vWF all along the plane of sectioning. Bars represent micrometers. (*From Reinders et al. 1988*).

a pro-polypeptide (741 amino acids, also known as von Willebrand antigen II) and mature vWF (2050 amino acid residues) (*Verwey et al., 1986; Fay et al., 1986*). Processing, assembly, storage and secretion of newly synthesized protein is thought to take place as follows (Figure 3). First, pro-vWF undergoes N-linked glycosylation and dimerization in the rough endoplasmic reticulum. Dimerization of the pro-vWF polypeptide occurs by disulfide bond formation mediated by cysteine residues located at the carboxy-terminal part of pro-vWF (*Fretto et al., 1986; Marti et al., 1987*). These dimers serve as protomers for multimerization, a process that occurs during the travel through the Golgi compartments (*Wagner and Marder, 1984*). During transfer through the Golgi, several other post-translational processes occur, including sulfation, processing of the mannose-type carbohydrate to the complex-type and proteolytic cleavage of the propolypeptide. Probably during the latter process multimers are formed. Co-expression of full-length vWF cDNA and cDNA of furin, a subtilisin- like proprotein processing enzyme with substrate specificity for paired basic amino acid residues, in Cos-1 cells resulted in complete processing of the vWF precursor into mature vWF (*van de Ven et al., 1990*). This observation suggests that also in endothelial cells a subtilisin-like protease is responsible for pro-peptide cleavage.

Also multimer assembly involves intermolecular disulfide bond formation. The cysteine residues involved in the latter process are located in the amino- terminal region of mature vWF (*Marti et al., 1987*). Only the largest and most biologically active vWF multimers are stored in the Weibel-Palade bodies. In contrast, all multimeric species are secreted in a constitutive way (*Sporn et al., 1986; Ewenstein et al., 1987*).

More than 90% of the pro-vWF molecule consists of four types of repeated domains, denoted A, B, C and D respectively (Figure 4, *Verwey et al., 1986; Titani et al., 1986; Bonthron et al., 1986; Shelton-Inloes et al., 1986*). Studies on the fate of pro-vWF and mutants that lack one or more of these domains, transiently expressed in monkey kidney cells, provided insights into the individual role of these domains in the assembly of vWF multimers. These studies revealed that multimerization of vWF is directed by the propolypeptide (domains D1 and D2) (*Verwey et al., 1987; Wise et al., 1988*). Upon deletion of pro-vWF, only dimers are formed. Also purified dimers that lack the pro-polypeptide are not able to multimerize (*Mayadas and Wagner, 1989*). As multimerization of vWF protomers involves the formation of intermolecular disulfide bonds in the region of the D3 domain (*Marti et al., 1987*), it seems possible that the propolypeptide serves to align the amino-terminal segments of vWF protomers to permit the formation of disulfide links. Both the amino-terminal and carboxy-terminal region of mature vWF serve an autonomous role in the assembly of multimers. A vWF deletion mutant lacking the A, B, C and D4 domains through the carboxy- terminus is able to form dimers (*Voorberg et al., 1990*). Similarly, a carboxy- terminal 151 amino acid residue may undergo dimerization (*Voorberg et al., 1991*). Taken together, these data suggest that dimerization and multimerization are processes that may occur independently.

THE PATHOPHYSIOLOGY OF VON WILLEBRAND FACTOR

Insights into the molecular basis of events associated with processing and secretion of von Willebrand factor by endothelial cells could provide a rationale for the understanding of diseases associated with either quantitative or qualitative abnormalities of circulating vWF. Disorders associated with apparently structurally defective vWF or disorders with either elevated or reduced or absent plasma levels are well-documented. It is beyond the scope of this chapter, however, to discuss the vast literature on the biology of these disorders in detail (for a review, see *Sadler and Davie, 1988*). It should be noted, however, that insights into the molecular basis underlying the pathophysiology is, with a few exceptions, rather limited. In general, one can only speculate about the cause of apparent aberrant synthesis of vWF.

The most common disorder associated with reduced plasma vWF levels or abnormal vWF protein synthesis is von Willebrand's disease, an autosomally inherited disorder with both

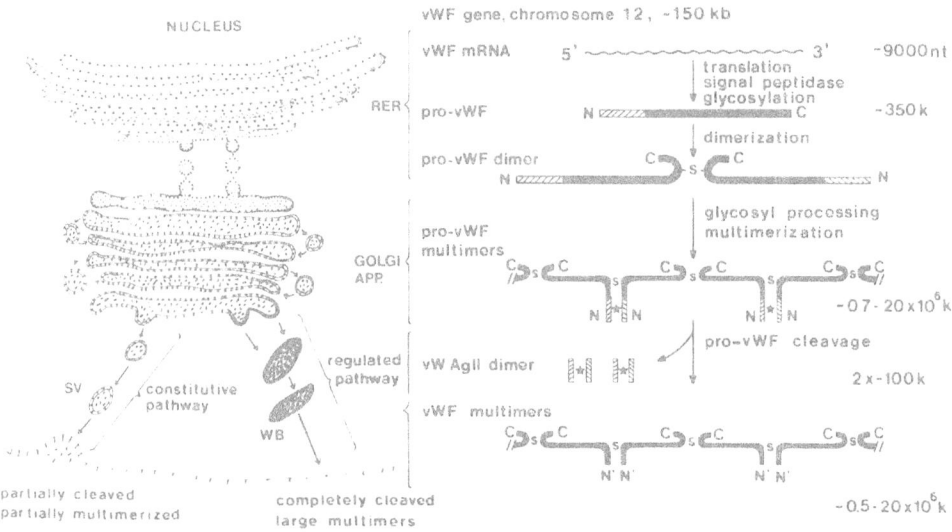

FIGURE 3: Schematic representation of the processing steps involved in the biosynthesis of vWF. The bars on the right represent the vWF precursor and the products derived from it. The hatched area represents the propeptide (vW Ag II) of vWF and the dark area mature vWF. WB = Weibel Palade body; SC = Secretory vesicles; * = non covalent interactions; S = disulfide bonds.

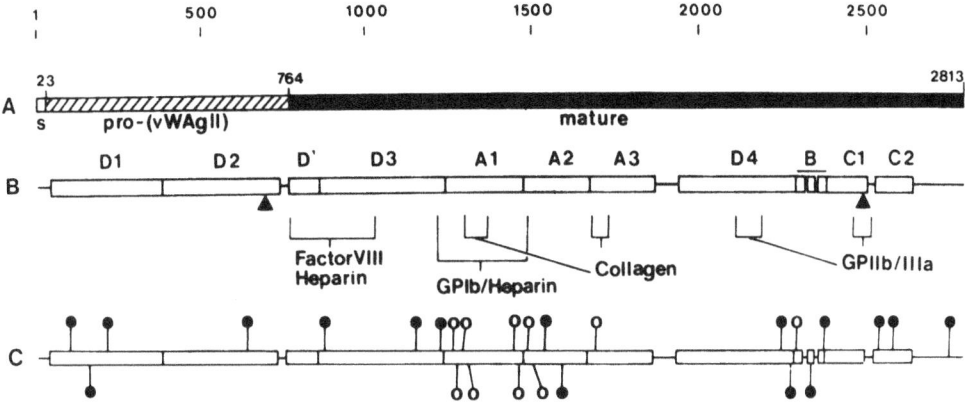

FIGURE 4: Structure of vWF precursor. A Precursor vWF includes a signal peptide (S), a pro-polypeptide (pro-vWAg II), and a mature subunit (mature). B Internal homologous domains and the structure-function relationship are depicted as are the positions of RGD tripeptides (triangles) known to be involved in cell attachment. GP = Platelet glycoproteins. C, sites of glycosylation in the pro- vWF subunit. N-linked and O-linked glycosylation sites are represented by closed and open symbols, respectively. The position of the four potential N-linked glycosylation sites in the pro-polypeptide is also indicated. (*From Verwey et al. 1988*).

dominant and recessive inheritance, and variable clinical expression. Clearly, von Willebrand's disease is a heterogeneous disorder, both from a clinical and genetic point of view. All cases described so far have in common that the bleeding diathesis is due to impairment of platelet adhesion to damaged vascular surfaces because of reduced or absent vWF levels or the production of functionally abnormal vWF molecules. Defective functioning could either be at the level of platelet binding or binding of vWF to the extracellular matrix.

The phenotypic classification scheme that is most widely employed today divides von Willebrand's disease into two broad categories according to the vWF multimer pattern (Figure 5). In one major subclass of von Willebrand's disease, all sizes of multimers are present. However, although the content of multimers is normal, the concentration of all multimers can be reduced to a variable extent. There are many ways to modify a gene, including gene deletions or defects in the regulatory elements that govern vWF expression, so that essentially no or little normal protein is derived from it. It is to be expected that in the near future, data will become available as to the genetic defects underlying aberrant synthesis of apparently normal vWF.

A second class of von Willebrand's disease comprises a bleeding disorder associated with normal or decreased protein vWF levels but abnormal multimeric composition. In general, in this subclass of von Willebrand's disease the larger multimers are absent (Figure 5). As the largest multimers, extending molecular weights greater than 10 million are the most biologically potent, a phenomenon that is probably related to the multivalency of these polymers, it is likely that defective polymerization is the cause of abnormal vWF function. As outlined above, several processes, including limited proteolysis and inter- and intra- molecular disulfide bond formation direct the vWF polymerization and probably correct sorting of vWF multimers within the endothelial cell.

Consequently, amino acid substitutions or deletions could result in defective multimer formation. Recently, a single amino acid substitution was characterized in a patient with defective polymerization that was held responsible for the defect (*Ginsburg et al., 1989*). The importance of vWF as a multifunctional protein is underscored by recent findings which show that abnormal binding of factor VIII to vWF may predispose to a bleeding tendency (*Nishino et al. 1989; Mazurier et al. 1990*).

VWF is frequently elevated in clinical conditions known to be associated with acute-phase-like responses of plasma proteins. In general, it is thought that vWF elevations seen under these conditions are without clinical consequences. Thrombotic thrombocytic purpura (TTP), a disorder of unknown cause, characterized by microangiopathic hemolytic anemia, thrombocytopenia, and wide- spread small vessel thrombosis, is probably an exception. In chronic relapsing TTP, the distribution of vWF multimers is skewed toward larger species than found in normal plasma (*Moake et al., 1982*). It is thought that in this variety of the disease

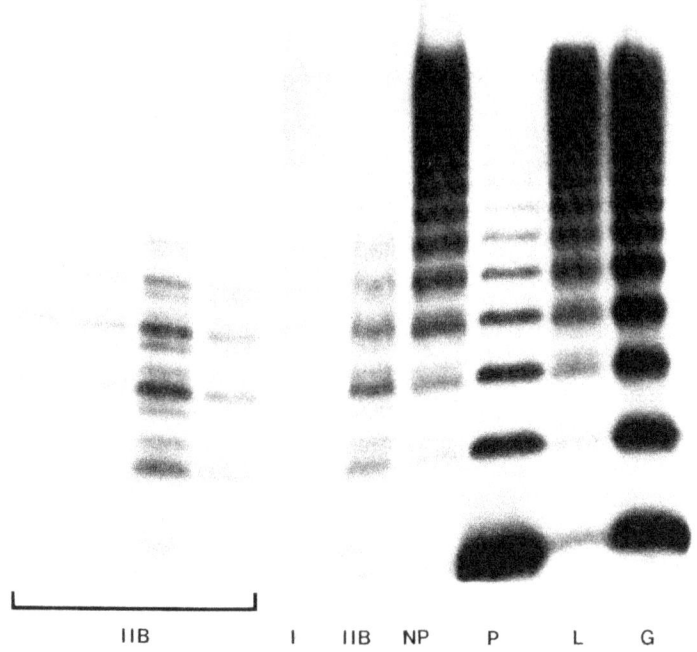

FIGURE 5: SDS-agarose gel electrophoresis of human vWF. Normal plasma (NP) and plasma from patients with von Willebrand's disease type I, IIB, and IIC (patient P and G; L is the father of P and G) were electrophoresed in agarose gels. The vWF multimers were labeled by incubating the gel with 125I anti-vWF monoclonal antibody and then visualized by autoradiography. In type I disease, the pattern is typically normal, but the quantity of vWF is reduced. Type IIB is relatively deficient in the largest multimers. Type IIC lack "satellite" bands as seen in normal plasma. In addition larger multimers are deficient.

abnormal large and potent vWF species are responsible for intravascular platelet agglutination. Similarly, in the hemolytic uremic syndrome, a disorder that resembles TTP, aberrant vWF multimerization could contribute to the pathogenesis of this disorder (*Moake et al., 1984*).

CONCLUDING REMARKS

More than 10 years after the discovery that endothelial cells produce vWF, it has become clear that these cells do not simply release this hemostatic component in a constitutive way, but, in addition, may release mature and biologically active vWF immediately upon exposure to stimuli that are generated in response to vascular injury. The ability to combine molecular biologic and cell biologic techniques has provided information regarding the biosynthesis and secretion of vWF by endothelial cells and possibly also by platelets. Similarly, these techniques have provided insights into the mode of action of vWF at the molecular level and it is anticipated that the molecular basis underlying disorders associated with abnormal vWF levels or function will be understood in the near future in greater detail.

REFERENCES

Bloom, A.L., Giddings, J.C. and Wilks, C.J. Factor VIII on the vascular intima: possible importance in hemostasis and thrombosis. *Nature New Biol.* 8:53, 1973.

Bonthron, D.T., Handin, R.I., Kaufman, R.J., Wasley, L.C., Orr, E.C., Mitsock, L.M., Ewenstein, B., Loscalzo, J., Ginsburg, D. and Orkin, S.H. Structure of pre- pro- von Willebrand factor and its expression in heterologous cells. *Nature* 324:270, 1986.

Brinkhous, K.M., Sandberg, H., Garvis, J.B., Mattson, C., Palm, M., Griggs, T. and Read, M.S. Purified human factor VIII procoagulant protein: comparative hemostatic response after infusion into hemophilic and von Willebrand disease dogs. *Proc. Natl. Acad. Sci. U.S.A.* 82:8752, 1985.

Cheresh, D.A. Human endothelial cells synthesize and express an Arg-Gly-Asp- directed adhesion receptor involved in attachment to fibrinogen and von Willebrand factor. *Proc. Natl. Acad. Sci. U.S.A.* 84:6471, 1987.

De Groot, Ph.G., Gonsalves, M.D., Loesberg, C., van Buul-Wortelboer, M.F., van Aken, W.G. and van Mourik, J.A. Thrombin-induced release of von Willebrand factor from endothelial cells is mediated by phospholipid methylation. Prostacyclin synthesis is independent of phospholipid methylation. *J. Biol. Chem.* 259:13329, 1984.

Ewenstein, N.M., Warhol, M.J., Handin, R.I. and Pober, J.S. Composition of the von Willebrand factor storage organelle (Weibel-Palade body) isolated from cultured human umbilical vein endothelial cells. *J. Cell Biol.* 104:1423, 1987.

Fay, P.J., Kawai, Y., Ginsburg, D., Bonthron, D., Ohlsson-Wilhelm, B.M., Chavin, S.I., Abraham, G.N., Handin, R.I., Orkin, S.H., Montgomery, R.R. and Marder, V.J. Propolypeptide of von Willebrand factor circulates in blood and is identical to von Willebrand antigen II. *Science* 232:995, 1986.

Folkman, J., Haudenschild, C.C. and Zetter, B.R. Long-term culture of capillary endothelial cells. *Proc. Natl. Acad. Sci. U.S.A.* 76:5217, 1979.

Fretto, L.J., Fowler, W.E., McCaslin, D.R., Erickson, H.P. and McKee, P.A. Substructure of human von Willebrand factor. *J. Biol. Chem.* 261:15679, 1986.

Ginsburg, D., Konkle, B.A., Cox, G.J., Montgomery, R.R., Bockenstedt, P.L., Johnson, T.A. and Yang, A.Y. Molecular basis of human von Willebrand's disease: Analysis of platelet von Willebrand factor mRNA. *Proc. Natl. Acads. Sci. U.S.A.* 86:3723, 1989.

Hoyer, L.W., de los Santos, R.P. and Hoyer, J.R. Antihemophilic factor antigen. Localization in endothelial cells by immunofluorescence microscopy. *J. Clin. Invest.* 52:2737, 1973.

Jaffe, E.A., Hoyer, L.W. and Nachman, R.L. Synthesis of von Willebrand factor by cultured human endothelial cells. *Proc. Natl. Acad. Sci. U.S.A.* 71:1906, 1974.

Kelly, R.B. Pathway of protein secretion in eukaryotes. *Science* 230:25, 1985.

Levine, J.D., Harlan, J.M. and Harker, L.A. Thrombin-mediated release of factor VIII antigen from umbilical vein endothelial cells in culture. *Blood* 60:531, 1982.

Loesberg, C., Gonsalves, M.D., Zandbergen, J., Willems, Ch., van Aken, W.G., Stel, H.V., van Mourik, J.A. and de Groot, P.G. The effect of calcium on the secretion of factor VIII-related antigen by cultured human endothelial cells. *Biochim. Biophys. Acta* 763:160, 1983.

Marti, T., Rosselet, S.J., Titani, K. and Walsh, K.A. Identification of disulfide-bridged substructures within human von Willebrand factor. *Biochemistry* 26:8099, 1987.

Mayadas, T. and Wagner, D.D. In vitro multimerization of von Willebrand factor is triggered by low pH. *J. Biol. Chem.* 264:13497, 1989.

Mazurier, C., Jorieux S., Delobel, J., Goudemand, M. (1990). A new von Willebrand factor (vWF) defect in a patient with factor VIII (FVIII) deficiency but with normal levels and multimeric patterns of both plasma and platelet vWF. Characterization of abnormal vWF/FVIII interaction. *Blood* 75:20, 1990.

McNiff, J.M. and Gil, J. Secretion of Weibel-Palade bodies observed in extra-alveolar vessels of rabbit lung. *J. Appl. Physiol. Respir. Environ. Exercise Physiol.* 54:1284, 1983.

Mertens, K, van Wijngaarden, A. and Bertina, R.M. The role of factor VIII in the activation of human blood coagulation factor X by activated factor IX. *Thromb. Haemostas.* 54:654, 1985.

Moake, J.L., Byrnes, J.J., Troll, J.H., Rudy, C.K., Weinstein, M.J., Colannino, N.M. and Hong, S.L. Abnormal VIII:von Willebrand factor patterns in the plasma of patients with the hemolytic-uremic syndrome. *Blood* 64:592, 1984.

Moake, J.L., Rudy, C.K., Troll, J.H., Weinstein, M.J., Colannino, N.N., Azocar, J., Seder, H.R., Hong, M.J. S.L. and Deykin, D. Unusually large plasma factor VIII: von Willebrand factor multimers in chronic relapsing thrombotic thrombocytopenic purpura. *N. Eng. J. Med.* 307:1432, 1982.

Nachman, R.L., Levine, R. and Jaffe, E.A. Synthesis of factor VIII antigen by cultured guinea pig megakaryocytes. *J. Clin. Invest.* 60:914, 1977.

Nishino M., Girma, J.P., Tothsdchild, C., Fressinaud, E. and Meyer, D. New variant of von Willebrand disease with defective binding to factor VIII. *Blood* 74:1591, 1989.

Reinders, J.H., de Groot, P.G., Gonsalves, M.D., Zandbergen, J., Loesberg, C. and van Mourik, J.A. Isolation of a storage and secretory organelle containing von Willebrand protein from cultured human endothelial cells. *Biochim. Biophys. Acta.* 804:361, 1984.

Reinders, J.H., de Groot, P.G., Sixma, J.J. and van Mourik, J.A. Storage and secretion of von Willebrand factor by endothelial cells. *Hemostasis* 18:246, 1988.

Reinders, J.H., de Groot, P.G., Daws, J., Hunter, N.R., van Heugten, H.A.A., Zandbergen, J., Gonsalves, M.D. and van Mourik, J.A. Comparison of secretion and subcellular localization of von Willebrand protein with that of thrombospondin and fibronectin in cultured human vascular endothelial cells. *Biochim. Biophys. Acta.* 884:306, 1985.

Sakariassen, K.S. Bolhuis, P.A. and Sixma, J.J. Human platelet adherence to artery subendothelium is mediated by factor VIII-von Willebrand factor bound to the subendothelium. *Nature* 279:636, 1979.

Shelton-Inloes, B.B., Titani, K. and Sadler, J.E. cDNA sequences for human von Willebrand factor reveal five types of repeated domains and five possible protein sequence polymorphisms. *Biochemistry* 25:3164, 1986.

Sporn, L.A., Chavin, S.I. and Marder, V.J. Biosynthesis of von Willebrand protein by human megakaryocytes. *J. Clin. Invest.* 76:1102, 1985.

Sporn, L.A., Marder, V.J. and Wagner, D.D. Inducible secretion of large biologically potent von Willebrand factor multimers. *Cell* 46:185, 1986.

Titani, K., Kumar, S., Takio, K., Ericsson, L.H., Wade, R.D. Ashida, K., Walsh, K.A., Chopek, M.W., Sadler, J.E.,and Fujikawa, K. Amino acid sequence of human von Willebrand factor. *Biochemistry* 25:3171, 1986.

Tschopp, T.B., Weiss, H.J. and Baumgartner H.J. Decreased adhesion of platelets to subendothelium in von Willebrand's disease. *J. Lab. Clin. Med.* 83:206, 1974.

Van Buul-Wortelboer, M.F., Brinkman, H.J.M., Reinders, J.H., van Aken, W.G. and van Mourik, J.A. Polar secretion of von Willebrand factor by endothelial cells. *Biochim. Biophys. Acta.* 1011:129, 1989.

Van de Ven, W.J.M., Voorberg, J., Fontijn, R., Pannekoek, H., Van den Ouweland, A.M.W., Van Duÿnhoven, H.L.P., roebroek, A.J.M. and Siezen, R.J> Furin is a subtilisin-like proprotein processing enzyme in higher enkaryotes. *Molec. Biol. Rep.* 14:265, 1990.

Van Dieijen, G., Tans, G., Rosing, J. and Hemker, H.C. The role of phospholipid and factor VIIIa in the activation of bovine factor X. *J. Biol. Chem.* 256:3433, 1981.

Verwey, C.L., Diergaarde, P., Hart, M. and Pannekoek, H. Full-length von Willebrand factor (vWF) encodes a highly repetitive protein, considerably larger than the mature vWF subunit. *EMBO J.* 5:1839, 1986.

Verwey, C.L., Hart, M. and Pannekoek, H. Expression of variant von Willebrand factor (vWF) cDNA in heterologous cells: requirement of the propolypeptide in vWF multimer assembly. *EMBO J.* 6:2885, 1987.

Verwey, C.L. Biosynthesis of human von Willebrand factor. *Hemostasis* 18:224, 1988.

Voorberg, J., Fontijn, R., van Mourik, J.A. and Pannekoek, H. Domains involved in multimer assembly of von Willebrand factor (vWF): multimerization is independent of dimerization. *EMBO J.* 9:797, 1990.

Voorberg, J., Fontijn, R., van Mourik, J.A. and Pannekoek, H. Assembly and routing of von Willebrand factor (vWF) variants: the requirements for disulfide-linked dimerization reside within the carboxy-terminal 151 amino-acids. *J. Cell Biol.* (in press).

Wagner, D.D., Oluisted, J.B. and Marder, V.J. Immunolocalization of von Willebrand factor protein in Weibel Palade bodies of human endothelial cells. *J. Cell Biol.* 95:355, 1982.

Wagner, D.D. and Marder, V.J. Biosynthesis of von Willebrand protein by human endothelial cells: processing steps and their intracellular localization. *J. Cell Biol.* 99:2123, 1984.

Weibel E.R. and Palade, G.E. New cytoplasmic components in arterial endothelia. *J. Cell Biol.* 23:101, 1964.

Weiss, H.J., Sussman, I.I., and Hoyer, L.W. Stabilization of factor VIII in plasma by the von Willebrand factor. Studies on posttransfusion and dissociated factor VIII and in patients with von Willebrand's disease. *J. Clin. Invest.* 60:390, 1977.

Wise, R.J., Pittman, D.D., Handin, R.I., Kaufman, R.J. and Orkin, S.H. The propolypeptide of von Willebrand factor independently mediates the assembly of von Willebrand multimers. *Cell* 52:229, 1988.

VII. MARKERS OF ENDOTHELIAL CELL INJURY AND REPAIR

MONITORING OF ENDOTHELIAL PLASMALEMMAL ECTOENZYME FUNCTION AS AN INDEX OF ENDOTHELIAL INJURY AND REPAIR

John D. Catravas

Department of Pharmacology and Toxicology
Medical College of Georgia
Augusta, Georgia 30912, U.S.A.

INTRODUCTION

A large number of physiologically important enzymes are distributed over the highly ciliated surface of endothelial cells and thus function as ectoenzymes. In this way, circulating substrates or inhibitors of the enzymes can interact without the expense in time and energy that would be required for the interaction with a cytoplasmic enzyme. For example, angiotensin converting enzyme (ACE) and 5′-nucleotidase (NCT) are distributed throughout the luminal endothelial cell surface, although not in identical fashion: ACE is expressed uniformly on the luminal endothelial surface, whereas NCT is primarily localized within the calveolae intracellularis (*Ryan et al., 1975; Ryan and Smith, 1971*). Nevertheless, both enzymes are present throughout the vasculature of the lung and can thus be considered as reflecting functions of the entire pulmonary circulation. ACE and NCT catalyze biologically important reactions. ACE catalyzes the conversion of the biologically inactive decapeptide angiotensin I to angiotensin II, an octapeptide with a wide spectrum of biological activity, including vasoconstriction, aldosterone release and stimulation of smooth muscle migration and proliferation. In addition, ACE deactivates the nonapeptide bradykinin, a compound with equally complex physiological actions, including vasodilation, prostaglandin release, participation in inflammatory processes, edema and pain. NCT catalyzes the dephosphorylation of 5′-AMP to adenosine, the latter possessing vasodilatory and antithrombogenic properties.

For several years, we and others have studied the function of these membrane-bound enzymes under physiologic conditions and have tried to learn how the behavior of these enzymes is affected by various lung pathologies. (*Catravas et al., 1988; Catravas and Gillis, 1981; Dobuler et al., 1982; Howell et al., 1988; Ryan and Catravas, 1991; Havill et al., 1989; McCormick et al., 1987; Riggs et al., 1988; Ryan, 1987*) Information to date suggests that endothelial ectoenzyme dysfunction is an early event in many types of endothelial and interstitial lung injury. Additional findings raise the intriguing hypothesis that monitoring of endothelial ectoenzyme function can be ultimately developed into a clinical procedure, useful in the early detection of vascular injury (*Morel et al., 1985; Gillis et al., 1986*).

METHODS

Many *in vivo* assays of endothelial ectoenzyme activity have employed modifications of the multiple indicator dilution technique, originally introduced for the measurement of cardiac output. Briefly, a radiolabelled substrate with or without a reference indicator of vascular space (e.g., indocyanine green, ^{125}I-albumin, etc.) is rapidly injected into a central vein. Simultaneously, arterial blood is withdrawn into a high-speed fraction collector (usually advancing at the rate of >1 sample tube/sec) for the duration of a single transpulmonary passage (usually around 10 sec). Later, the radioactivity associated with the surviving substrate, formed product and reference indicator (or optical density in the case of a dye like indocyanine green) is quantified in each sample. This methodology allows for monitoring of very rapid interactions between the substrate and the endothelial - bound enzyme, thus minimizing the contribution from the corresponding soluble (plasma) enzyme, provided of course, that the serum enzyme is in low concentrations relative to the bound enzyme. This is the case for ACE (*Catravas and Gillis, 1981*), NCT (*Catravas and White, 1984; Ryan et al., 1971*) and aminopeptidase P (*Chen et al., 1991*). Detailed descriptions of this procedure have been published previously (*Catravas, 1986; Catravas and White, 1984*).

In the middle 1970s, specific synthetic radiolabelled tripeptide substrates of ACE were first introduced and were quickly employed in assays *in vivo*, because, unlike their natural counterparts - angiotensin I and bradykinin - they could be quantified easily and rapidly (*Chung et al., 1986*). Furthermore, their affinity for ACE was such that measurable hydrolysis occurred during a single transpulmonary passage, *in vivo*. The first and by far the most frequently used synthetic ACE substrate is ^{3}H-benzoyl-Phe-Ala-Pro; others include benzoyl-Ala-Gly-Pro (*Catravas and Ryan, 1986*), benzoyl-Phe-Gly-Pro and benzoyl-Phe-His Leu. There is a wide range of reactivity among the four substrates. However, they are all highly specific for ACE and their rates of hydrolysis range from 20-85% during a single passage through the pulmonary circulation of various laboratory animals. The radiolabelled substrate benzoyl-Gly-His-Leu (or Hip-His-Leu) has been widely utilized in *in vitro* assays, however, because of its relatively low reactivity with ACE (less than 5% conversion during a transpulmonary passage), it is of limited usefulness in *in vivo* assays.

One or more ectoenzymes can be assayed simultaneously. For example, both ACE and NCT activities can be measured with a single procedure, as described above, utilizing ^{14}C labelled 5´-AMP, the natural substrate of NCT, and a tritiated ACE substrate, with or without an intravascular indicator (*Catravas and White, 1984*). The inclusion of the intravascular reference indicator is optional, since, during a single passage through the pulmonary circulation, the substrates themselves are restricted within the vascular spaces and thus the arterial blood concentration of total ^{3}H or ^{14}C can be used for the calculation of pulmonary blood flow (*Catravas and Gillis, 1981; Catravas and White, 1984*).

In addition to - or instead of - substrate utilization measurements, transpulmonary binding of radiolabelled ectoenzyme inhibitors can also provide useful information. The ACE inhibitors captopril and RAC-X-65 have been studied thus far. Binding of the inhibitor to the endothelial ectoenzyme is inferred from the difference in the effluent plasma concentrations between the inhibitor and the reference indicator (*Catravas et al., 1990; Ryan and Catravas, 1991; Turrin et al., 1986*).

When trace amounts of substrate or inhibitor are injected, so that hydrolysis or binding, respectively, proceeds under first order reaction conditions (*Segel, 1975*), two parameters of enzyme function can be calculated, as follows.

For substrate hydrolysis, the first order reaction parameter is:

$$A_{max}/K_m = \dot{Q} * ln\,([S_o]/[S])$$

where $[S_o]$ and $[S]$ are the initial and final concentrations of the substrate in the arterial effluent plasma and $\dot{Q}$ is pulmonary plasma flow. Substrate concentrations are expressed in dpm/ml plasma or mol/ml plasma based on the specific radioactivity (dpm/mol) of each compound. $[S_o]$ is calculated as the total radioactivity per ml plasma (i.e., radioactivity associated with both surviving substrate and product), whereas $[S]$ is computed from the radioactivity associated with surviving substrate only. By definition, A_{max} is the product of available enzyme mass (E) and the constant of product formation, k_{cat}, as well as the product of V_{max} and microvascular plasma volume, with units of mass over time (e.g., mol/min) (*Catravas and White, 1984*).

Additionally, substrate utilization is expressed either as $ln([S_o]/[S])$ or as percent metabolism (%M) calculated as:

$$\%M = 100 * ([S_o]-[S])/[S_o]$$

For inhibitor binding, the second order reaction parameter is:

$$B_{max} = \dot{Q} * ln\,([I_o]/[I])$$

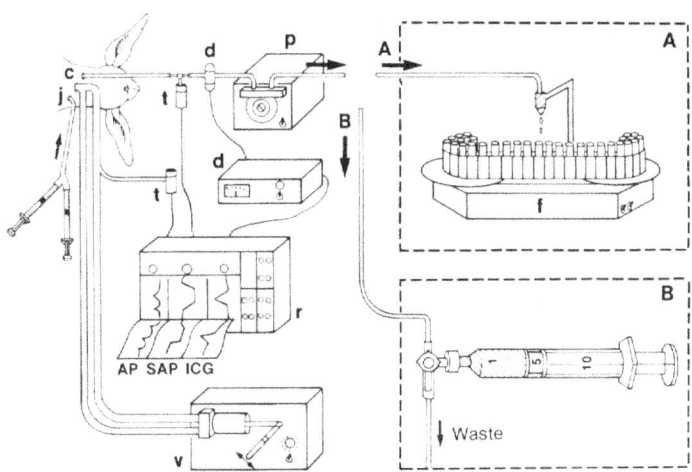

FIGURE 1: Diagram of the multiple sample and single sample methods used in determining endothelial enzyme function, *in vivo*. A bolus containing radiolabelled substrates and dye was injected into the central venous circulation, and blood samples were withdrawn from the arterial side with a peristaltic pump either into a fraction collector (A) or into a single disposable syringe (B). Blood flow was measured via a densitometer. The rest of the equipment is for the anesthesia and monitoring of vital signs. c = carotid artery catheter; j = jugular vein catheter; t = pressure transducer; d = densitometer cuvette and densitometer; p = peristaltic withdrawal pump; r = recorder; AP = airway pressure tracing; SAP = systemic arterial pressure tracing; ICG = indocyanine green (Cardiogreen) tracing; v = ventilator. (*With permission from Toivonen et al., 1988*)

where $[I_o]$ and $[I]$ are the initial and final concentrations of the free inhibitor in the arterial effluent plasma, calculated and expressed as described above for substrates. B_{max} is by definition the product of available enzyme mass (E) and enzyme - inhibitor association constant, k_1 (*Catravas et al., 1990*).

Similarly, inhibitor binding is expressed either as $ln\ ([I_o]/[I])$ or as percent binding (%B) calculated as:

$$\%B = 100 * ([I_o]-[I])/[I_o]$$

For both substrates and inhibitors, "available enzyme" reflects the amount of enzyme in the *perfused* microvasculature available to interact with the circulating substrates. Available enzyme is thus proportional to the perfused microvascular endothelial (luminal) surface area. Consequently, changes in A_{max}/K_m or B_{max} values reflect changes in microvascular surface area, as, for example, during microvascular recruitment or derecruitment, restrictive vascular disease, postnatal pulmonary development, etc. (*Catravas et al., 1990; Fanburg and Glazier, 1973; Harris et al., 1987; Pitt and Lister, 1984; Ryan, 1987; Toivonen and Catravas, 1987; Toivonen and Catravas, 1986; Toivonen and Catravas, in press*).

RESULTS

A schematic diagram of the experimental setup used to measure pulmonary microvascular ectoenzyme function in rabbits is shown in Figure 1. Panel A depicts the classical procedure, while in Panel B the modification to a single sample method is shown. The latter was attempted in an effort to simplify the technique by eliminating the requirement for a fraction collector as well as the processing of multiple blood samples, and thus making it more adaptable for use in the clinic. Comparison of data obtained by the two methods indicate similar accuracy and sensitivity to pathophysiologic stimuli (*Toivonen, et al., 1988*).

Examples of results obtained using the multiple sample method to evaluate angiotensin converting enzyme function in control rabbits are shown in Figure 2. In panels A and B, two

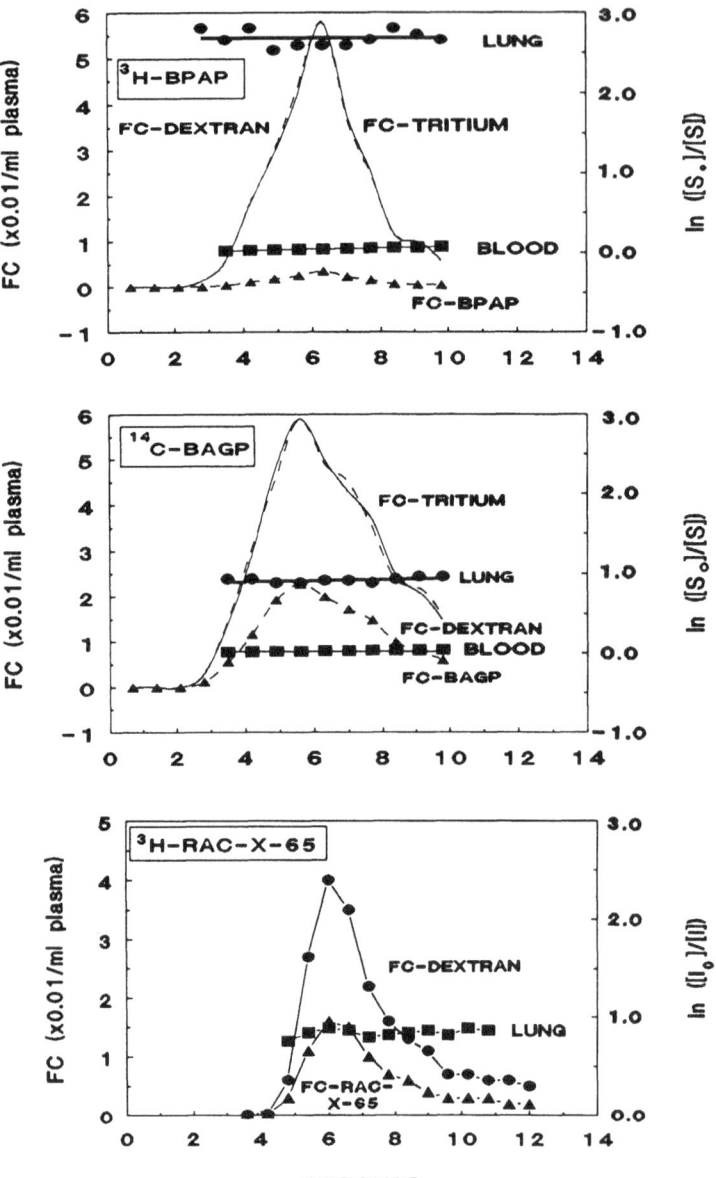

FIGURE 2: Single-pass transpulmonary metabolism of two ACE substrates and binding of an ACE inhibitor under first order reaction conditions in an anesthetized, rabbit. Top: A mixture of ^{3}H-benzoyl-Phe-Ala-Pro (BPAP, 0.1 nmol) and ^{14}C-dextran (intravascular indicator) was injected into the right atrium. Arterial plasma outflow concentration curves of the two isotopes (solid and dashed, respectively) are congruent indicating that the ACE substrate remained within the vascular space. In a separate experiment *in vitro*, BPAP hydrolysis in blood (solid squares) was measured and found to be minimal. Thus, most of the observed substrate metabolism was due to endothelial-bound (lung) ACE. Middle: Similar experiment with a different ACE substrate, ^{14}C-benzoyl-Ala-Gly-Pro (BAGP) and ^{3}H-dextran. Notice the lower affinity of BAGP for ACE, compared to BPAP, as reflected in lower $\ln([S_o]/[S])$ values (filled circles) and higher amounts of substrate surviving a transpulmonary passage (filled triangles). Bottom: A similar experiment in which a specific ACE inhibitor, ^{3}H-RAC-X-65 (0.08 nmol) and ^{14}C-dextran were injected. This dose of inhibitor binds less than 2% of pulmonary ACE and does not significantly reduce substrate metabolism. (*With permission from Ryan and Catravas, 1991*)

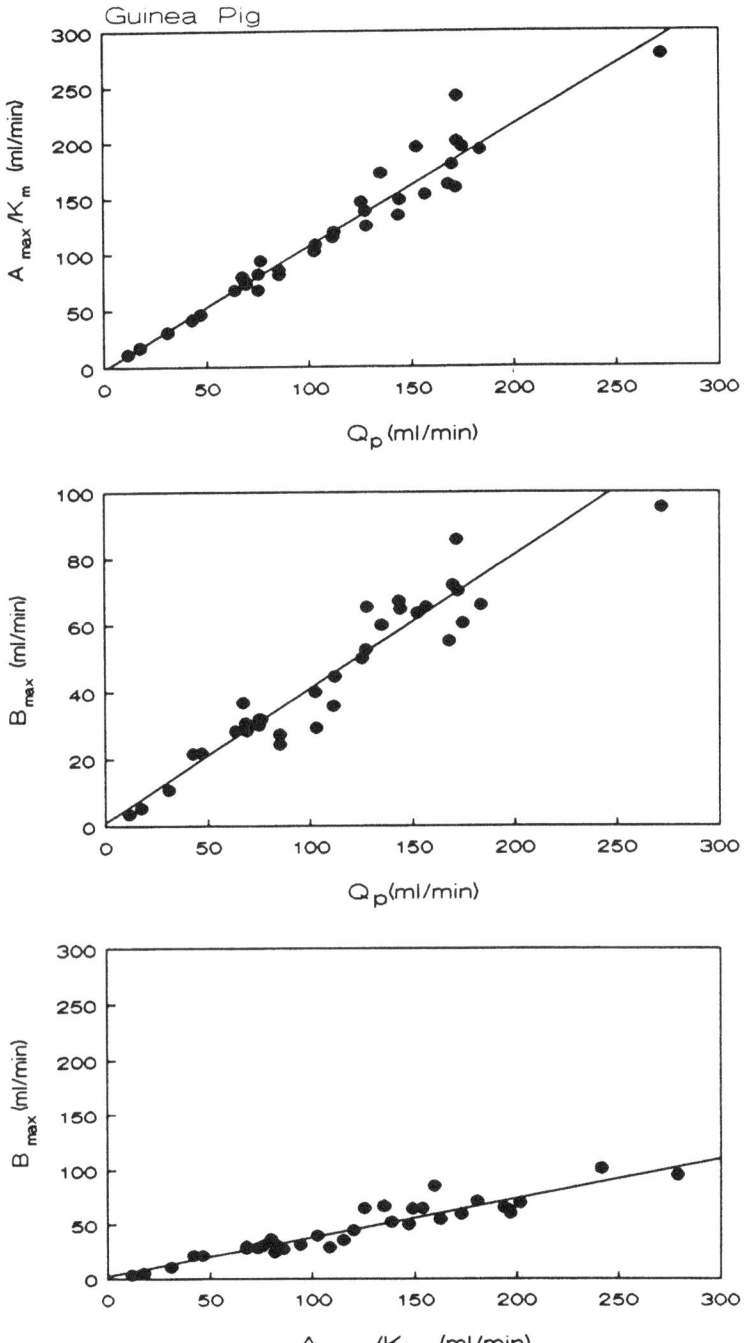

FIGURE 3: Positive correlation of A_{max}/K_m, calculated from BAGP hydrolysis, or B_{max}, calculated from RAC-X-65 binding, (both parameters proportional to endothelial ACE mass and hence to perfused microvascular surface area) to each other and to pulmonary blood or plasma flow. Results of individual experiments using anesthetized guinea pigs in which blood flow varied spontaneously. Linear regression analysis (n = 34): Q_p v. A_{max}/K_m, y = 1.089(x) + 0.01, r = 0.975; Q_p v. B_{max}, y = 0.40(x) + 0.02, r = 0.93; B_{max} v. A_{max}/K_m, y = 0.359(x) + 0.046, r = 0.937; $p < 0.001$ for each comparison. (*With permission from Ryan and Catravas, 1991*)

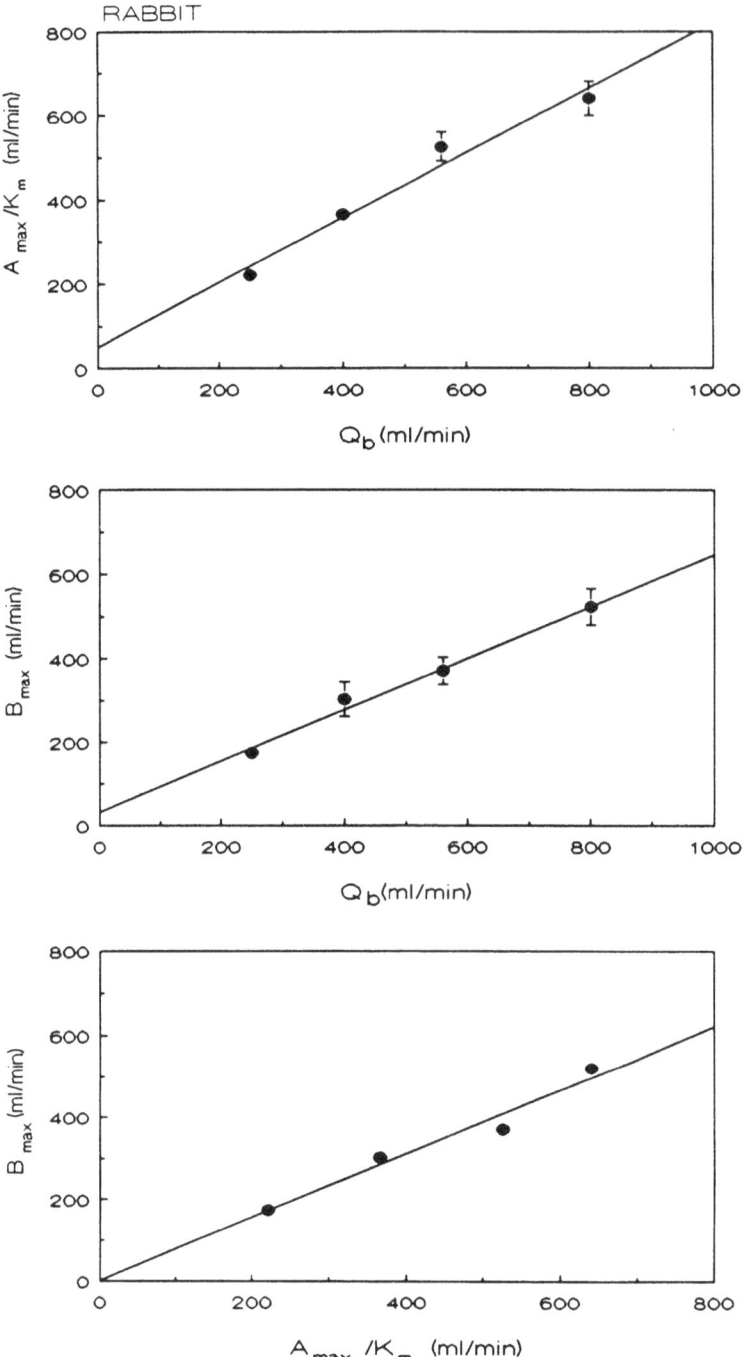

FIGURE 4: Similar studies as in Figure 3. Results are (mean ± S.E.) from 8 experiments in anesthetized rabbits placed on total heart bypass and tested at four fixed blood flow values. Linear regression analysis (n=4): Q_b v. A_{max}/K_m, y=0.768(x) + 52.34, r=0.98; Q_b v. B_{max}, y=0.614(x) + 34.33, 4=0.99; B_{max} v. A_{max}/K_m, y=0.776(x) + 2.44, r + 0.981; p<0.001 for each comparison (*With permission from Ryan and Catravas, 1991*).

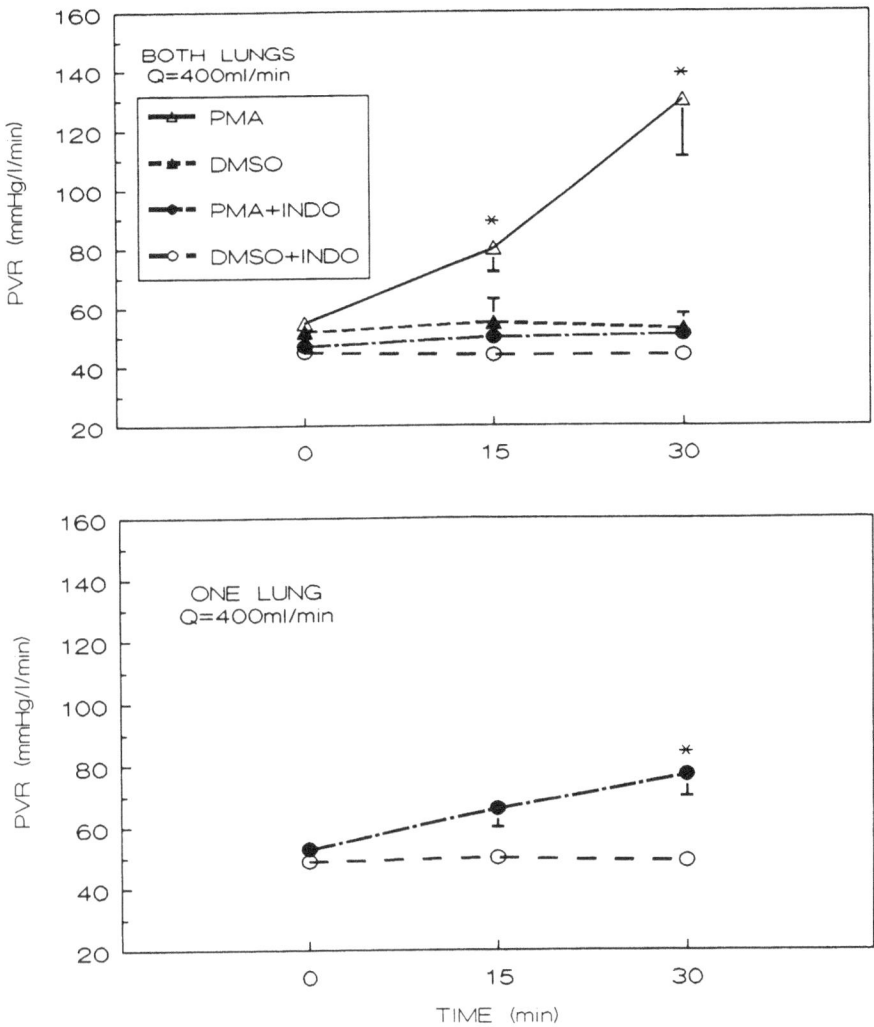

FIGURE 5: Effects of indomethacin (1.5 mg/kg) on the pulmonary vascular resistance response to phorbol myristate acetate (PMA 15μg) in rabbits with partially recruited lungs (Q =400ml/min through the entire lung; upper panel) and with fully recruited lungs (Q =640ml/min through the left lobe only; lower panel). (*From Chen and Catravas, In Press*)

different ACE substrates, BPAP (^{3}H-benzoyl-Phe-Ala-Pro) and BAGP (^{14}C-benzoyl-Ala-Gly-Pro), were utilized, respectively. The most remarkable difference between these two substrates is their relative reactivity: BAGP is about half as active as BPAP and consequently less susceptible to computational errors, as previously reported (*Toivonen and Catravas, 1987*). Under first order reaction conditions, single transpulmonary metabolism of BPAP by ACE is approximately 85%, whereas that for BAGP is 60%. Since %M values are not directly proportional to enzyme activity, this illustrates why the difference in %M is much smaller than the difference in the reactivity between the two substrates, when expressed, for example, as $ln\ [S_o]/[S]$ (see Figure 2 and Table 2). Interestingly, %M of e.g., BAPGP is very similar among dogs, rabbits, mini pigs and cats (*Chen et al., 1991*), suggesting that capillaries of different size animals may be of comparable size and length. This figure also shows the relatively insignificant contribution of plasma ACE to the transpulmonary metabolism of the two compounds: in separate experiments, substrate hydrolysis in blood, *in vitro*, over comparable times was only a small fraction of the metabolism observed *in vivo*. This is also the case for NCT and its substrates and reflects the low concentration of the plasma enzymes relative to the endothelium-bound ACE and NCT. For ACE, for example, the membrane bound enzyme is estimated to be 80-100 times more concentrated than that in plasma (*Ryan, 1983*).

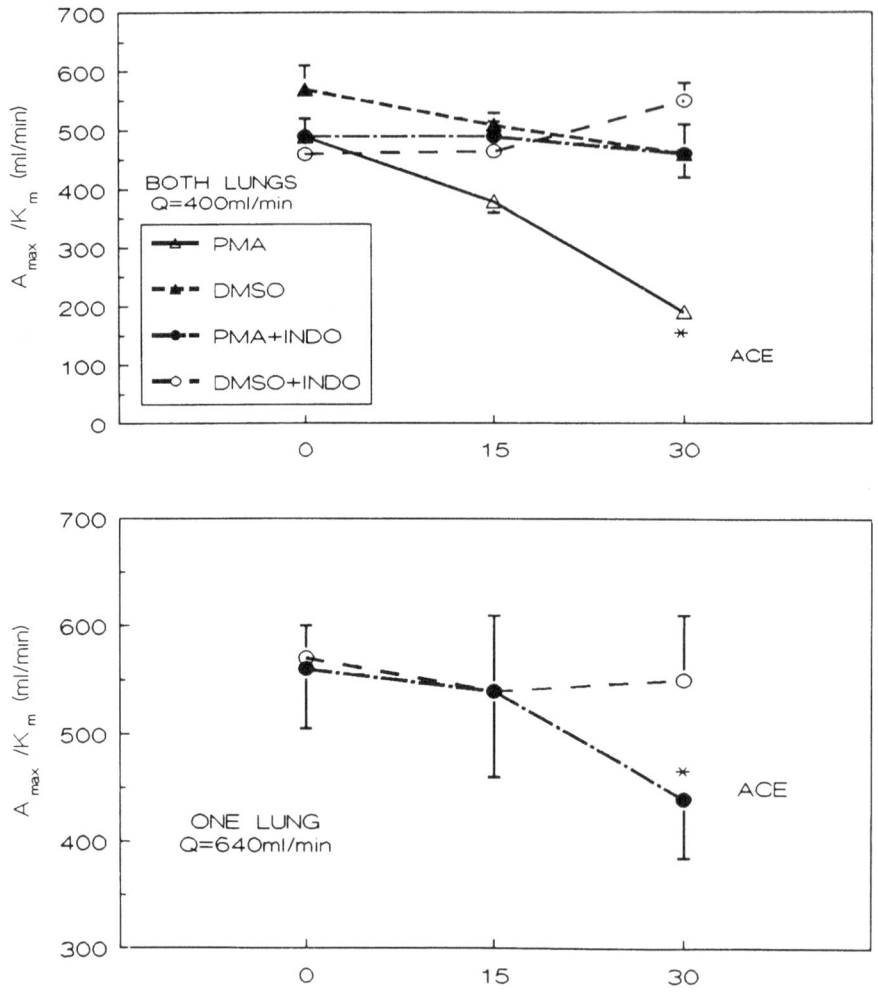

FIGURE 6: Effects of indomethacin (1.5 mg/kg) on PMA-induced changes in the modified first order parameter, A_{max}/K_m of ACE for BPAP in rabbits with partially recruited lungs (Q_b = 400 ml/min through the entire lung; upper panel), or fully recruited lungs (Q_b = 640 ml/min through the left lobe only; lower panel). *(From Chen and Catravas, In Press)*

To confirm experimentally that, since A_{max}/K_m and B_{max} are directly proportional to enzyme mass (E), they do indeed reflect the surface area of the perfused microvascular endothelium, we investigated the changes in these parameters in response to increasing pulmonary blood flow (*Catravas and Ryan, 1986; Toivonen and Catravas, in press*). As shown in Figure 3, in anesthetized guinea pigs and in Figure 4 in anesthetized rabbits, A_{max}/K_m of BAGP for ACE and B_{max} of RAC-X-65 for ACE increased linearly with increasing blood flow, suggesting recruitment of new microvessels. This was further confirmed in additional experiments which measured single pass transpulmonary hydrolysis of BPAP by ACE in rabbits placed on total heart bypass so that pulmonary blood flow was controlled and was stepwise elevated between 50% and 250% of normal cardiac output (*Toivonen and Catravas, 1991*).

To examine whether agents or modalities which cause lung microvascular injury also affect ectoenzyme function, anesthetized rabbits placed on total heart bypass were administered phorbol myristate acetate, an agent believed to act by activating circulating neutrophils which subsequently are sequestered in the microvasculature and release toxic free radicals. As shown in Figure 5 (top panel), at pulmonary blood flow of 400 ml/min (approximately the normal cardiac output of a rabbit which corresponds to about 40% recruitment of the pulmonary microvasculature (*Toivonen and Catravas, in press*), PMA caused a time-dependent increase in pulmonary vascular resistance which was not observed in vehicle (DMSO) - treated controls and was prevented by prior administration of indomethacin. However, when full pulmonary microvascular recruitment was achieved by perfusing only the left lung at 640 ml/min blood flow (*Toivonen and Catravas, in press*; lower panel) indomethacin was no longer able to prevent the PMA-induced increase in PVR. Qualitatively similar effects of PMA on ACE function are presented in Figure 6, where in partly recruited lungs (top panel), PMA caused a decrease in the modified first order parameter, A_{max}/K_m, while no changes were observed in DMSO - treated controls or in following treatment with indomethacin. Again, when microvessels were fully recruited, indomethacin was unable to prevent the reduction in ACE activity.

DISCUSSION

Several laboratories have utilized procedures similar to those described herein in various animal models (*Dawson et al., 1989; Howell et al., 1988; Pitt and Lister, 1984; Ryan, 1987*) to demonstrate that routine *in vivo* assays are feasible, that enzyme function is altered early on under many, but not all (*Howell et al., 1988*), instances of pulmonary injury, and that certain parameters of ectoenzyme activity can provide useful information on the normal physiology of the pulmonary circulation, particularly on the relative - and perhaps even absolute - estimation of perfused microvascular lumenal surface area (*Toivonen and Catravas, in press*).

We have presented here evidence of decreased ACE activity in one model of lung injury: PMA, a model of acute lung injury commonly used to study the pathogenesis of the adult respiratory distress syndrome (ARDS). Other models studied by us and other investigators include hyperoxia (*Dobuler et al., 1982; Howell et al., 1988*), bleomycin (*Catravas, 1988*), irradiation to the chest, (*Catravas et al., 1988*) anaphylaxis (*Ryan, 1987*), etc. In most of these studies, enzyme dysfunction occurred early in time and suggested, if not a causal relationship with subsequent interstitial injury, at least the possibility that changes in endothelial ectoenzyme function could be utilized as early indices of the subsequent, and frequently irreversible, lung damage. In the past, others have reported promising results utilizing changes in endothelial transport functions as early indicators of pulmonary injury, employing similar techniques (*Morel, et al., 1985; Gillis et al., 1986*). Thus, it is conceivable that approaches such as these presented here could develop into clinically useful procedures which will contribute to the early diagnosis of vascular disease. This may be of considerable importance, as there is increasing awareness that a greater than previously thought number of diseases (e.g. pulmonary and systemic hypertension, atherosclerosis, scleroderma, etc.) or side effects of therapeutic procedures (e.g. cardiopulmonary bypass; irradiation; bleomycin) may involve endothelial dysfunction or injury.

In addition to studies of pathologic influences on endothelial enzyme function, *in vivo* assays under normal physiologic conditions have recently provided evidence (*Catravas and Ryan, 1986; Harris et al., 1987; Pitt and Lister, 1984; Pitt et al., 1987; Ryan, 1983; Toivonen and Catravas, 1987*) confirming earlier hypotheses (*Fanburg and Glazier, 1973*) that such measurements could be used to obtain at least relative estimations of perfused microvascular surface area. Also, as our data with PMA suggest, quantification of perfused microvascular surface area could help elucidate the mechanism of action of certain drugs. In the rabbit model presented here (Figures 5 and 6), indomethacin prevents the PMA-induced increase in pulmonary vascular resistance and decrease in A_{max}/K_m values, in partially recruited lungs only. This suggests that the "protective" action of indomethacin may involve diverting blood flow to previously unperfused (and hence uninjured) areas, rather than preventing or repairing cell dysfunction.

While most of the studies to date have focused on the hydrolysis of ACE substrates, other endothelial ectoenzymes, such as 5´-nucleotidase (*Catravas and White, 1984*), aminopeptidase P (*Chen et al., 1991*) or carbonic anhydrase (*Effros et al., 1980*) have been used with promising results. Simultaneous assay of more than one enzyme (see Methods) could help differentiate between specific and diffuse endothelial dysfunction. A still different approach involves estimating the single pass transpulmonary binding of specific enzyme inhibitors. It appears that the parameters %B, $ln \, ([I_o]/[I])$ and B_{max} provide qualitatively similar information with their corresponding terms for substrate hydrolysis (%M, $ln \, ([S_o]/[S])$ and A_{max}/K_m, respectively), with the potentially important difference that measurements of substrate hydrolysis under first order reaction conditions involve estimates of two rate constants: substrate-enzyme affinity (K_m) and product formation (k_{cat}), whereas inhibitor binding is limited to estimates of inhibitor-enzyme affinity (k_1). It is unclear at present whether pathophysiologic influences can affect the binding vs. the catalytic subunit of an endothelial ectoenzyme differently.

To date, significant advances have been achieved in the routine performance of *in vivo* assays of endothelial ectoenzyme function as well as in the interpretation of experimental results. During the next few years, two relevant areas will be most likely addressed: a) testing this knowledge in the clinic in order to find out whether it can be useful in the early diagnosis of vascular disease, and b) applying these experiences in gaining new insights into the physiology of the pulmonary circulation, in health and disease.

REFERENCES

Catravas, J.D. Michaelis-Menten kinetics of pulmonary endothelial-bound angiotensin converting enzyme in conscious rabbit. *Adv. Exp. Med.* 198:445, 1986.

Catravas, J.D. Pulmonary toxicity of anticancer drugs: alterations in endothelial cell function, in *Organ directed Toxicity of Anticancer Drugs*, Hackler, Lazo, and Tritton, Nijoh, Eds., 118:1988.

Catravas, J.D., Burch, S.E., Spurlock, B., and Mills, L.R. Early effects of ionizing radiation on pulmonary endothelial angiotensin converting enzyme and 5´-nucleotidase, *in vivo*. *Toxicol. Appl. Pharmacol.* 94:342, 1988.

Catravas, J.D. and Gillis, C.N. Metabolism of ^{3}H-Benzoyl-Phe-Ala-Pro by pulmonary angiotensin converting enzyme *in vivo*, effects of bradykinin, SQ14225 or acute hypoxia, *J. Pharmacol. Exp. Ther.*, 217:263, 1981.

Catravas, J.D. and Ryan, J.W. ^{3}H-Benzoyl-ala-gly-pro: a new synthetic substrate of pulmonary endothelial angiotensin converting enzyme *in vivo*. *Fed. Proc.* 45:554, 1986.

Catravas, J.D., Ryan, J.W., Quinn, N.E. and Anthony, B.L. *In vivo*, determination of lung kininase II inhibition by angiotensin converting enzyme inhibitors. *Br. J. Pharmacol.* 101:121-127, 1990.

Catravas, J.D. and White, R.E. Kinetics of pulmonary angiotensin converting enzyme and 5´-nucleotidase, *in vivo*. *J. Appl. Physiol.* 57:1173, 1984.

Chen, X-L., Orfanos, S.E. and Catravas, J.D. Effects of indomethacin on PMA-induced endothelial enzyme dysfunction, *in vivo*. *Amer. J. Physiol.* (In Press)

Chen, X-L., Orphanos, S.E., Ryan, J.W., Chung, A.Y.K. and Catravas, J.D. Species variation in pulmonary endothelial aminopeptidase P activity. *J.Pharmacol. Exp. Ther.*, 1991.

Chung, A.Y.K., Ryan, J.W., Ryan, J.P.A., and Ryan, U.S. Radiolabelled substrates for angiotensin converting enzyme. *Adv. Exp. Med. Biol.* 198A:427, 1986.

Dawson, C.A., Bongard, R.D., Rickaby, D.A., Linehan, J.H. and Roerig, D.L. Effect of transit time on metabolism on a pulmonary endothelial enzyme substrate. *Am. J. Physiol.* 257:H853, 1989.

Dobuler, K.J., Catravas, J.D. and Gillis, C.N. Early detection of oxygen-induced lung injury in conscious rabbits: reduced activity of angiotensin converting enzyme and removal of 5-hydroxytrptamine. *Am. Rev. Respir. Dis.*, 126:534, 1982.

Effros, R.M., Shapiro, L. and Silverman, P. Carbonic enhydrase activity of rabbit lungs. *J. Appl. Physiol. 49:589, 1980.*

Fanburg, B.L. and Glazier, J.B. Conversion of angiotensin I to angiotensin II in the isolated perfused dog lung. *J. Appl. Physiol.* 35:325, 1973.

Gillis, C.N., Pitt, B.R., Wiedemann, H.P., Hammond, G.L. Depressed prostaglandin E_1 and 5-hydroxytryptamine removal in patients with adult respiratory distress syndrome. *Am Rev. Respir. Dis.* 134:739-744, 1986.

Harris, T.R., Pou, N.A., Wright, P.E. and Bernard, G.R. Norepinephrine uptake measures changes in lung capillary surface during E coli endotoxemia in sheep. *Fed. Proc.* 45:1099, 1987.

Havill, A.M., Riggs, D., Pitt, B.R. and Gillis, C.N. Resolution of impaired pulmonary function and pulmonary hypertension after phorbol ester administration in rabbits. *Am. Rev. Respir. Dis.* 140:782, 1989.

Howell, R.W., Hansen-Flaschen, J.H. and Wheeldon, E.B. Pulmonary angiotensin-converting enzyme activity in the oxygen-toxic sheep. *Am. Rev. Respir. Dis.* 138:160, 1988.

McCormick, J.R., Chrzanowski, R., Andrieni, J. and Catravas, J.D. Early pulmonary endothelial dysfunction after intravenous administration of phorbol myristate acetate (PMA) in rabbits. *J. Appl. Physiol.* 63:1972, 1987.

Morel, D.L., Dargent, F., Bachmann, M., Suter, B.M. and Junod, A.F. Pulmonary extraction of serotonin and propranolol in patients with adult respiratory distress syndrome. *Am. Rev. Respir. Dis.* 132:479, 1985.

Pitt, B.R. and Lister, G. Kinetics of pulmonary angiotensin converting enzyme activity in conscious developing lambs. *J. Appl. Physiol.* 57:1158, 1984.

Pitt, B.R., Lister, G., Davies, P. and Reid, L. Correlation of pulmonary ACE activity and capillary surface area during postnatal development. *J. Appl. Physiol.* 62:2031-2041, 1987.

Riggs, D., Havill, A.M., Pitt, B.R. and Gillis, C.N. Pulmonary angiotensin-converting enzyme kinetics after acute lung injury in the rabbit. *J. Appl. Physiol.* 64:2508, 1988.

Ryan, J.W. Assay of peptidase and protease enzyme, *in vivo. Biochem. Pharmacol.* 32:2127, 1983.

Ryan, J.W. Assay of pulmonary endothelial surface enzymes *in vivo*, in *Lung Biology in Health and Disease*. Lenfant, C., E., Marcel Dekker, New York 161, 1987.

Ryan, J.W. and Catravas, J.D. Angiotensin converting enzyme as an indicator of pulmonary microvascular function. *In Focus on pulmonary pharmacology and toxicology,* ed. by M. Hollinger, CRC Press Inc., Boca Raton, pp. 183-210, 1991.

Ryan, J.W., Ryan, U.S., Schultz, D.R., Whitaker, C., Chung, A., and Dorer, F.E. Subcellular localization of pulmonary angiotensin converting enzyme (kininase II). *Biochem. J.*, 146:497, 1975.

Ryan, J.W. and Smith, U.S. Metabolism of adenosine-5´-monophosphate during circulation through the lungs. *Trans. Assoc. Am. Physicians* 84:297-306, 1971.

Segel, U.G., *Enzyme Kinetics*, John Wiley & Sons, New York. 54 and 474, 1975.

Toivonen, H.J. and Catravas, J.D. Effect of acid-base imbalance on pulmonary angiotensin converting enzyme, *in vivo. J. Appl. Physiol.* 63:1629, 1987.

Toivonen, H.J. and Catravas, J.D. Effects of airway pressure on pulmonary endothelial angiotensin converting enzyme function, *in vivo. J. Appl. Physiol.* 61:1041, 1986.

Toivonen, H.J. and Catravas, J.D. Effects of lung blood flow on converting enzyme kinetics: evidence for microvascular recruitment. *J. Appl. Physiol. (In Press)*.

Toivonen, H.J. Makari, N. and Catravas, J.D. Monitoring of pulmonary enzyme function *in vivo*: an animal model for a simplified procedure. *Anesthesiology*, 68:44, 1988.

Turrin, M., Pitt, B.R., Ryan, J.W., Chung, A.Y.K., Clark, M.B. and Gillis, C.N. Uptake of N[1(S)-carboxy-(4-OH-3-^{125}I-phenyl)ethyl]-L-Ala-L-Pro, an inhibitor of angiotensin-converting enzyme by rabbit lungs *in situ. J. Pharmacol. Exp. Ther.* 238:14, 1986.

PATHOPHYSIOLOGICAL SIGNIFICANCE OF ENDOTHELIAL CELL INTEGRINS

Jan A. van Mourik, Jacques G. Giltay
and Albert E.G. Kr. von dem Borne

Central Laboratory of the Netherlands Red Cross
Blood Transfusion Service
Amsterdam, The Netherlands

INTRODUCTION

At the outer surface of the endothelial cell plasma membrane a number of structurally closely related glycoprotein complexes ("integrins") are exposed that serve an important role in mediating the anchorage of the endothelium to extracellular matrix proteins. These surface receptors have also in common that they serve as recognition sites for multivalent matrix proteins such as fibronectin, vitronectin, collagens or laminin. (*Ruoslahti and Pierschbacher, 1987; Ruoshlati, 1988*). Defective functioning of these surface receptors, e.g. due to structural defects or to antibody-mediated dysfunction, may, therefore, affect the integrity of the vessel wall. Although the structure and function of integrins produced by endothelial cells and a variety of other cell types have now been described in detail and their similarities in terms of structure and mode of action have been appreciated (*Buck and Horwitz, 1987; Ruoshlati, 1988*), insight into the pathophysiology of disorders associated with molecular defects or dysfunction of integrins is limited. Because of the wide cellular distribution and similarities in both structure and function, one would expect that a genetically determined defect of those integrins that are under the same genetic control affects the integrity of a variety of cell types. So far this has not been found. Only when the expression of certain integrins is cell-specific, as is the case with the platelet integrin glycoprotein (GP) IIb/IIIa, or the surface receptors LFA-1, Mac-1, and p150/95, (members of the integrin family which are expressed by leukocytes only), isolated, cell-specific disorders associated with a defective integrin function might be expected. Indeed, many cases of isolated platelet- and leukocyte disorders due to integrin deficiency have been identified. (*Anderson and Springer, 1987, Clemetson and Luscher, 1988*). Acquired integrin dysfunction, for instance, due to the formation of integrin- specific antibodies could be of broader clinical significance. For instance, several members of the integrin family show marked antigenic polymorphism, and one might expect that an immunogenic response elicited by these polymorphic determinants will lead to systemic integrin dysfunction. However, similar to the genetically determined integrin defects, such immune-mediated disorders seem in general restricted to a single cell-type (e.g. platelets; *von dem Borne and Ouwehand, 1989*, or leukocytes; *Pischel et al., 1987*). Here we wish to provide a picture of our current knowledge of genetically determined and acquired functional abnormalities of integrins expressed by vascular endothelial cells.

STRUCTURAL AND FUNCTIONAL FEATURES OF INTEGRINS

Integrins comprise a family of structurally homologous surface receptors, consisting of non-covalently linked alpha- and beta subunits, which mediate cell-cell and cell-matrix interactions. The integrin family is divided into three subfamilies; the VLA protein family

TABLE I: Related integrins on endothelial cells and platelets

Subfamily	Endothelial cells		Platelets		Ligands
	Member	subunits	Member	Subunits	
VLA	VLA-2	a2/b1	GP Ia/IIa	a2/b1	Collagens
	VLA-5	a5/b1	GP Ic*/IIa	a5/b1	Fibronectin
	VLA-6	a6/b1	GP Ic/IIa	a6/b1	Laminin
Cytoadhesin	Vitronectin receptor	aV/b3	Vitronectin receptor	aV/b3	Vitronectin
			GP IIb/IIIa	aIIb/IIIa	Fibrinogen Fibronectin v Willebrand factor

* The alpha-5 and alpha-6 subunit differ slightly in electrophoretic mobility

(Hemler et al., 1987), the Leu-Cam proteins (Springer et al., 1987), and the cytoadhesins (Ginsberg et al., 1988). These subfamilies are characterized by a common beta subunit (beta-1, beta-2 and beta-3). Each beta subunit can associate with a number of structurally related alpha subunits. Recently, several other beta subunits have been identified that may form complexes with previously described alpha subunits. So far, eleven alpha subunits and seven beta subunits have been reported and may combine to form sixteen different complexes since the same alpha subunit may associate with more than one beta subunit (Ruoslahti and Giancotti, 1989). Many integrins bind to multivalent extracellular matrix proteins, including collagens, fibronectin, laminin and vitronectin, and thereby mediate cell- extracellular matrix interactions. Some integrins, such as LFA-1 mediate cell- cell interactions through the binding of cellular ligands such as the endothelial membrane molecules ICAM-1 or ICAM-2 (Staunton et al., 1989).

CELLULAR DISTRIBUTION OF INTEGRINS

The pattern of integrin expression varies between cell types. Most cells express at least two distinct integrin subfamilies. Some integrins are clearly cell-specific. For instance, the cytoadhesin GP IIb/IIIa (a beta-3 integrin) is only expressed by platelets and megakaryocytes (Ginsberg et al., 1988), whereas the leu-CAM subfamily is exclusively expressed by leukocytes. On the other hand, the fibronectin- and vitronectin receptor (membrane receptors belonging to the VLA- subfamily) are widely distributed among cell types. Two of the integrin subfamilies have been localized on endothelial cells which are also shared by blood platelets (Table I). Thus far, four members of these subfamilies have been identified on endothelial cells. Three of them belong to the VLA-subfamily, the other integrin, the vitronectin receptor, is a member of the cytoadhesin subfamily.

POLYMORPHISM OF ENDOTHELIAL AND PLATELET INTEGRINS

Several integrins are genetically polymorphic glycoproteins and can express one or more alloantigens. These alloantigenic structures are integral parts of the constituent subunits of the integrins and may be immunogenic. The majority of alloantigens carried by integrins have been identified on platelets. For instance, both the platelet GP IIb/IIIa- and the GP Ia/IIa

TABLE II: Distribution and molecular basis of integrin-associated alloantigens

Antigen	Antigen location		Polymorphism		Frequency (%)
	Platelets	Endothelial Cells	Nucleotide	amino acid	
Zwa (PlA1)	GP IIIa	beta-3*	T (CTG)	leu 33	97
Zwb (PlA2)	GP IIIa	beta-3	C (CCG)	Pro 33	27
Baka (Leka)	GP IIb	ND#	T (ATC)	Ile 843	91
Bakb (Lekb)	GP IIb	ND	G (AGC)	Ser 843	60
Yuka (Pena)	GP IIIa	beta-3			02
Yukb (Penb)	GP IIIa	beta-3			99
Bra	GP Ia/IIa	VLA-2			20
Brb	GP Ia/IIa	VLA-2			99

* Also detected on fibroblasts and smooth muscle cells.
# Not detectable.

complex are polymorphic in nature (reviewed by *von dem Borne and Ouwehand, 1989*). Both the GP IIb- and GP IIIa subunit of the GP IIb/IIIa complex are known to bear a number of clinically important alloantigenic determinants that are responsible for eliciting the immune response in two clinical syndromes, neonatal alloimmune thrombocytopenic purpura and posttransfusion purpura (*reviewed by Shulman and Jordan, 1987*). In posttransfusion purpura, antibody formation occurs following transfusion and exposure to antigenic structures not present on the platelets of the recipient. Similarly, in neonatal alloimmune thrombocytopenic purpura, maternal sensitization occurs following the exposure of paternal antigens on the platelets of the fetus. In both cases, antibody formation may lead to severe thrombocytopenia or platelet dysfunction. The alloantigen system most frequently implicated in these disorders is Zw (or PlA). There are two serologically defined allelic forms of the Zw alloantigen, Zwa (PlA1) and Zwb (or PlA2), both of which have been localized to GP IIIa, the beta-subunit of the GP IIb/IIIa complex. Similarly, the alloantigen system Bak (or Lek) is associated with GP IIb and the Br system located to the platelet GP Ia/IIa complex (Table II).

Recent studies on the structural features of the GP IIIa molecule responsible for Zwa (PlA) polymorphism revealed a C - T polymorphism at base 196 of the GP IIIa cDNA (*Newman et al., 1989*). This single base change results in a leucine/proline polymorphism at amino acid residue 33 from the amino-terminus of GP IIIa and is likely to impart significant differences in the conformation of these two allelic forms of the GP IIIa molecule and, hence, the expression of the alloantigenic determinant (*Kolodziej et al., 1990*). Similarly, the alloantigen system Bak is associated with a isoleucine/serine polymorphism at residue 843 of (mature) GP IIb (*Lyman et al., 1989*).

The demonstration that certain alloantigens are carried by integrins and the discovery of the structural homology of the platelet integrins with endothelial cell adhesion receptors led to the observation that endothelial cells also express alloantigens (*Leeksma et al., 1987; Giltay et al., 1988*). Immunochemical analysis revealed that only the alloantigens that are carried by GP IIIa, are also carried by the endothelial counterpart (i.e. the beta-chain of the vitronectin receptor). Similarly, other cell types, such as smooth muscle cells and fibroblasts, express alloantigens associated with the beta-chain of their cytoadhesin (*Giltay et al. 1989*). Alloantigens known to reside on platelet GP IIb (e.g. Bak or Lek), were not found on endothelial cells (*Giltay et al., 1988*), most likely because of substantial structural differences between platelet GP IIb and the alpha-chain of the vitronectin receptor.

It is clear now that the cytoadhesins expressed by various cell types share distinct genetic markers of the beta-chain of the cytoadhesins, such as the Zw- system and also the Yuk system, another alloantigen system located in the beta- chain of the cytoadhesins (*Giltay et al., 1988*). Recently, we have shown that an alloantigen carried by the platelet GP Ia/IIa complex (or VLA-2), the Br- antigen, is expressed by the endothelial VLA-2 receptor as well (*Giltay et al., 1990*). Similarly, an antigenic polymorphism of the alpha-chain has been reported (*Pischel et al., 1987*).

As briefly outlined above, the polymorphic nature of integrins is also of clinical significance. For instance, it is well documented that platelet alloimmunization may result in severe thrombocytopenia. Since alloantibodies directed to platelet integrins may crossreact with related surface receptors expressed by other cell types, one might expect that platelet alloimmunization may affect the function and integrity of these cells, including vascular endothelial cells, as well. So far there is no evidence in support of this view. It has been shown that the expression of certain alloantigens is conformation- induced (*Flug et al., 1990*). It seems likely that the conformation may vary between cell types and, hence, the antibody binding capacity of the putative alloantigenic site.

MOLECULAR BASIS OF ABNORMAL SYNTHESIS OF INTEGRINS

The alpha- and beta subunits of the integrins are encoded by separate genes and transport of a newly synthesized receptor to the cell surface takes place only after the subunits have combined. Evidence with variant cell lines has suggested that deficiency in expression or processing of the alpha- and beta subunit may lead to a failure to process and express the other constituent unit of the complex (*Kishimoto et al., 1987; Cheresh and Spiro, 1987; O'Toole et al., 1989*). This suggests that mutations which affect expression of either the alpha- and beta subunit alone may affect expression of both subunits. This view is clearly supported by recent studies on the identification of gene abnormalities in Glanzmann's thrombasthenia, an autosomal recessive bleeding disorder most commonly characterized by a markedly reduced expression of the GP IIb/IIIa complex. For instance, it has been shown that GP IIb/IIIa deficiency can be caused by an isolated defect in the GP IIIa gene, leading to reduced or absent synthesis of GP IIIa (*Bray and Shuman, 1990*). As the counterpart of GP IIIa (the beta chain of the vitronectin receptor) is also present in vascular endothelial cells, and these subunits are products of only one gene (*Bray et al.; 1988; Rosa et al., 1988*), one would expect that in the patients studied the expression of the vitronectin receptor in endothelial cells (and other cells that express the vitronectin receptor) is abnormal as well. So far, there is no evidence in support of this view. We have directly shown that in a patient with Glanzmann's thrombasthenia, the vitronectin receptor is normally produced by both endothelial cells and smooth muscle cells (*Giltay et al., 1987*). The cause of the defective expression of GP IIb/IIIa in this patient has not been established (either GP IIb or IIIa deficiency or, less likely, both). Clearly an intriguing question remains to be answered: If indeed, in certain cases of Glanzmann's thrombasthenia, the common beta chain of the vitronectin receptor and the platelet GP IIb/IIIa complex are deficient (*Bray et al., 1990*), why then is the pathophysiological consequence of the beta-3 deficiency only manifest at the platelet level? These and other questions raised above, will certainly be the focus of future studies.

REFERENCES

Anderson, D.C. and Springer, T.A. Leukocyte adhesion deficiency: an inherited defect in the Mac-1, LFA-1, and p150,95 glycoproteins. *Ann. Rev. Med.* 38:175, 1987.

Bray, P.F., Barsh, G., Rosa, J.-P, Luo, X.Y., Magaenis, E. and Shuman, M.A. Physical linkage of the genes for platelet membrane glycoproteins IIb and IIIa. *Proc. Natl. Acad. Sci. U.S.A.* 85:8683, 1988.

Bray, P.F. and Shuman, M.A. Identification of an abnormal gene for the GPIIIa subunit of the platelet fibrinogen receptor resulting in Glanzmann's thrombasthenia. *Blood* 75:881, 1990.

Buck, C.A. and Horwitz, A.F. Cell surface receptors for extracellular matrix molecules. *Annu. Rev. Cell Biol.* 3:179, 1987.

Cheresh, D.A. and Spiro, R.C. Biosynthetic and functional properties of an Arg-Gly-Asp-directed receptor involved in human melanoma cell attachment to vitronectin, fibrinogen, and von Willebrand factor. *J. Biol. Chem.* 262:17703, 1987.

Clemetson, K.J. and Luscher, E.F. Membrane glycoprotein abnormalities in pathological platelets. *Biochim. Biophys. Acta.* 947:53, 1988.

Flug, F., Espinola, R., Liu, L.-X. and Karpatkin, S. A 33-mer peptide spanning the 33rd amino acid polymorphism leucine/proline of platelet GP IIIa is not the PLA1/A2 epitope. *Clin. Res.* 38:425A, 1990 (Abstract).

Giltay, J.C, Leeksma, O.C., Breederveld, C. and van Mourik, J.A. Normal synthesis and expression of endothelial IIb/IIIa in Glanzmann's thrombasthenia. *Blood* 69:809, 1987.

Giltay, J.C., Leeksma, O.C., von dem Borne, A.E.G.Kr. and van Mourik, J.A. Alloantigenic composition of the endothelial vitronectin receptor. *Blood* 72:230, 1988.

Giltay, J.C., Brinkman, H.J.M., von dem Borne, A.E.G.Kr. and van Mourik, J.A. Expression of the alloantigen Zwa on human vascular smooth muscle cells and foreskin fibroblasts. A study on normal individuals and a patient with Glanzmann's thrombasthenia. *Blood* 74:965, 1989.

Giltay, J.C., Brinkman, H.J.M., Vlekke, A., Kiefel, V., van Mourik, J.A. and von dem Borne, A.E.G.Kr. The platelet glycoprotein Ia-IIa-associated Br-alloantigen system is expressed by cultured endothelial cells. *Br. J. Haematol.* 75:557, 1990.

Ginsberg, M.H., Loftus, J.C. and Plow, E.F. Cytoadhesins, integrins and platelets. *Thrombos. Haemostas.* 59:1, 1988.

Hemler, M.E., Huang, C. and Schwarz, L. The VLA protein family. Characterization of five distinct cell surface heterodimers each with a common 130,000 molecular weight beta subunit. *J. Biol. Chem.* 262:3300, 1987.

Hynes, R.O. Integrins: a family of cell surface receptors. *Cell* 48:549, 1987.

Kishimoto, T.K., Hollander, N., Roberts, T.M., Anderson, D.C. and Springer, T.A. Heterogeneous mutations in the beta subunit common to the LFA-1, Mac-1, and p150,95 glycoproteins cause leukocyte adhesion deficiency. *Cell* 50:193, 1987.

Kolodziej, M., Goldberger, A., Poncz, M., Newman, P.J. and Bennett, J.S. Evidence that a GP IIIa polymorphism is responsible for the PlA1 and PlA2 alloantigens by heterologous expression of the platelet GP IIIa. *Clin. Res.* 38:425A, 1990 (Abstract).

Leeksma, O.C., Giltay, J.C., Zandbergen-Spaargaren, J., Modderman, P.W., van Mourik, J.A. and von dem Borne, A.E.G.Kr. The platelet alloantigen Zwa or PlA1 is expressed by cultured endothelial cells. *Br. J. Haematol.* 66:369, 1987.

Lyman, S., Aster, R.H. and Newman, P.J. Polymorphism of human platelet membrane glycoprotein IIb associated with the Bak a/Bak b alloantigen system. *Blood* 58a, 1989 (Abstract).

Newman, P.J., Derbes, R.S. and Aster, R.H. The human platelet alloantigens, PlA1 and PlA2, are associated with a leucine 33/proline 33 amino acid polymorphism in membrane glycoprotein IIIa, and are distinguishable by DNA typing. *J. Clin. Invest.* 83:1778, 1989.

O'Toole, T.E., Loftus, J.C., Plow, E.F., Glass, A.A., Harper, J.R. and Ginsberg, M.H. Efficient surface expression of platelet GPIIb-IIIa requires both subunits. *Blood* 74:14, 1989.

Pischel, K.D., Marlin, S.D., Springer, T.A., Woods, Jr., V.L. and Bluestein, H.G. Polymorphism of lymphocyte function-associated antigen-1 demonstrated by a lupus patient's alloantiserum. *J. Clin. Invest.* 79:1607, 1987.

Rosa, J.-P., Bray, P.F., Gayet, O., Johnston, G.I., Cook, R.G., Jackson, K.W., Shuman, M.A. and McEver, R.P. Cloning of glycoprotein IIIa cDNA from human erythroleukemia cells and localization of the gene to chromosome 17. *Blood* 72:593, 1988.

Ruoslahti, E. and Pierschbacher, M.D. New perspectives in cell adhesion: RGD and integrins. *Science* 238:491, 1987.

Ruoslahti, E. Fibronectin and its receptors. *Annu. Rev. Biochem.* 57:375, 1988.

Ruoslahti, E. and Giancotti, F.G. Integrins and tumor cell dissemination. *Cancer Cells* 1:119, 1989.

Shulman, N.R. and Jordan Jr., J.V. Platelet immunology. In: *Hemostasis and Thrombosis, Basic Principles and Clinical Practice*. Ed by J. Hirsh, V.J. Marder and E.W. Salzman. Lippincott Company, Philadelphia, pp. 452-529, 1987.

Springer, T.A., Dustin, M.L., Kishimoto, T.K. and Martin, S.D. The lymphocyte function associated LFA-1, CD2 and LFA-3 molecules; cell adhesion receptors of the immune system. *Annu. Rev. Immunol.* 5:223, 1987.

Staunton, D.E., Dustin, M.L. and Springer, T.A. Functional cloning of ICAM-2, a cell adhesion ligand for LFA-1 homologous to ICAM-1. *Nature* 339:61, 1989.

von dem Borne, A.E.G.Kr. and Ouwehand, W.H. Immunology of platelet disorders. In: *Bailliere's Clinical Haematology*, (Ed. Caen, J.P.), Tindall, London, pp 749-781, 1989.

VIII. EPILOGUE

VASCULAR ENDOTHELIUM: PHYSIOLOGICAL BASIS
OF CLINICAL PROBLEMS

Kenneth L. Brigham

Center for Lung Research and Division of Pulmonary
and Critical Care Medicine
Vanderbilt University
Nashville, Tennessee 37232, U.S.A.

MEETING SUMMARY

Perspective

It is virtually certain that if you plan a meeting in a place as exotic and delightful as Corfu it will be easy to attract a distinguished audience regardless of the topic. It is equally certain that one would be inclined to use whatever mental gymnastics necessary to fit one's area of expertise to the chosen topic in order to justify an invitation to participate. Thus the choice of the topic of this conference assured that the endothelium would receive the credit and the blame for a broad spectrum of physiologic and pathologic events relevant to clinical medicine. An analogy could be drawn to the apocryphal story about the blind man and the elephant with the conclusion, illustrated in Figure 1, that the endothelium is very like a tree. Another quote comes to mind, the origin of which I am not sure but I first heard it from Joseph Milic-Emili, "To a man with a hammer, everything looks like a nail".

Human diseases which were attributed to endothelial dysfunction at this conference include: the adult respiratory distress syndrome, sepsis syndrome, atherosclerosis, renal failure, ischemia; stroke, coagulopathies, pulmonary or systemic hypertension, Alzheimer's disease, cancer and snake bite. In each case, data were presented which supported the contention that abnormalities of endothelial structure or function were important, although, to be fair, the magnitude of the role of the endothelial dysfunction in the clinical disease was often not clear.

As with diseases, it is also possible to catalogue some endothelial functions which could be related to diseases. These functions include: a physical barrier, a biochemical processor, a modulator of inflammation, immunity and coagulation, a regulator of vascular tone and a modulator of organ structure (e.g. angiogenesis). Situated at the interface between circulating blood and tissue and comprising an enormous surface area, endothelium is particularly suited to performing many of these functions. It is not difficult to imagine connections between endothelial dysfunction and clinical disease.

Clinical Relevance of Endothelial Dysfunction

As summarized above, it is possible and, in some cases, even likely, that human disease is either a direct or an indirect result of dysfunction of the endothelium. Some of the ways in which this could be the case will be discussed below.

There are other implications for endothelial functions in human disease. For example, detection of changes in some functions of endothelium could be useful in detecting disease early or in following the course of disease and the response to therapy. This approach might not require that the specific endothelial function being measured was relevant to the disease of interest, as long as the function accurately marked the disease and accurately reflected its progress. Either relevant or irrelevant endothelial dysfunction might thus be useful to measure as a practical matter. Since some therapies are aimed directly at improving specific endothelial functions, assessments of such therapies require that the specific function can be measured, but functions not related to the pathology of interest (a yes, yes and unrelated option) may also be useful.

Understanding how endothelial cells work and how to manipulate them may have therapeutic implications in human disease as well. Obviously, if malfunction of the endothelium were responsible for a disease, and a drug could be given which would correct the malfunction, the disease ought to be cured. However, less direct rationales might also be applicable. For example, although endothelial function might be perfectly normal in some diseases, treatments which affect endothelial functions might be beneficial. An example might be inhibition of angiotensin converting enzyme as treatment for (systematic) hypertension.

Recent research suggests even more novel uses of endothelium in the therapy of human diseases. Endothelial cells, including those derived from humans, can be grown in culture, and using current techniques for manipulating DNA, genes encoding virtually any protein of choice can be introduced into the genome of these cultured cells. The transformed cells can then be implanted in one of several ways into intact animals where they produce the new protein for which they were genetically engineered. In theory, this approach could be used to supply deficient secreted proteins. It may also be possible to introduce functioning foreign DNA into endothelial cells *in vivo* using liposome vehicles (or other approaches) so that implantation of cells may not be necessary.

How is Endothelium Involved in Human Disease?

Because the bewildering array of human diseases attributed to endothelial dysfunction at this conference is difficult to assimilate, and because of our general proclivity for classifying things, I suggest that there are three general ways in which endothelium may involve itself in the pathogenesis of human disease. These are illustrated in Figure 2.

Type I pathogenesis would be the simple relationship where a dysfunction of endothelium is directly responsible for the clinical disease. An example here might be coagulopathies where

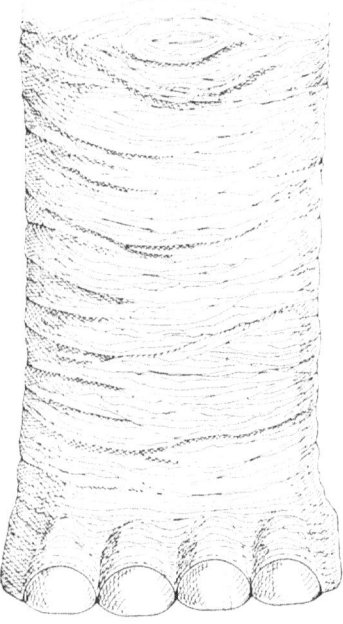

FIGURE 1: The endothelium is like a tree

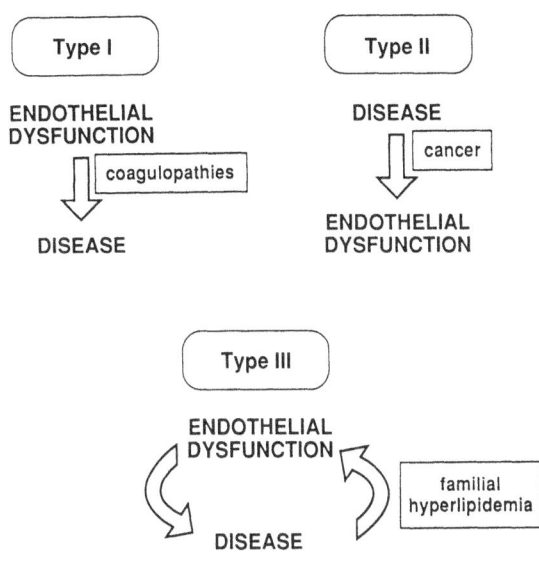

FIGURE 2: Pathogenesis

deficiency of a protein essential for proper coagulation and normally produced primarily by endothelial cells results in a clinical bleeding diathesis.

Type II pathogenesis would be where the disease was not primary in the endothelium, but where endothelial function is disrupted by the basic disease with the result that the disease is compounded. The example I have chosen here is cancer where actual organ structure is altered in part because of neoangiogenesis, the abnormal proliferation of endothelium presumably caused by the primary disease.

Most complex of all is type III pathogenesis where there is a primary abnormality of endothelial function or structure causing a disease which further compromises endothelial function and exaggerates the clinical disease. This vicious cycle might be exemplified by some familial hyperlipidemias where there is a congenital deficiency of endothelial receptors for LDL resulting in hyperlipidemia, and the high concentrations of lipids in the circulating blood further alter endothelial function.

In an effort to lighten the intellectual burden imposed by such a classification, I offer the following three short poems:

A Short Poem for Type I Pathogenesis

And should I bleed
I could not stop
Until I bled
The final drop
Because a fact-
Or does not come
Out from my en-
Dothelium.

A Shorter Poem for Type II Pathogenesis

No need to rave
Pointless to scold
They're only doing
What they're told.

Bad Poem for Type III Pathogenesis

Poor little cell
It could not make
Receptors which
Would let it take

Out from the blood
The LDL
Which then destroyed
The little cell.

Some Questions Not Completely Answered in Corfu

1. What diseases are primarily diseases of the endothelium?

Much of the material presented at the conference associated abnormalities of endothelial structure or function with diseases, but it is still not easy to establish, in many cases, specific dysfunctions of the endothelial abnormalities as the pathognomonic lesion in human diseases.

2. Which endothelial functions are universal?

This is a lumper versus splitter question. Should we view endothelium, wherever it occurs, as a single organ or is there sufficient organ, species and other regional specialization of function that the site at which endothelial cells exist is more important than their morphology and embryonic derivation?

3. Is endothelium an organ?

Related to Question 2, but broader, more philosophical.

4. Are cultured endothelial cells functionally relevant to endothelial cells *in vivo*?

This deals with the recurring problem in biomedical investigation of whether it is more valuable to use a controllable system or to use a "physiologic" system. The short answer to that question is both, but that begs the larger question which can only be answered in the long run by more research in a spectrum of experimental preparations.

5. What is EDRF? Does it matter?

As reflected in the other parts of this volume, the answer to the first part of this question is not entirely resolved. The answer to the second part is even more obscure at the moment as it relates to human disease (the topic of the conference).

6. What does endothelin really do? Does it matter?

What endothelin is is pretty clear, but exactly what it does (apart from vasoconstrict) and the role in human disease is only beginning to be identified.

7. What diseases could be cured by therapy aimed directly at endothelium?

In the interest of symmetry, this question **follows** from Question 1. However, it is conceivable that therapy aimed specifically at endothelium might cure a disease which was not a direct consequence of endothelium dysfunction.

ACKNOWLEDGEMENTS

This work was supported by NIH, National Heart, Lung and Blood Institute, Grants: HL 19153, (SCOR in Pulmonary Vascular Diseases); HL 07123 (NRSA Institutional Training

Grant in Multidisciplinary Respiratory Diseases); ROI HL 34208. Additional support was from the John W. Cooke and Laura W. Cooke Foundation; the Bernard Werthan, Sr. Foundation; the Harry H. and Martha W. Straus Foundation; and the Upjohn Company.

IX.   ABSTRACTS OF ORAL AND POSTER PRESENTATIONS

EFFECTS OF INTERLEUKIN-1 α/β, TUMOR NECROSIS FACTOR-α AND INTERFERON-γ ON HUMAN ENDOTHELIAL CELL PERMEABILITY

*,**Anne Burke-Gaffney, *Deirdre Cooney,
*E. Bruce Mitchell and **Alan Keenan

*Children's Research Centre,
Our Lady's Hospital for Sick Children
Crumlin, Dublin 12, Ireland

**Department of Pharmacology
University College Dublin
Dublin, Ireland

Cytokine-activated endothelial cells can play a number of pivotal roles in the development of cell-mediated immune responses, and as such, play a beneficial role in host defence. Endothelial activation can also contribute to disease processes such as vascular leak syndrome, following Interleukin-2 therapy for solid tumors and Kawasaki Syndrome, a childhood panvasculitis.

We have used an *in vitro* model system, cultured human umbilical vein endothelial cells (HUVEC) grown on Transwell-COL membrane assemblies, to investigate the direct effect of interleukin-1 α/β (IL-1 α/β), tumor necrosis factor-α (TNF-α) and interferon-γ (IFN-γ) on permeability to ^{125}I labelled bovine serum albumin (BSA). We also investigated the effect of conditioned medium obtained from PPD stimulated human lymphocytes on endothelial monolayer permeability. This conditioned medium mimics the inflammatory infiltrate that characterizes the T-cell mediated response to antigen (e.g. in experimental delayed-type hypersensitivity) and bathes the post capillary venules of the inflamed tissue. It may be possible, therefore, to use this model to identify cytokines/mediators responsible for augmenting vascular permeability in cell mediated immune responses.

HUVEC were isolated from umbilical vein by collagenase digestion and grown to confluence in 25cm^2 tissue culture flasks. Confluent cells (4-7 days) were trypsinized (0.05% trypsin and 0.02% EDTA) and seeded at a density of 2.5 - 3.0 x 10^5 cells/10µl, onto Costar Transwell - COL membrane assemblies (6.5mm diameter, 0.4µm pore size). One to two days after seeding onto membranes, the monolayers were stimulated with cytokine, heat inactivated cytokine (87^{0}C, 30 min), cytokine in the presence of polymyxin B (10µg/ml) or conditioned medium. Permeability was determined by incubating (1 hr, 37^{0}C, shaking water bath), monolayers with ^{125}I labelled BSA in RPMI 1640 containing 10% FCS. ^{125}I labelled BSA was added to the upper compartment of the membrane assembly and 100µl aliquots were sampled from the lower compartment after 1 hr. Radioactivity was determined by gamma counting.

A 2-3 fold increase in permeability of HUVEC monolayers was seen with IL-1 α/β, TNF-α, IFN-γ. Heat inactivated cytokine did not cause an increase in permeability. Cytokine in the presence of polymyxin B (10µg/ml) gave a permeability increment similar to the increase observed with cytokine alone. The permeability change caused by each cytokine was shown to be dose and time dependent. Supernatants from PPD stimulated lymphocytes (collected on day 7) gave a 4.6 fold increase in permeability which was significantly greater than that produced by control supernatants (P<0.002).

A marked increase in permeability of HUVEC monolayers was observed with all 3 cytokines used in the assay. The variability in the permeability change after 48 hr stimulation with IL-1α/β and TNF-α may be due to inactivation of these cytokines or to a certain degree of recovery from the cytokine induced injury. Control studies using heat inactivated cytokine and cytokine in the presence of polymyxin B suggest that the permeability increase is in fact attributable to the cytokine and not to contamination by endotoxin. The increase in permeability caused by supernatants from PPD stimulated lymphocytes may be due to synergistic effects of a number of cytokines including IL-1 α/β, TNF, IFN-γ and will be investigated further using specific cytokine neutralizing antibodies.

This work was funded by EOLAS (The Irish Science and Technology Agency) and the Children's Research Centre.

BLOOD-BRAIN BARRIER *IN VITRO*: PHARMACEUTICAL APPLICATIONS

*,**Romeo Cecchelli, *,**M.P. Dehouck, **P. Delorme,
*J.C. Fruchart

*SERLIA - INSERM U 325
Institut Pasteur
59019 Lille, France

**Laboratoire de Neurobiologie Fonctionnelle
Universite de Lille I
59650 Villenduve D´ Ascq
Lille, France

The passage of substances across the blood-brain barrier is regulated by cerebral capillaries which possess certain distinctly different morphological and enzymatic properties compared to capillaries of other organs. Investigations of the functional characteristics of brain capillaries have been facilitated by the use of cultured brain endothelial cells, but in most studies some characteristics of the *in vivo* system are lost.

To provide an *in vitro* system for studying brain capillary function, we have developed a process of coculture that closely mimics the *in vivo* situation by culturing brain capillary endothelial cells on one side of a filter and astrocytes on the other. In these conditions, endothelial cells retain all the endothelial cell markers and the characteristics of the blood-brain barrier including tight junctions and γ-glutamyl transpeptidase activity. The average electric resistance for the monolayers was 661 $\Omega \bullet cm_2$. The system is impermeable to inuline and sucrose, but allows the transport of leucine. Arabinose treatment increases transcellular transport flux by 70%. The relative ease with which such monolayers can be produced in large quantities would facilitate the *in vitro* study of brain capillary functions.

NEUTROPHILS ARE REQUIRED FOR PMA-INDUCED ENDOTHELIAL ECTOENZYME DYSFUNCTION: STUDY IN CULTURED ENDOTHELIAL CELLS

Xilin Chen, *M. Tzanela, M. Baumgartner, *J.R. McCormick and J.D. Catravas

Department of Pharmacology and Toxicology
Medical College of Georgia
Augusta, Georgia 30912-2300, U.S.A.

*Veterans Administration Medical Center
Augusta, Georgia 30912, U.S.A.

We have studied the role of neutrophils (PMN) in phorbol myristate acetate (PMA)-induced endothelial ectoenzyme (angiotensin converting enzyme, ACE and 5´-nucleotidase, NCT) dysfunction in cultured rabbit aortic endothelial cells (EC), using [^{3}H]-Benzoyl-Phe-Ala-Pro and [^{14}C]-5´-AMP as substrates, respectively, under first order reaction conditions. PMA alone (1-1000 ng/ml) or PMN in the absence of PMA did not affect ACE activity. When PMA was incubated together with PMN (PMN:EC ratio=1.25:1) for 4 h in Earl's salts, a dose dependent decrease in ACE activity was observed: threshold PMA concentration was 2 ng/ml and at 10 ng/ml, ACE activity was totally inhibited. This decrease in ACE activity was also dependent on PMN concentration and was detectable at as low as 1.25:10 PMN:EC ratio. The inhibition of ACE activity was also time-dependent, occurring as early as 1 h after incubation with PMN and PMA (10 ng/ml) and reaching maximum at 4 h. When incubated alone with EC for 4 h, PMA caused a small but significant increase in NCT activity (12-29%), which was PMA dose independent from 2-1,000 ng/ml. In the presence of PMN (1.25:1, PMN:EC ratio), PMA produced a significant decrease in NCT activity (20-26%) which, however, was dose-independent from 2-10 ng/ml. Pretreatment of EC with PMA (10 ng/ml) for 12 h, followed by incubation of EC with PMN in the absence of PMA for 4 h, did not affect ACE activity. Pretreatment of PMN (1:1, PMN:EC ratio), with PMA (10 ng/ml) for 1 h, followed by incubation of EC with activated PMN (after washing out the PMA) for 4 h, produced a significant decrease in ACE activity, whereas, the supernatant (containing PMA) from activated PMN had no effect on ACE activity. PMA also increased PMN adherence to the endothelial monolayer (measured using ^{51}Cr-labelled PMN) from 16% to 86% after 10ng/ml PMA for 1 h. Pretreatment of PMN with anti CDw18, monoclonal antibody 60.3 or physical separation of PMN from EC with a 2µ filter prevented PMA-induced ACE dysfunction. These results suggest that 1) by activating PMN, PMA produces endothelial ectoenzyme dysfunction in cultured EC; 2) EC ectoenzyme dysfunction occurs earlier in time and at lower PMN:EC ratios than reported for alteration in EC structural integrity. Our findings are consistent with the hypothesis that PMA induced endothelial ectoenzyme dysfunction is mediated by neutrophils as reported for PMA-induced increase in cell permeability and cell lysis, and provide a model for further study of the neutrophil mediated early endothelial cell injury. (*Supported by HL31422*).

NEUTROPHILS ARE REQUIRED FOR PMA-INDUCED ENDOTHELIAL ECTOENZYME DYSFUNCTION: STUDY IN PERFUSED LUNG PREPARATION

Xilin Chen, *M. Tzanela, *J.R. McCormick and J.D. Catravas

Department of Pharmacology and Toxicology
Medical College of Georgia
Augusta, Georgia 30912-2300, U.S.A.

*Veterans Administration Medical Center
Augusta, Georgia 30912, U.S.A.

We have studied the effects of PMA (15µg) on pulmonary endothelial ectoenzyme (angiotensin converting enzyme, ACE and 5´-nucleotidase, NCT) function in isolated rabbit lungs perfused in situ with platelet poor plasma (PPP), platelet rich plasma (PRP), with and without rabbit peritoneal neutrophils (PMN) or with PMN treated with anti CDw18 monoclonal antibody 60.3 (MoAb-PMN). Enzyme activities were estimated from the hydrolysis of substrates [^{3}H]-Benzoyl-Phe-Ala-Pro, ([^{3}H]-BPAP, ACE) and [^{14}C]-5´-AMP (NCT) during a single transpulmonary passage, using indicator dilution techniques. In all treatment groups, PMA produced a delayed increase in pulmonary vascular resistance to about 3x the control values. In PPP (n=4) and PRP (n=6) groups, PMA produced no changes in enzyme activity. In the presence of PMN (n=5), percent metabolism (%M) of [^{3}H]-BPAP decreased from 87±3.2% to 77±3.7% (15 min post PMA) and 75±6.2% (30 min post PMA) as did the apparent first order rate constant (A_{max}/K_m) for ACE from 821±114 to 617±59 (15 min post PMA) and 613±61 ml/min (30 min post PMA). K_m values of ACE for BPAP and NCT for AMP were elevated from 9.2±2.2 to 19.3±3 µM and 6.7±1.2 to 15.1±3.6 µM, respectively, 30 min after PMA, while A_{max} values (product of enzyme mass and rate of product formation) did not change. Pretreatment of PMN with MoAb 60.3, which inhibits PMN adherence to endothelial cells, (n=5), prevented the PMA- induced change in %M, A_{max}/K_m and K_m for ACE and NCT. These results indicate that 1)PMA induces a delayed pulmonary vasoconstriction which is independent of blood cell elements; 2)PMA-induced endothelial enzyme dysfunction is platelet independent and PMN dependent; 3)PMN adherence to endothelial cell is required for the PMA•mediated lung endothelial ectoenzyme dysfunction. (*Supported by HL31422*)

ENDOTHELIAL-DEPENDENT RELAXATIONS IN ATHEROSCLEROTIC HUMAN CORONARY ARTERIES

Adrian H. Chester, G.S. O'Neil and M.H. Yacoub

National Heart and Lung Institute
Harefield Hospital
London, United Kingdom

Nitric oxide (NO) has recently been implicated to be the mediator of endothelium-dependent vasorelaxation in animal tissues. We have studied whether this mediator plays a role in endothelial function in the human coronary circulation. In addition, we have examined if atherosclerosis affects the ability of the coronary endothelium to release NO under either stimulated or basal conditions.

189 ring segments of human epicardial coronary artery were removed from explanted hearts of 36 patients (2-64 years) undergoing transplantation for ischemic and non-ischemic heart disease (IHD and non-IHD), yielding atherosclerotic and non-atherosclerotic tissue. Vessel segments were placed in 5ml organ baths in gassed Tyrodes solution and maintained at 37^0C while changes in vessel tension were measured.

Substance P caused a maximum dilatation of $89.1 \pm 8.5\%$ and $60.1 \pm 6.4\%$ in non-IHD and IHD segments, respectively. The diseased vessels were 10 fold less sensitive to the effect of substance P. In normal and diseased vessels, L-NG-monomethyl-arginine (L-NMMA) (an inhibitor of NO formation) blocked the relaxation induced by substance P to $34.1 \pm 10.5\%$ and $22.0 \pm 9.7\%$, respectively.

The maximal response to U46619 was increased on both addition of L-NMMA and removal of the endothelium, indicating that NO opposes constriction by U46619. However, there was a significant decrease in magnitude of the tension induced by L-NMMA in atherosclerotic arteries, indicating that a basal secretion of NO is attenuated in diseased vessels.

We conclude that the protective capacity of the endothelium, which is mediated by the release of endogenous NO, is attenuated under basal and stimulated conditions in human coronary arteries affected with atherosclerosis. The arterial wall may then be exposed to direct vasoconstrictor effects of factors that mediate vasospasm and increase the risk of thrombosis.

USE OF *IN VITRO* AUTORADIOGRAPHY TO LOCALIZE BINDING SITES FOR PUTATIVE ENDOTHELIUM-DERIVED CONSTRICTOR/DILATOR SUBSTANCES IN HUMAN VASCULAR TISSUE

Michael R. Dashwood

National Heart and Lung Institute
Harefield Hospital
Middlesex, United Kingdom

In recent years, there has been increasing interest in the role of vascular endothelial cells in the control of blood vessel reactivity. It is now well established that many agents produce vasodilatation via an endothelium-dependent relaxing factor (EDRF). In 1988, Yanagisawa and his colleagues described the isolation of endothelin, an extremely potent vasoconstrictor peptide, which was released from cultured porcine endothelial cells. We have been using high and low resolution *in vitro* autoradiography (essentially as described by *Young and Kuhar, 1979*) to localize binding sites (putative receptors) for vasoactive compounds in human coronary tissue from patients undergoing bypass grafting or cardiac transplantation.

Tissue removed during surgery is frozen as soon as possible in liquid nitrogen. Slide-mounted, cryostat-cut sections are then incubated in buffer containing radiolabelled ligand, chosen for its selectivity for the receptor under investigation (eg. [^{125}I] substance P, endothelin, beta/alpha adrenoceptor ligands etc.), the degree of non-specific binding being established by incubating paired slides in the presence of excess concentrations of unlabelled ligand. After incubation, washing, and drying, sections are either apposed to Hyperfilm ^{3}H or dipped in liquid nuclear emulsion for low and high resolution autoradiography, respectively. After processing, autoradiographic images are photographed and densitometric analysis performed where appropriate. Underlying or adjacent sections are stained for histology and, in certain cases, the immunocytochemical identification of endothelial cells or transmitter presence performed.

Using autoradiography, we have identified binding sites for substance P that are associated with endothelial cells of the coronary arteries and internal mammary artery, whereas, binding sites for endothelin are associated with the smooth muscle and perivascular structures of these vessels. Adrenergic receptors exhibit different distributions with alpha-and beta-adrenoceptor subtypes both being associated with the vascular smooth muscle, yet beta adrenoceptors also showing strong luminal binding, presumably to the endothelium. In order to show that these "binding sites" identified autoradiographically, are indeed putative "receptors" functional studies are carried out in parallel wherever possible.

Using such an approach, we have described the presence of an endothelium dependent binding of [^{125}I] substance P to human epicardial coronary arteries and have shown in angiographic studies that substance P produces a concentration dependent dilatation of human coronary arteries *in vivo* (*Crossman et al, 1988*). Similarly, we have described dense binding of [^{125}I] endothelin to human coronary arteries and shown endothelin to produce a concentration-dependent contraction of human coronary artery segments *in vitro* (*Chester et al, 1989*). Finally, we have shown that [^{125}I] endothelin binding sites are present on human cardiomyocytes, and that endothelin has positive inotropic activity on isolated human cardiomyocytes *in vitro* (*Moody et al, 1990*).

In vitro autoradiography is a useful technique for establishing the localization and distribution of "receptors" in coronary and vascular tissue. The application of densitometric analysis to autoradiographic images allows estimates of receptor number to be made in tissue and also comparisons to be made, for example, between normal and pathological tissue. It is extremely important, where possible, to complement receptor mapping data with functional studies.

REFERENCES

Chester, A.H. et al. *Am. J. Cardiol.* 63:1395-1398, 1989.
Crossman, D.C. et al. *J. Physiol. (London)* 407:15, 1988.
Moody, C.J. et al. *Circ. Res.* in press.
Yanagisawa, M. et al. *Nature* 332:411-415, 1988.
Young, W.S. and Kuhar, M.J. *Brain Res.* 179:255-270, 1979.

ANGIOGENIN ACTS AS A DIRECT MITOGEN ON BOVINE BRAIN CAPILLARY ENDOTHELIAL CELLS

*,**M.P. Dehouck, **M. Chamoux, *J.C. Fruchart,
**G. Spik, **J. Montreuil and *R. Cecchelli

*SERLIA et INSERM U 325
Institut Pasteur
Lille, France

** UMR-CNRS 1 1 1,
Universite des Sciences et Techniques de Lille I
59650 Vileneuve D'Ascq,
Lille, France

Angiogenin is one of the potent inducers of neovascularization when compared to other angiogenic polypeptides recently described, such as acidic and basic fibroblast growth factor, transforming growth factor (α and β) and tumor necrosis factor α. First characterized by Vallee's group in conditioned medium from human colon adenocarcinoma cells HT 29, angiogenin has also recently been isolated from human plasma and bovine milk. Angiogenin stimulates angiogenesis *in vivo* when tested on the developing vascular system of the chick chorioallantoic membrane. However, no direct causal relationship between angiogenin and mitogenesis has yet been firmly established. Furthermore, it has been reported that angiogenin does not appear to be a growth factor for endothelial cells. Here, we report that bovine milk angiogenin is mitogenic *in vitro* for bovine brain capillary endothelial cells but not for aortic endothelial cells. This difference in the behavior of large blood vessels and capillary derived endothelial cells could be due to the occurrence of angiogenin binding sites, which have been found only on capillary endothelial cells.

PLATELET FUNCTION IN PATIENTS WITH VASCULITIS SECONDARY TO BEHCET'S DISEASE

F. Ferkan Demircioglu, T. Gursel, M.A. Gurer, V. Sepici, M. Bozkurt and A. Gulekon

Gazi University School of Medicine
Ankara, Turkey

In this study, platelet aggregation was studied in patients with vasculitis secondary to Behcet's Disease. The platelet aggregation responses to adenosine diphosphate, adrenalin and collagen were carried out in seven patients with Behcet's disease prior to drug treatment, in seven patients receiving colchicine treatment, and in a control group. In untreated patients, the aggregation response to these three agents exhibited no significant difference in either maximal light transmission or initial angle values. In patients with colchicine treatment, however, the aggregation response to these three agents decreased significantly in maximal light transmission, while no significant difference was noted in initial angle values. This finding suggests that colchicine treatment decreases the ability of platelets to aggregate and may also prevent thromboembolic phenomena.

REFERENCES

Demircioglu, F., Boke E., Demircin, M. et al. Abdominal aortic aneurysm with inferior vena cava obstruction. *Angiology* 40:227-232, 1989.

Demircioglu, F., Boke E., Demircin, M. et al. Aortic aneurysm due to Behcet's disease. *Deri Hast Frengi Ars.* 20:175-181, 1986.

Demircioglu, F., Komsuoglu, B., Dundar S. Echocardiographic evaluation of left ventriculum in Behcet's disease. *Second Behcet's Day Istanbul Univ.* Pub. 72-82, 1984.

Dundars, S., Demircioglu, F., Ozerkan, K. Familial Behcet's Syndrome. *GATA Mil. Med. Acad. Bul.* 21:287-292, 1979.

RESISTANCE VESSEL ENDOTHELIUM: ISOLATION, CLONING AND CHARACTERIZATION STUDIES

Clement A. Diglio, P. Grammas, M. Kumar,
F. Giacomelli and J. Wiener

Wayne State University School of Medicine
Detroit, Michigan 48201, U.S.A.

Vascular endothelial cells are not homogeneous throughout the body but have unique functional properties depending upon organ and species. Considerable evidence also suggests functional differences between endothelial cells derived from micro and macrovascular origin. In this report, we have successfully isolated and characterized endothelial cells derived from enriched preparations of rat cerebral resistance vessels. Previous work from our laboratory demonstrated that cerebromicrovascular smooth muscle cells derived from resistance vessels grew at a slower rate when compared to large vessel (aorta) derived smooth muscle cells and failed to tolerate standard freezing procedures at low passage levels (*J. Cell Physiol.129:131, 1986*). Using collagenase dissociation, we were able to significantly enhance the presence of resistance vessel endothelial cells. Once established and cloned, these cultures express Factor VIII antigen and exhibit phenotypic characteristics of cultured endothelial cells. In addition, these cells, under normal culture conditions, spontaneously form "tube-like" arrays. The presence of the *mas* oncogene (putative AII receptor) mRNA was found in subconfluent and confluent resistance vessel endothelial cell cultures using a probe (pRM I) that contains the gene for the rat *mas* oncogene. These results demonstrate that resistance vessel derived endothelial cells can easily be isolated and subcultured to provide an *in vitro* cell system to study cerebral angiogenesis. In addition, the availability of both smooth muscle and endothelial cell population from the same microvascular source will provide a unique system to study their cell-cell interactions *in vitro*.

BIOCHEMICAL ABNORMALITIES OF THE BLOOD-BRAIN BARRIER IN ALZHEIMER'S DISEASE

Paula Grammas, *A.E. Roher, **M.J. Ball

Department of Pathology
Wayne State University
Detroit, Michigan 48201, U.S.A.

*Anatomy and Cell Biology
Wayne State University
Detroit, Michigan 48201, U.S.A.

**University of Western Ontario
London, Canada

Alzheimer's disease (AD) is a progressive degenerative disorder characterized by widespread functional disturbances of human brain involving loss of memory reasoning and judgement. The role of the cerebral circulation in the pathogenesis of this disorder has not been well delineated, although both structural and functional abnormalities of the blood-brain barrier (BBB) in AD have been documented. Our laboratory has previously shown that important biochemical functions of the BBB are modulated by adrenergic receptors (*J. Neurochem. 4-4:1732, 1985*). The purpose of this study was to examine whether adrenergic control of BBB function, at the level of adrenergic receptors, is altered in AD. Cerebral microvessels were isolated from autopsy material from AD patients and non-demented elderly controls by a modification of our procedure for rat microvessel isolation. Alpha and β-adrenergic receptors were characterized using the specific antagonists [^{3}H] prazosin (PZ) and ^{125}I-iodocyanopindolol (ICYP), respectively. Examination of PZ binding indicated a significant ($P < 0.001$) decrease in α-receptor binding at each ligand concentration (0.05 - 0.3nM) and a two-fold decrease in maximum binding capacity [B_{max} in microvessels from AD patients with no change in receptor affinity (K_D)]. Interestingly, β-adrenergic receptor binding parameters (B_{max} and K_D) were comparable in control and AD microvessels. These data demonstrate a selective decrease in α-adrenergic receptors in cerebral microvessels in AD and suggest that this abnormality may alter BBB functions, such as permeability, that are thought to be modulated by α-receptors. Thus, this *in vitro* preparation can be used to unravel the biochemical basis of cerebral endothelial cell dysfunction in AD that may contribute to the multistep pathogenesis of cell death in AD. [*Supported in part by a grant from the American Health Assistance Foundation (PG) and the Atkinson Charitable Foundation of Toronto (MJB)*].

MACROPHAGE ACTIVATION IN BIORESORBABLE VASCULAR GRAFTS

Howard P. Greisler

Department of Surgery
Loyola University Medical Center
Maywood, Illinois 60153, U.S.A.

The biochemical and biomechanical characteristics of prosthetic materials may affect the physiology of cells in the local microenvironment and thereby modulate their expression of growth factors,y which may affect re-endothelialization and anastomotic pseudointimal hyperplasia. Our work examines how various parameters of prosthetic materials may differentially induce growth factor production generally and by the macrophage specifically. The implantation of lactide/glycolide copolymeric bioresorbable vascular grafts into animal models results in macrophage phagocytosis of the material followed shortly by a rapid transinterstitial ingrowth of myofibroblasts, capillaries, and endothelial cells. Such tissue ingrowth is not seen in response to similarly woven Dacron or ePTFE grafts. The rate of tissue ingrowth parallels the kinetics of macrophage mediated prosthetic resorption. Autoradiographic analyses have demonstrated mitotic indices of regenerating myofibroblasts to be 27% three weeks after implantation of polyglactin 910 (PG910) prostheses with a diminution to 1% after three months. By contrast, mitotic index following implantation of Dacron prostheses never exceeds 1.5%. The stimulus for induced cell proliferation appears to be a humoral factor. Freshly explanted grafts incubated for two hours in a physiologic salt solution release into that solution mitogenic activity capable of stimulating DNA synthesis in BALB/C3T3 mouse fibroblasts and rabbit aortic smooth muscle cells, the magnitude of this response correlating strongly with the *in vivo* mitotic index data. These PBS eluates tested negatively for PDGF by RIA, and other growth factors are likely involved.

Peritoneal macrophages were harvested from both normal New Zealand White rabbits and diet induced atherosclerotic NZW rabbits and were cultured in media supplemented with either PG910 or Dacron. Normal rabbit macrophages exposed to PG910 released into their media significantly more mitogenic activity for LE-II endothelial cells. Western blot analysis of this conditioned media revealed immunoreactivity with a basic FGF antibody. This antibody did not immunoreact with the media from macrophages grown in the absence of material or in the media of macrophages harvested from the atherosclerotic rabbits and grown in the presence of PG910. Western blotting was negative using antibodies against acidic FGF and PDGF. Macrophages harvested from atherosclerotic rabbits released significantly more mitogenic activity under basal conditions (in the absence of prosthetic material) when tested against BALB/C3T3 cells, aortic smooth muscle cells, or LE-II cells. Macrophages from normal rabbits were stimulated by PG910 to release more mitogen for all three cell lines. The macrophages from the atherosclerotic rabbits, however, released minimal additional mitogen in response to PG910 and in response to Dacron, inhibited LE-II proliferation, suggesting the release of a growth inhibitor by "atherosclerotic macrophages" in response to Dacron. Attempts to amplify mRNA for TGF-β using 30 cycles of sequential reverse transcriptase and polymerase chain reaction did not reveal the presence of TGF-β as the possible growth inhibitor.

In additional studies, we have demonstrated that HBGF-1, a potent endothelial cell mitogen, can be applied to biomaterial surfaces using the sequential application to that surface of fibronectin, heparin, and HBGF-1. Using ^{125}I HBGF-1, we found the retention of 44% of applied growth factor to the surface following seven days in circulation. After this time, the applied HBGF-1 was eluted from explanted grafts and shown to be intact by SDS PAGE, and eluted growth factor was then added to quiescent LE-II cells and demonstrated retention of mitogenic activity. However, an *in vivo* enhancement of endothelialization by this technique has not been achieved. This was shown by *in vitro* studies to be due to the loss of mitogenic activity until the bond to immobilized heparin is broken. Further modifications of these techniques are currently being studied.

It is anticipated that the application of specific growth factors to a surface or the stimulation of growth factor production by macrophages in response to an implanted biomaterial will lead to a new generation of more clinically efficacious small diameter vascular grafts. (*Supported by a grant from The National Institutes of Health, number RO1 HL41272*).

A DRUG THAT INCREASES cGMP BLOCKS THROMBOXANE INDUCED PULMONARY HYPERTENSION AND REDUCES EDEMA FORMATION IN OXIDANT LUNG INJURY

M.H. Jafri, Jr. and Gail H. Gurtner

New York Medical College
Valhalla, New York 10595, U.S.A.

Oxidants such as t-butyl-hydroperoxide (t-bu-OOH) cause vascular injury via lipid peroxidation and the consequent release of thromboxane and the peptide leukotrienes. It has been shown that agents which increase cAMP ameliorate lung (J. Appl Physiol. 62(1): 47-54, 1987). In this study, we sought to determine if agents which increase cGMP may play a similar role. We investigated the modulation of t-bu-OOH induced injury by sodium nitroprusside (SNP) in the isolated rabbit lung perfused with Krebs-Henseleit buffer and ventilated with 5% CO_2 in air. The perfusate was pretreated with 1mM SNP or no pharmacological intervention for twenty minutes before serial one minute challenges with t-bu-OOH. The challenges occurred in ten minute intervals with [t-bu-OOH] = 300µM in the first four challenges, 600 µM in the fifth challenge, and 900 µM in the sixth challenge. In the untreated group, infusion of t-bu-OOH produces an acute increase in peak PPa, airway pressure, and lung weight gain. SNP treated lungs exhibited a peak PPa of 10.55 ± .64 vs. 30.86 ± 6.83 mm Hg in untreated lungs ($\bar{X}$ ±SEM). The mean airway pressure of treated lungs was 9.43 ± .35 vs. 19.97 ± 3.66 mm Hg in controls. Increases in lung weight gain were significantly reduced by treatment with SNP: .2 ± .09 vs. 1.08 ± .56 grams/minute ($p<.05$). In other experiments, the left atrial pressure of the lung preparation was raised to 5, 10, and 15 mm Hg in an attempt to further characterize the protective effect of SNP. There was no significant difference in lung weight gain at elevated left atrial pressure between the SNP treated lungs and control. These findings suggest that the protective effect of SNP in the perfused rabbit lung may be due to its action on microvascular pressure. (*Supported by PHS Grant HL35483.*)

MAG-FURA-2 ELICITED FLUORESCENT RESPONSE OF SUBCELLULAR MAGNESIUM IN ENDOTHELIAL CELLS: A SHARP CONTRAST WITH CALCIUM

Brendan A. Hayes, *P.V. Avdonin and U.S. Ryan

Department of Surgery
Washington University School of Medicine
St. Louis, Missouri, U.S.A.

*Institute of Experimental Cardiology
Moscow, U.S.S.R.

The intracellular role of Mg^{2+} relative to Ca^{2+} is unclear, but it is clear that free Mg^{2+} ion plays an important role in regulation of cell growth, homeostasis and response to extracellular agonists. The spectral response of the Mg^{2+} indicator MAG-FURA-2 is similar to the Ca^{2+} indicators except that it occurs near the 0.1-5.0 mM range of Mg^{2+} that is normally found within cells. Mg^{2+} selectivity versus Ca^{2+} is sufficiently high ($Kd > 10.0$ mM) that typical ranges of Ca^{2+} (10-10000 nM) do not interfere with Mg^{2+} (Molecular Probes Handbook 1989-91).

We have found that bovine pulmonary artery endothelial cells (BPAEC) predictably load and fluoresce with the Mg^{2+} binding dye MAG-FURA-2. However, in marked contrast with Ca^{2+} behavior subcellular Mg^{2+} levels do not display rapid fluctuations in response to exogenous stimulatory agents: rapid and transient agonist-induced Ca^{2+} concentration changes occur on a timescale of seconds to minutes and are terminated equally quickly (*Avdonin et al., 1989*). In sharp contrast ATP, thrombin and bradykinin do not have any perceptible affect on subcellular free Mg^{2+}: while increasing agonist concentrations and exposure times of up to 15 minutes were used the excitation and emission spectra of MAG-FURA-2 loaded BPAEC's exhibit little difference. These results indicate that intracellular Mg^{2+} exchanges far slower than Ca^{2+} and that changes in Mg^{2+} flux are prolonged and could provide a different signal type to the cell (*Maguire, 1988*). These data support the theory (*Maquire, 1984; Grubs et al., 1987*) that magnesium is a "chronic" regulatory ligand whose intracellular concentration orchestrates the magnitude over which other agents, such as Ca^{2+}, are allowed to act. There are abundant examples of other important cellular processes activated or inhibited specifically by free Mg^{2+}: examples include Ca^{2+} transport ATPase of sarcoplasmic reticulum, (*Epstein et al., 1976; Jones, 1979; Martonosi, 1969*) acetyl CoA synthetase (*Guynn et al., 1974*) and phosphorylase kinase (*King et al., 1981*).

The agonist induced spiking reported (*Jacob et al., 1988*) for calcium is mirrored by the somewhat analogous behavior of magnesium. MAG-FURA-2 loaded BPAEC's exhibit an unsolicited and asynchronous spiking: the cause of this phenomenon is unknown.

REFERENCES

Avdonin, P.V., et al. *Tissue and Cell* 21:171-178, 1989.
Epstein, M., Kuriki, Y., Biltonen, R.L. and Racker, E. *Biochem.* 19:5564-5568, 1976.
Grubs, R.D. and Maguire, M.E. *Magnesium* 6:113-127, 1987.

Guynn, R.W., Webster, J.L., Jr. and Veech, R.L. *J. Biol. Chem.* 249:3248-3254, 1974.
Jacob, R., Merritt, J.E. and Hallam, T.J. *Nature* 335:40-45, 1988.
Jones, L.R. *Biochem. Biophys. Acta.* 557:230-242, 1979.
King, M.M. and Carlson, G.M. *J. Biol. Chem.* 256:11058-11064, 1981.
Maguire, M.E. *Ann. N.Y. Acad. Sci.* 551:201-217, 1988.
Maguire, M.E. *Trends Pharmacol. Sci.* 5:73-77, 1984.
Martonosi, A.J. *Biol. Chem.* 244:613-620, 1969.
Molecular Probes Handbook of Fluorescent Probes. 96-97:219-220, 1989-1991.

A DECREASE IN PLASMINOGEN ACTIVATOR INHIBITOR-1 ACTIVITY AFTER SUCCESSFUL PTCA IS ASSOCIATED WITH A SIGNIFICANTLY REDUCED RISK FOR DEVELOPMENT OF CORONARY RESTENOSIS

Kurt Huber, *M. Jorg, P. Probst, F. Kaindl and *B.R. Binder

Department of Cardiology
University of Vienna
Vienna, Austria

*Laboratory of Clinical Experiments
Department of Physiology
University of Vienna
Vienna, Austria

To determine a possible relation of changes in plasma levels of plasminogen activator inhibitor 1 (PAI-1) and tissue plasminogen activator (t-PA) to the development of coronary restenosis after successful coronary angioplasty (PTCA), we followed 104 patients for a period of 12 months after their first PTCA. PAI-1 plasma levels (functional activity) and t-PA antigen were determined before PTCA and 33 days, 3 months and 6 months, thereafter.

Thirty-four patients (32.69%) developed angiographically proven coronary restenosis within 4-48 weeks (median 12.5 weeks) after PTCA. Before PTCA, no significant differences could be demonstrated in t-PA antigen or PAI-1 plasma levels between the groups of patients with or without restenosis. During the whole observation period, t-PA plasma levels were not significantly different between the two groups; however, PAI-1 plasma levels were significantly higher at all time points after PTCA in patients who developed coronary restenosis ($p<0.005$).

When the pattern of PAI-1 plasma levels over time (increase or decrease between two consecutive time points of blood collection) was used to discriminate between the two study groups only, 3.5% - 18% of patients with a decrease in PAI-1 developed coronary restenosis within the following observation time as compared to up to 58% of patients exhibiting a PAI-1 increase ($p<0.05$ - $p<0.0005$).

The study demonstrates for the first time a strong relation between development of coronary restenosis and PAI-1 elevation in selected patients.

ENDOTHELIAL CELL HEMOSTATIC FUNCTION
AFTER HEART TRANSPLANTATION

Beverley J. Hunt, H. Segal and M. Yacoub

Department of Research Haematology
Harefield Hospital
Harefield, United Kingdom

We hypothesize that derangement of endothelial cell hemostatic function may be involved in the development of accelerated atherosclerosis (AA), (which is the major cause of mortality in heart transplant recipients (HTR) after their first year). Endothelial injury can alter hemostatic function, and endothelial injury from chronic rejection has been implicated as an etiological factor in AA. Changes in endothelial cell hemostatic function have also been implicated in the development of naturally- occurring atheroma which has similarities with AA.

Plasma levels of ATIII (chromogenic), vWF antigen (ELISA), tissue-plasminogen activator antigen (t-PA) (ELISA), PAI activity (chromogenic) and euglobulin clot lysis times (ELCT) were assessed in 115 steady state HTRs who had received a heart transplant more than one year previously and who received continuous immunosuppression of cyclosporin and azathioprine ± prednisolone. They were compared with 40 age-matched controls and 20 with ischemic heart disease (IHD).

ATIII levels were higher ($p<.001$) in HTR ($1.14 \pm .31$ IU/ml v controls $.94 \pm .21$ IU/ml), the greatest changes were in those transplanted for IHD ($1.16 \pm .26$ IU/ml). vWF levels were also raised ($p<0.001$) in the HTRs ($1.94 \pm .96$ IU/ml v controls $.97 \pm .27$ IU/ml). There was a correlation ($p=.004$) between vWF levels and creatinine levels, and vWF levels tended to be higher in those receiving prednisolone.

Fibrinolysis was significantly poor ($p<0.002$) in HTRs (median ECLT >360 min v controls of 265 min). However, on analysis of the HTRs according to their indication for transplantation, poor fibrinolysis was confined to those transplanted for IHD, and their ECLT times were not significantly different from the non- transplanted group with IHD. PAI levels were higher ($p<0.05$) in those HTRs transplanted for IHD (11.7 ± 6 AU/ml) than HTR without IHD (9.8 ± 8.8 AU/ml), those with IHD (10.2 ± 6.8 AU/ml) and the controls (8.7 ± 5 AU/ml). There was a strong relationship of prolongation in ECLT with PAI levels (Spearman's correlation $p<0.001$). There was a trend for the patients receiving prednisolone to have higher PAI levels. T-PA levels were higher ($p<0.001$) in those with IHD (8 ± 3.4 ng/ml) and those transplanted for IHD (8.5 ± 3.5 ng/ml) against the controls (5.1 ± 2.1 ng/ml) and the HTR without previous IHD (94.9 ± 1.9 ng/ml).

Twenty-three patients had AA. When compared to the other HTRs they had higher levels of vWF (2.2 ± 1.1 IU/ml v $1.86 \pm .89$ IU/ml) and t-PA levels were lower (7.0 ± 2.5 v 8.5 ± 3.7 ng/ml) although these changes did not reach significance.

In conclusion, endothelial hemostatic function is perturbed in HTRs. All HTRs had increased levels of plasma vWF and ATIII. Those transplanted for IHD have poor fibrinolysis, similar to non-transplanted patients with IHD, suggesting some of these changes were present prior to transplantation. Some hemostatic changes may be related to the use of cyclosporin and prednisolone. The relationship with the development of AA requires further study.

RESPONSE OF VASCULAR CELLS TO HERPES SIMPLEX VIRUS (HSV) INFECTION

Nicholas A. Kefalides, J.M. Brinker and Z. Ziaie

Connective Tissue Research Institute and Department of Medicine
University of Pennsylvania
Philadelphia, Pennsylvania 19104, U.S.A.

We have demonstrated that endothelial (EC) and smooth muscle cells (SMC) support replication of several human viruses, including HSV-1 and HSV-2 (*Ziaie, et al., 1986; Lashgari, et al., 1987*). HSV infection of EC leads to induction of Fc and C3b receptors and to increased granulocyte adherence to EC. Infection of EC or SMC with HSV-1 or HSV-2 suppresses host-cell protein and proteoglycan synthesis (*Ziaie, et al., 1986; Lashgari, et al., 1987; Kaner, et al., 1990*). The rate of suppression (2-6 hrs. post-infection) is virus dose dependent and varies with the specific host protein: fibronectin > types I, III and IV collagen > von Willebrand Factor > thrombospondin > actin, and tubulin. The reduction in specific host protein synthesis correlates with reduced steady-state levels of mRNA in infected cells (*Brinker, et al., 1990; London, et al., 1990*). The early phase (2-4 hrs. post-infection) of suppression of host-cell protein synthesis is virion dependent and independent of new viral protein synthesis. Clinical studies have shown that lithium salts prevent recurrence of HSV infections in humans. *In vitro* studies have suggested that lithium prevents viral DNA replication. Since early suppression of host-cell protein synthesis is mediated by a virion-associated function, we studied the effect of lithium chloride (LiCl) on protein synthesis and host, as well as viral mRNA levels, in HSV-1 infected EC. In the presence of LiCl, host-cell protein synthesis was maintained for a number of host proteins compared to cultures without LiCl. The ability of LiCl to maintain host protein synthesis in HSV-infected EC depended on virus dose, LiCl concentration (30mM > 20mM > 10mM) and varied with the host protein, i.e. the synthesis was more pronounced for thrombospondin and plasminogen activator inhibitor-1 than for fibronectin or collagen type IV (*Ziaie and Kefalides, 1989*). LiCl was must effective when added from 0-3 hrs. post-infection. The degree of synthesis of a given protein correlated with an increase in the level of its corresponding mRNA. Although immediate-early proteins of HSV were synthesized, the synthesis of early and late HSV peptides was suppressed. Both the synthesis of HSV DNA polymerase protein and its corresponding mRNA level were totally suppressed in the infected EC treated with LiCl. This resulted in complete suppression of production of infectious HSV particles. (*Supported by NIH Grants AR20553, HL29492 and AR07490*).

REFERENCES

Brinker, J.M., Ziaie, Z. and Kefalides, N.A. Virus Research (In Press) 1990.
Kaner, R.J., Iozzo, R. V. Ziaie, Z. and Kefalides, N.A. *Am. J. Resp. Cell & Mol. Biol.* 2:423-431, 1990.
Lashgari, M.S., Friedman, H.M. and Kefalides, N.A. *Biochem. Biophys. Res. Commun.* 143:145-151, 1987.

London, F.S., Brinker, J.M., Ziaie, Z. and Kefalides, N.A. *Lab. Invest* 62:189-195, 1990.
Ziaie, Z. and Kefalides, N.A. *Biochem. Biophys. Res. Commun.* 160:1073-1078, 1989.
Ziaie, Z., Friedman, H. M. and Kefalides, N. A. *Collagen Rel. Res.* 6:333-350, 1986.

ENDOTHELIAL CYCLIC NUCLEOTIDES ARE UNLIKELY TO REGULATE EDRF RELEASE

M. Kuhn, A. Otten, J.C. Frolich and *U. Forstermann

Department of Clinical Pharmacology
Hannover Medical School
Germany

*Abbott Laboratories
Chicago, Illinois, U.S.A.

Among the various vasoactive substances released by endothelial cells, are two powerful vasodilators: endothelium-derived relaxing factor (EDRF) and prostacyclin. Many of the stimuli of EDRF release lead to the simultaneous formation of prostacyclin (*Busse, et al., 1985*). The mechanism of action of the two autacoids is completely different: EDRF stimulates soluble guanylate cyclase, whereas, prostacyclin activates adenylate cyclase in vascular smooth muscle cells. Endothelial cells also possess adenylate cyclase and soluble as well as particulate guanylate cyclase (*Dembinska-Kiec, et al., 1980; Ganz, et al., 1986*). This prompted us to investigate in cultured bovine aortic endothelial cells (BAE cells) whether or not: 1) EDRF and prostacyclin increase the cyclic GMP and/or cyclic AMP content of the producing cells themselves and 2) increase in endothelial cyclic nucleotides influences the release of EDRF.

EDRF activity was measured using rat fetal lung fibroblasts (RFL-6, ATCC, Rockville) as detectors. These cells contain considerable amounts of soluble guanylate cyclase. Therefore, the increase in cyclic GMP in RFL-6 cells can be used as a bioassay to detect EDRF formation (*Ischii, et al., 1989*). Co-incubation of RFL-6 cells with the conditioned media of BAE cells stimulated with bradykinin (10 nM) or Ca^{2+}-ionophore A23187 (1 µM) markedly enhanced the cyclic GMP content of RFL-6 cells by 9.3 fold and 9.9 fold, respectively. Pretreatment of BAE cells with N^{ω}-nitro-L-arginine (0. 1 mM), which inhibits EDRF synthesis, completely abolished the cyclic GMP response in RFL-6 cells.

Bradykinin- and A23187- stimulated EDRF formation was accompanied by 2 fold increases in the cyclic GMP content of the producing BAE cells themselves. Pretreatment of BAE cells with atrial natriuretic peptide (0.1 µM) or sodium nitroprusside (10 µM) strongly enhanced the cyclic GMP content of BAE cells by 6.5 fold and 4.1 fold, respectively. These increases in endothelial cyclic GMP had no effect on bradykinin- and A23187- stimulated release of EDRF.

Bradykinin (10 nM) and A23187 (1 µM) also stimulated endothelial prostacyclin production by 2.4 fold and 5.6 fold, respectively. Endothelial cyclic AMP levels remained unchanged during the release of prostacyclin. Moreover, a high concentration of exogenous prostacyclin (3 µM) also was without effect on cyclic AMP content of BAE cells. Pretreatment of BAE cells with the selective inhibitor of cyclic AMP phosphodiesterase AH 21-132 (0.1 mM), increased the cyclic AMP content of BAE cells 3.7 fold. This increase had no effect on the release of EDRF by bradykinin and A23187.

In summary, our study shows that in BAE cells, EDRF produces only a modest increase in cyclic GMP and that prostacyclin has no effect on cyclic AMP. Marked increases in levels of either cyclic nucleotide produced by other substances do not alter EDRF production. We conclude that there is no feedback mechanism of cyclic nucleotides on EDRF formation in these cells.

REFERENCES

Busse, R. et al., *Basic Res. Cardiol.* 80:475-90, 1985.
Dembinska-Kiec, A. et al., Naunyn-Schmiedeberg's Arch. *Pharmacol.* 311:67-70, 1980.
Ganz, P. et al., *Proc. Natl. Acad. Sci. U.S.A.* 83:3552-3556, 1986.
Ischii, K. et al., *J. Appl. Cardiol.* 4:505-12, 1989.

THE HUMAN LUNG AS AN ENDOTHELIN CLEARANCE ORGAN: NORMAL FUNCTION AND DERANGEMENT IN PULMONARY HYPERTENSION

David Langleben, R.D. Levy, P. Cernacek and D.J. Stewart

Royal Victoria and Sir Mortimer B. Davis-Jewish General Hospitals
McGill University
Montreal, Canada

Endothelin (ET) is a recently described circulating peptide released by vascular endothelial cells. It has potent vasoconstrictor and smooth muscle proliferative properties. In most models, ET is a pulmonary vasoconstrictor. Animal studies suggest that the lung acts as a net clearance organ for ET and normally reduces circulating ET levels, but this function has not been examined in humans. The effects of pulmonary hypertension on this clearance function are unknown, and it is also unknown if altered ET clearance (or excessive release) is an initiating step in the development of pulmonary hypertension.

We, therefore, measured plasma immunoreactive ET (irET) levels in 16 control subjects: 8 normal volunteers and 8 patients with coronary disease but no pulmonary hypertension; and also in 27 patients with pulmonary hypertension: 7 primary, 3 with lung disease, 4 with congenital heart disease, 4 with autoimmune or collagenvascular disease, 5 with thromboembolic disease and 5 with valvular heart disease. To obtain an index of net clearance of ET across the lung, systemic venous blood, representing pulmonary inflow blood, and systemic arterial blood, representing pulmonary outflow, were sampled in each subject. Net clearance should yield an arteriovenous (A:V) ratio of less than unity. The irET levels in extracted plasma were determined by use of a specific and sensitive radioimmunoassay, using an antibody to ET-1.

In normal, and coronary disease patients, the mean A:V ratios were less than 1.0 (0.59 ± 0.35 and 0.53 ± 0.64, respectively), consistent with considerable clearance of ET in the lung. In patients with secondary pulmonary hypertension of all causes, the mean A:V ratio was 0.99 ± 0.43. However, patients with primary pulmonary hypertension had a mean A:V ratio of 2.21 ± 0.73, which was significantly different from both controls and patients with secondary pulmonary hypertension ($p < 0.001$) and from unity ($p < 0.01$). Venous levels of irET were elevated in both primary and secondary pulmonary hypertension, as compared to controls.

Thus, the normal human lung may act as a net endothelin clearance organ. This function is somewhat altered in secondary pulmonary hypertension. In contrast, the A:V ratio is greatly changed in primary pulmonary hypertension, indicating a large net release of endothelin in most patients. Abnormal ET clearance by the lung may lead to the increased circulating levels seen in pulmonary hypertension. Furthermore, given endothelin's biologic actions, its increased production and release in the lung may be contributing to the pathogenesis of primary pulmonary hypertension. (*Support: Medical Research Council of Canada, Heart and Stroke Foundation of Canada and Quebec, Quebec Lung Association*).

SECRETION OF ENDOTHELIAL-LEUKOCYTE ADHESION MOLECULE (ELAM-1) UPON DELETION OF PORTIONS OF ITS CARBOXYL TERMINUS

T.J. Ahern, M.A. Shaffer, D.S. Sako and
Glenn R. Larsen

Genetics Institute
87 Cambridge Park Drive
Cambridge, Massachusetts 02140, U.S.A.

The carboxyl terminus of endothelial-leukocyte adhesion molecule (ELAM-1) consists of a 22-amino acid, hydrophobic domain followed by a 32-amino acid, hydrophilic domain. In an effort to identify the portions of the carboxyl terminus necessary for efficient subcellular transport and anchoring of ELAM-1 in the endothelial cell membrane, we have altered these domains by site-directed mutagenesis of the molecularly cloned sequence of ELAM-1. Expression of mature ELAM-1 in mammalian cells produced no detectable ELAM-1 in the conditioned media, as determined by pulse-chase labeling with ^{35}S-methionine. Modifications within the carboxyl terminus of ELAM-1 resulted in expression and secretion of the variant forms. Unlike cells expressing full-length ELAM-1, none of the cells expressing secreted mutant forms bound either neutrophils or HL-60 cells (a promyelocytic precursor of neutrophils), indicating that functional ELAM-1 was not associated with the cell. We conclude that the carboxyl terminal region is required for anchoring ELAM-1 into the cell membrane.

BASEMENT MEMBRANE BIOSYNTHESIS AS A TARGET FOR DEVELOPING INHIBITORS OF ANGIOGENESIS WITH ANTITUMOR ACTIVITY

Michael E. Maragoudakis, E. Missirlis, G. Karakiulakis,
M. Bastaki and N. Tsopanoglou

Department of Pharmacology
University of Patras Medical School
Patras, Greece

A method was developed for assessing angiogenesis using the rate of basement membrane (BM) collagen biosynthesis as a biochemical index. In the chick chorioallantoic membrane (CAM) system, which is used as a model for studying angiogenesis by many investigators, the rate of radiolabeled proline incorporation into collagenous proteins was 11-fold higher at day 10, the stage of maximum angiogenesis, than at day 15 of chick embryo development, when angiogenesis had reached a plateau. In addition, it was shown that inhibition of BM collagen biosynthesis prevented angiogenesis (*Maragoudakis, et al., 1988*).

Using this methodology, we have initiated a search for specific inhibitors of BM biosynthesis and evaluated their effects on angiogenesis in the CAM system and their antitumor effects in rats bearing a Walker 256 carcinosarcoma.

There are many enzymatic steps in the biosynthesis and assembly of BM by the proliferating endothelial cells of the new vessel, where inhibition can lead to anti-angiogenic effects. 8,9-dihydroxy-7-methyl-benzo(b)quinolizinium bromide (GPA 1734) inhibited BM synthesis by interfering with the hydroxylation reactions of proline and lysine in the formation of collagen type IV. GPA 1734 prevented angiogenesis in the CAM, while a closely related analog 9,10-dihydroxy-7-methyl-benzo(b)quinolizinium bromide (GPA 1967), which was without effect on BM biosynthesis, had no effect on angiogenesis. Tricyclodecan-9-yl-xanthate (D609), which inhibited BM synthesis by an as yet unknown mechanism, also prevented angiogenesis. The structurally related analogs, ethyl and propyl xanthates, had no effect on BM synthesis and angiogenesis. This suggested a specificity of BM synthesis inhibitors in preventing angiogenesis.

It has been established that solid tumors are angiogenesis dependent, and that inhibitors of angiogenesis, such as protamine or the combination of heparin with cortisone, prevent tumor growth and metastasis (*Folkman, 1985*). The aforementioned inhibitors of angiogenesis, GPA 1734 and D609, were also shown to have antitumor effects in rats bearing Walker 256 carcinosarcoma. Implantation of 10 mg of Walker tumor ip in rats caused a growth of tumor mass 10-15 g within 10 days. Treatment of the rats with GPA 1734 and D609 (50 mg/kg/day, ip) caused a 65% and 85% reduction in tumor growth, respectively. The antitumor effects of these compounds were dose dependent and related to their inhibitory effect on angiogenesis, since the structurally related analogs of ethyl and propyl xanthates, which did not inhibit angiogenesis, were without effect on tumor growth. These compounds were not cytotoxic in cell cultures of Walker 256 carcinoma *in vitro* at concentrations up to 0.1 mM and did not cause any obvious toxic effects to the treated animals.

The results suggest that a search for compounds that interfere specifically with BM biosynthesis by the proliferating endothelial cells may provide novel anti-angiogenic agents with potential application in tumor chemotherapy. Such inhibitors are not expected to have serious side effects for short-term treatment since both BM biosynthesis and angiogenesis are extremely slow processes under physiological conditions.

REFERENCES

Folkman, J. *Ad. Cancer Res.* 43:175, 1985.
Maragoudakis, M.E. et al., *J. Pharm. & Exper. Therap.* 244:729, 1988.

MODULATION OF CYCLIC NUCLEOTIDE ACCUMULATION
IN CULTURED VASCULAR CELLS

Nandor Marczin, *U.S.Ryan and John D. Catravas

Department of Pharmacology and Toxicology
Medical College of Georgia
Augusta, Georgia 30912, U.S.A.

*Monsanto Company
800 N. Lindbergh Boulevard
St. Louis, Missouri 63110, U.S.A.

Vascular endothelium actively contribute to vasomotion by releasing vasoactive substances and modulating the effects of various agents on the tone of the underlying smooth muscle.

Cyclic nucleotides seem to be involved in the vasodilation elicited by EDRF. Stimulation of the soluble guanylate cyclase within the smooth muscle leads to relaxation. A feed back mechanism in the endothelial cells was proposed involving soluble guanylate cyclase. Endothelial cells release EDRF in response to acetylcholine, however, the signal transduction mechanism is not clear. In other cell types, muscarinic receptor stimulation is associated with the inhibition of adenylate cyclase.

Endothelial cells possess a controlled contractile machinery which is thought to be the source of both the relaxant response to certain physiological stimuli and of endothelial retraction during pathological conditions. Modulation of cyclic nucleotide levels within the endothelial cells may affect the contractile mechanism and interfere with the pathologic process.

We have studied the effects of several vasodilators on the adenylate and guanylate cyclase systems in long term cultured rabbit and calf endothelial (EC) and smooth muscle cells (SMC). EC were mechanically harvested from calf pulmonary artery (CPAE), aorta (CAE) or rabbit aorta (RAE). SMC were similarly obtained from rabbit pulmonary artery (RPASM). Confluent cell monolayers (2 cm^2 wells) were incubated in medium alone or in the presence of cyclic nucleotide phosphodiesterase inhibitor (1 mM IBMX) and then exposed to isoproterenol (Iso), sodium nitroprusside (SNP), atriopeptin II (ANP) or acetylcholine (ACh). Cyclic nucleotide (cAMP and cGMP) levels were measured by radioimmunoassay.

Basal cAMP levels were higher in RPASM than in EC. IBMX alone increased basal cAMP levels significantly. Iso produced a further dose- and time-dependent increase in cAMP content in all cultures, although the magnitude and rate of cAMP accumulation varied considerably among different cell types.

In the presence of Iso, ACh inhibited cAMP accumulation in a dose dependent manner in calf EC (CAE and CPAE) only, but not in RAE and RPASM. IBMX alone exerted no significant effect on cGMP levels. SNP produced a dose- and time dependent increase in cGMP content of smooth muscle cultures only. IBMX facilitated this response.

Stimulation of EC by ANP resulted in a 2-3 fold increase in cGMP levels under basal conditions; in the presence of IBMX, the increase was 50-100 fold. RPASM exhibited a much less pronounced response to ANP, and only in the presence of IBMX (2 fold).

Basal cGMP levels were higher in cocultures of EC and RPASM than in either RPAS or EC alone. In these cocultures, in the absence of IBMX, ACh had no effect on cGMP accumulation, however, bradykinin produced a 50% increase in cGMP levels.

POLYMORPHONUCLEAR LEUKOCYTES (PMNL)-ENDOTHELIUM INTERACTION *IN VITRO*: EFFECTS OF ISOLATION TECHNIQUES OF PMNL

Gary J. Morrissey, B. Rogers, D. McIver and J. Powe

Department of Nuclear Medicine
Victoria Hospital
London, Ontario, Canada

Radiopharmaceutical Development Group
Department of Pharmacology and Toxicology
University of Western Ontario
Ontario, Canada

The hypothesis that polymorphonuclear leukocytes (PMNL) play a major role in the pathogenesis of vascular endothelial injury necessitates the need for techniques to isolate PMNL for subsequent experimental investigations. It has been suggested that various isolation protocols may alter the functional integrity of the isolated white blood cells. Extrapolation of this concept to the investigation of their role and mechanisms of interaction with vascular endothelium led this laboratory to investigate what effect in vitro isolation techniques had on aggregation response, oxygen radical chemiluminescence and the pattern of intracellular calcium response of stimulated PMNL.

PMNL were isolated by 3 different protocols. Whole blood was collected from a single donor with an ACD anticoagulant, (7:1) blood:ACD. The Percoll (Pharmacia) protocol involved hetastarch sedimentation of the RBC. Leukocyte rich plasma (LRP) was layered on top of a 3 phase Percoll/plasma discontinuous gradient. Separation of the granulocyte and agranulocyte fractions through centrifugation yielded granulocyte populations greater than 98% pure. The Ficol protocol involved hetastarch sedimentation of the RBC. LRP was layered on top of a 2 phase discontinuous gradient of Ficol Hypaque (Sigma) and separation of the granulocyte and agranulocyte fractions through centrifugation yielded granulocyte populations greater than 95% pure. The ammonium chloride (NH_4Cl) protocol involved hetastarch sedimentation of the RBC. LRP was mixed with 8 mls of .87% NH_4Cl for 2 minutes to rid the cell suspension of contaminating RBC. The cells were centrifuged and resuspended in 3 mL of PBS. The leukocyte suspension was layered on top of a 2 phase Ficol Hypaque discontinuous gradient. Separation of the granulocyte and agranulocyte fractions through centrifugation yielded granulocyte populations of greater than 95%. All cell preparations were diluted to 1×10^7 cells/mL with 0.5% BSA. Aggregometry and oxygen radical chemiluminescence measurements were made in response to $10^{-6}, 10^{-7}, 10^{-8}$ fMLP and 10^{-7} and 10^{-8} M PMA. Intracellular calcium measurements utilizing Indo-1 were made in response to $10^{-6}, 10^{-7}$ and 10^{-8} fMLP.

The aggregometry results displayed similar dose response patterns to fMLP for all 3 isolation techniques. The oxygen radical chemiluminescence responses were calculated as normalized percent dose response and mean maximum chemiluminescence of PMNL to varying doses of fMLP. Both representations of chemiluminescence displayed dose response patterns, although the cells isolated by the NH_4Cl protocol displayed a noticeably enhanced response compared to cells isolated by the other protocols. Intracellular calcium measurements were calculated as mean and maximum x-fold increases of intracellular calcium over baseline for PMNL in response to varying doses of fMLP isolated by the 3 different protocols. PMNL isolated by the Ficol and NH_4Cl protocols displayed patterns suggestive of threshold responses. The PMNL isolated by the Percoll technique did not display this pattern. Aggregation and oxygen radical chemiluminescence responses of PMNL to varying doses of PMA were calculated as latent times to aggregation and the slope of the response curve of aggregation and chemiluminescence for varying doses of PMA. PMNL isolated by the Ficol protocol displayed an extended latent time to aggregation as well as an enhanced slope of the response curve for chemiluminescence to PMA.

These results suggest that different isolation techniques yield PMNL populations that demonstrate variations in certain responses to different stimuli. The isolation technique utilized for harvesting PMNL for investigation of roles and mechanisms of endothelial-PMNL interaction may play a pivotal role in interpreting and comparing experimental data.

CHEMOTAXINS INHIBIT NEUTROPHIL ADHESION TO AND TRANSMIGRATION OF CYTOKINE-ACTIVATED ENDOTHELIUM BY A CD 11/18-INDEPENDENT MECHANISM

Rene Moser and Jorg Fehr

Department of Internal Medicine
University Hospital
Zurich, Switzerland

Evidence is growing that the CD 11/18 (Mac-1) adhesion complex on polymorphonuclear neutrophils (PMN) is not the only surface structure to interact with monokine-activated endothelium resulting in attachment and rapid penetration of such layers by PMN. Moreover, little is known about mechanisms counteracting this extravasation process. Within this scenario, we observed strong inhibition of endothelium-governed PMN adhesion transmigration by chemotoxins. Culminating at maximal chemotactic concentrations, formylated chemotactic peptides (fMLP) and C5a rapidly inhibit PMN adhesion to endothelial layers activated by interleukin- 1 (IL-1), tumor necrosis factor (TNF) or endotoxin; whereas, lower inhibitory activity could be ascribed to IL-8, leukotriene B_4 and the spreading factors endotoxin and TNF were inactive. Such inhibition was limited to the initiation phase of PMN/endothelium interaction. Focussing on the inhibitory capacity of anti-CD18 antibodies 60.3 and IB-4 (M. Patarroyo), we found only incomplete inhibition of IL-1-provoked PMN-endothelium interaction [IB-4: $76 \pm 10\%$, n=8; 60.3: $77 \pm 11\%$, n=8; cf. control W6/32: $12 \pm 4\%$). However, inhibition by these two antibodies was greatly enhanced ($90 \pm 4\%$ and $92 \pm 3\%$, respectively) when PMN were pre-activated (Mac- 1 upregulated) with fMLP (10 nM) or C5a (100 ng/ml), clearly delineating a Mac-1-dependent from a Mac-1-independent interaction. Moreover, Mac-1-independent residual adherence of non-activated PMN in the presence of IB-4/60.3 was completely (100%/97%) eliminated by subsequent PMN activation with fMLP or C5a. These results suggest (with respect to the Mac-1 complex) an inversely regulated adhesive PMN surface structure that inhibits cytokine-provoked PMN extravasation, particularly in non-inflamed regions of the body. [*Abstract also submitted for presentation at the Federation Meetings (American Federation for Clinical Research/American Society for Clinical Investigation) in May 1990 in Washington, D.C.*]

COMBINED CYTOTOXIC EFFECTS OF BACTERIAL SHIGA TOXIN, TNF, IL-1 AND LPS ON VASCULAR ENDOTHELIAL CELLS, *IN VITRO*

Tom G. Obrig, C.B. Nelson and *P.J. Del Vecchio

University of Rochester Medical Center
Rochester, New York 14642, U.S.A.

*Albany Medical College
Albany, New York 12208, U.S.A.

To better understand the development of the altered coagulation state seen in kidney glomeruli from hemolytic uremic syndrome (HUS) patients, our laboratory is continuing to investigate the effects of bacterial Shiga toxin (ST) on human umbilical vein endothelial cells (HUVEC) and human kidney cells (HKC) isolated from glomerular remnants. The present study is an initial survey of the cytotoxic effects of the 65kD ST combined with either LPS, hrIL-1 or hrTNF on both HUVEC and HKC. The HKC (mesangial) and HUVEC expressed approx. 10^7 and 10^6 Shiga toxin receptors per cell, respectively.

Cytotoxicity LD_{50} values indicated that HKC were 50 to 500-times more sensitive to Shiga toxin than were HUVEC. Combinations of ST (lpM) and E. coli LPS (10ng/ml) exhibited synergistic cytotoxic activity vs. HUVEC. IL-1 (10ng/ml) alone was not cytotoxic to either HUVEC or HKC, but IL-1 increased the cytotoxicity of ST towards HUVEC. In contrast, TNF (10ng/ml) alone was toxic to HUVEC and this effect was additive in the presence of Shiga toxin. Prostacyclin release from HUVEC was enhanced by Shiga toxin (lnM) at 4hr and 12hr to approximately 135% of control values. It is concluded that cytokines, in specific combinations, may influence Shiga toxin action on vascular endothelial cells *in vitro*. This appears not to be the case with renal mesangial cells, *in vitro*. [*This study was supported in part by USPHS grant AI-24431 (to TGO)*].

APPLICATION OF AN OPIOID AFFINITY LABEL
FOR QUALITATIVE ANALYSIS OF MU-RECEPTORS

Huseyin A. Oktem, J. Moitra, S. Benyhe, M. Szucs,
I. Lengyel and A. Borsodi

Department of Biology
Middle East Technical University
Ankara, Turkey

Institute of Biochemistry and Biophysics
Biological Research Center
Hungarian Academy of Sciences
Szeged, Hungary

Potency of an opioid affinity label for identification of mu-opioid receptors in neuronal and non-neuronal membrane preparations of different species, ages and fractions were studied.

The irreversible ligand, Tyr-D-Ala-Phe(Me)-Gly-chloromethyl ketone (DAMCK), was able to label all membrane fractions prepared from brain tissues of different species. In rat brain and rabbit cerebellum membranes, four major bands were observed at 58, 45, 38 and 25 KDa, when analyzed by SDS-polyacrylamide gel electrophoresis. In guinea pig cerebellum and frog brain membranes, the labelled band at 38 KDa was absent.

When membranes prepared from young and adult chicken brains were labelled with the ligand, almost similar labelling profiles were observed. The major label was at 58 KDa. In adult animals, an additional labelled protein was observed at 70 KDa.

^{3}H-DAMCK was also used for identification of opioid receptors located on intracellular membranes. For this purpose, synaptic plasma (SPM) and microsomal membrane fractions from rat brain were used. In both fractions, labelled proteins were at around 50 and 30 KDa. In SPM fraction, an additional labelled protein was observed at 20 KDa.

Labelling studies on rat brain microvascular membranes revealed the presence of several labelled proteins. The major labelled band was around 55 KDa with additional minorly labelled bands around 30 and 20 KDa, respectively.

These results suggest that ^{3}H-DAMCK can be efficiently used for identification of mu-opioid receptors independently of tissue source and species.

PULMONARY ENDOTHELIAL ENZYME FUNCTION IN VIVO, AFTER CHEST IRRADIATION

S.E. Orfanos[*], X-L Chen[*], S.E. Burch[**],
J.W. Ryan[***] and J.D. Catravas[*]

[*]Department of Pharmacology and Toxicology
Medical College of Georgia
Augusta, Georgia 30912, U.S.A.

[**]Department of Radiology
Medical College of Georgia
Augusta, Georgia 30912, U.S.A.

[***]Department of Medicine
University of Miami School of Medicine
Miami, Florida 33125, U.S.A.

We investigated the early effects of ionizing radiation on pulmonary endothelial functions in rabbits exposed to a single dose (30 Gy) of ionizing radiation to the chest, with or without post-radiation indomethacin administration. Utilizing multiple indicator-dilution techniques, we measured metabolism (M) of ^{3}H-benzoyl-Phe-Ala-Pro (BPAP) and ^{14}C-benzoyl-Ala-Gly-Pro (BAGP) by endothelial bound angiotensin converting enzyme (ACE), M of ^{14}C-5´-AMP by 5´-nucleotidase (NCT) and binding (B) of the specific ACE inhibitor ^{3}H-RAC-X-65 (RAC) during a single transpulmonary passage in anesthetised, artificially ventilated, open-chest rabbits in which both systemic and pulmonary circulations were fully supported by an extracorporeal pediatric pump. Experiments were performed 7-8 hours post irradiation or sham at constant blood flows (Q_b) of 250, 400, 560, 800 ml/min and, after removal of the right lung (in the indomethacin-treated animals), at 640 ml/min, a Q_b sufficient to achieve full microvascular recruitment. We also calculated A_{max}/K_m and B_{max} ($=k_1 \bullet E$), the modified reaction constants of a substrate and inhibitor respectively, and the apparent constants K_m and A_{max} ($=E \bullet k_{cat}$).

We found that: 1) Increasing pulmonary blood flow levels increased perfused microvascular surfase area, as reflected in the proportionally increasing A_{max}/K_m or B_{max} values for all substrates and inhibitor, respectively, for both ACE and NCT, in control and irradiated animals. 2) Ionizing radiation to the chest produced endothelial ectoenzyme dysfunction, as reflected in the altered enzyme kinetics and substrate metabolism and inhibitor binding, which were present within a wide range of physiologic pulmonary blood flow values. 3) Post-radiation administration of indomethacin corrected or prevented most evidence of radiation-induced endothelial dysfunction, but failed to reverse the radiation-induced depression of the binding of ACE with its inhibitor (RAC-X-65). 4) The corrective or protective actions of indomethacin did not appear to involve shunting of the blood through normal or less injured vessels. (*Supplied by HL31422 and HL22087*)

ENDOTHELIAL FUNCTION OF VESSELS USED AS CORONARY ARTERY BYPASS GRAFTS

Greg S. O'Neil, A.H. Chester, T.N. Luu,
S.P. Allen and M.H. Yacoub

National Heart and Lung Institute
Harefield Hospital
Harefield
London, United Kingdom

Endothelial function has been implicated to be an important mechanism which may contribute to the patency of coronary artery bypass grafts. We have examined the capacity of the human saphenous vein, internal mammary artery (IMA), and the human gastroepiploic artery (GEA) to generate cyclic GMP, the second messenger that translates EDRF release into smooth muscle relaxation.

Segments of native saphenous vein (SV_N), surgically distended (SV_D), IMA, and GEA were removed from multi-organ donors or from patients undergoing coronary artery surgery. A total of 287 vessel segments were utilized from 28 patients (SV_N-55 segments, 6 patients; SV_D-58 segments, 6 patients; IMA-96 segments 10 patients; GEA 78 segments, 6 patients). Segments were challenged with either substance P (SP) (10^{-8}M), bradykinin (BK) (10^{-6}M), acetylcholine (ACH) (10^{-6}M), glyceryl trinitrate (GTN) (1 µg/ml), or vehicle for a period of 45 seconds. In some segments, the endothelium was removed by gentle rubbing. After stimulation, segments were flash frozen, crushed and the nucleotide extracted and assayed by specific radioimmunoassay. A Lowry technique was used to normalize all data to pmol/mg protein.

In the IMA and SV, all agonists (SP, ACH, BK) produced an increase in cyclic GMP formation compared to control. This effect was abolished by removal of the endothelium. Levels of cGMP were always lower in distended segments. In both SV_N and SV_D, increases were significantly less than in the IMA. The GEA showed a similar profile of cyclic GMP increases in response to SP. However, all parameters measured (control, SP and GTN) were approximately 10 fold greater than in the IMA.

This data suggests that the activity of the guanylate cyclase system or the ability of the smooth muscle to respond to cyclic GMP, may play a crucial role in determining the relaxatory capacity of blood vessels. Furthermore, surgical preparation of the SV may exacerbate an already compromised ability to respond to releasers of EDRF.

PATHOGENESIS OF HEMORRHAGE INDUCED BY RATTLESNAKE VENOMS AND THEIR PURIFIED HEMORRHAGIC TOXINS

Charlotte L. Ownby

Department of Physiological Sciences
Oklahoma State University
Stillwater, Oklahoma U.S.A.

One of the most striking results of rattlesnake venom poisoning and of experimental injection of rattlesnake venom is local hemorrhage. The extreme rapidity of the development of the hemorrhagic lesion makes it very difficult to treat. One antivenom is made commercially in the United States for treatment of rattlesnake bites, and it has been shown experimentally to have poor neutralization capacity when used after injection of venom. We have been investigating the pathogenesis of hemorrhage induced by rattlesnake venoms and their purified hemorrhagic components in an effort to improve the treatment of snake bite induced hemorrhagic damage.

So far, we have studied the pathogenesis of hemorrhage induced by the crude venom of the Western Diamondback Rattlesnake (Crotalus atrox) and three of its purified hemorrhagic toxins (HTa, HTb and HTe); the crude venom of the Timber Rattlesnake (Crotalus horridus) and its purified hemorrhagic toxin; HPIV; and bilitoxin, isolated from the venom of the Common Cantil (Agkistrodon bilineatus). Electron microscopy of skeletal muscle tissue taken at various times after the intramuscular injection of either venom or toxin showed that in all cases the pathogenesis of hemorrhage was "per rhexis", a process in which endothelial cells lining the walls of capillaries are destroyed, allowing blood cells and plasma to escape into the connective tissue.

Initially, endothelial cells become swollen, sometimes form blebs of their plasma membranes, and eventually rupture. Intercellular junctions remain structurally intact at all times, even when the endothelial cells have lysed. Fibrin deposition was observed in some vessels, sometimes both intravascularly and extravascularly, and platelet aggregations were common. Although the basal lamina was sometimes disrupted, it was equally common to find an intact basal lamina beneath a ruptured endothelium.

The exact mechanism of action of hemorrhagic toxins on the endothelium is not known, but our studies have shown that endothelial cells can be severely damaged while the basal lamina and intercellular junctions remain intact. It is possible that these toxins act directly on endothelial cells.

ANTIGENIC HETEROGENEITY OF VASCULAR ENDOTHELIUM

Chris Page, M. Rose and M. Yacoub

National Heart and Lung Institute
Harefield Hospital
Harefield, United Kingdom

In view of the participation of vascular endothelial cells (EC) in a wide range of normal and pathological conditions, phenotypic heterogeneity of EC is an important concept for consideration: different sub-populations of EC may play differing biological roles in normal and patho-physiological states.

We investigated the antigenic status of vascular endothelium from different sites of the human cardiovascular system and umbilical cord. The presence of endothelium was defined by immunoreactivity with a pan-endothelial cell (EC) marker EN4. This being established, the endothelium was studied using a panel of monoclonal antibodies. These included other recognized EC markers (FVIII-RAg, Pal-E and 44G4) and markers of immune activation (MHC I, MHC II, ICAM-1 and OKM5). With reference to the table below, it can be seen that a distinct heterogeneity of antigenic expression exists between large vessel (Aorta, Pulmonary artery, Coronary artery and Umbilical vessels) and small vessel (myocardial capillary) endothelium.

TABLE: Phenotypic status of *in situ* EC examined

TISSUE Vessels n=	PA (5)	A (5)	Adult Heart CA (3)	E (4)	AV	C	Fetal Heart E	C (1)	Hu. Umbil. Cord Vein (4)	Artery (4)
MABs										
EN4	3	3	3	3	3	3	3	3	3	3
FVIII	3/4	3/4	3/4	3/4	3/4	1	1	1	4	3
PalE	2	2/1	2/1	3	3/2	3	3	3	3	3
44G4	2/1	3/2	3/2	3	3	3	4	3	3/4	3/4
MHC I	2	2	3/2	2	3	3	3/2	3/2	2/0	2/0
HLA-DR	0	0	0	0	3	3	0	0	0	0
HLA-DP	0	0	0	nd	2	3	0	0	0	0
HLA-DQ	0	0	0	nd	1	0	0	0	0	0
ICAM-1	2	2	3/2	2	3	3	0	2	3/2	3
OKM5	0	0	nd	0	3	3	0	3	0	0
OKMI	0	0	nd	0	0	0	0	0	0	0

Key 0= undetectable, 1 to 4 = increasing intensity of stain. PA.= Pulmonary artery, A= Aorta, CA= Coronary Artery, E= Endocardium, AV= Arterioles/Venules, C= Capillary

Capillary endothelium exhibited strong expression of immunological markers (MHC I and II, ICAM-1 and OKM5) with infrequent FVIII-Rag binding. Conversely, large vessel endothelium exhibited strong reactivity with FVIII-RAg, weak expression of immunological markers MHC I/ICAM-1, with MHC II and OKM5 being undetectable.

Furthermore, endocardial EC were found to be phenotypically similar to large vessel endothelium.

The question arises as to whether the heterogenous expression of EC antigens reflects a permanent specialized state for specific functions or is an induced response maintained by factors in the immediate environment.

SECRETION OF MONOCYTE CHEMOTACTIC ACTIVITY BY BLEOMYCIN TREATED RAT PULMONARY ARTERY ENDOTHELIAL CELLS

Sem H. Phan, F. Wolber and *U.S. Ryan

Department of Pathology
University of Michigan Medical School
Ann Arbor, Michigan U.S.A.

*Monsanto Company
St. Louis, Missouri 63167, U.S.A.

Increased lung mononuclear phagocyte numbers accompany many forms of lung injury, including those associated with fibrosis. In previous studies, bleomycin-treated alveolar macrophages have been shown to secrete monocyte chemotactic activity which could contribute to the increase in lung mononuclear phagocytes. In this study, rat pulmonary artery endothelial cells were examined for their ability to secrete similar activity in response to bleomycin, since the lung endothelial cell is also among the first cells to be exposed to bleomycin and is known to be an important *in vivo* source of cytokines.

Confluent cultures of rat pulmonary artery endothelial cells were exposed to various doses of bleomycin in serum free medium for 2-24 hours. The conditioned media were harvested, dialyzed exhaustively and tested for monocyte chemotactic activity using modified Boyden blindwell chambers.

The results show that bleomycin stimulated monocyte chemotactic activity secretion in a dose-dependent fashion, reaching maximal response at about 50-100 ng/ml. Kinetic studies show that >6-8 hours are required for full activity to be expressed. Gel filtration analysis by high performance liquid chromatography revealed the activity to be heterogeneous with major peaks of activity having molecular weights between 5-15 kD. In preliminary studies, this monocyte chemotactic activity was neutralized by antisera to murine JE, but not by antibody to transforming growth factor-β, and thus may correspond to the recently described monocyte chemotactic protein variously referred to as monocyte chemotactic protein-1 (MCP-1) or monocyte chemotactic activating factor (MCAF), and found to be secreted by human monocytes, fibroblasts and endothelial cells upon stimulation by cytokines. Human MCP-1 has an estimated molecular weight of approximately 13 kD, and due to significant sequence homology between MCP-1 and murine JE, it is now considered to be the human counterpart of murine JE. Hence, in early bleomycin-induced lung injury, direct stimulation of monocyte chemotactic activity secretion by endothelial cells may play an important role in the recruitment of peripheral blood monocytes into the lung interstitium and alveolar space. This increase in lung mononuclear phagocyte numbers is likely to promote fibrosis due to their ability to secrete fibrogenic cytokines. [*Part of this work was supported by NIH grants HL28737, HL31963, HL39925 and DK38149, and was undertaken during the tenure of an established investigatorship (SHP) from the American Heart Association*].

PROTEIN KINASE C MODULATION OF SEROTONIN TRANSPORT BY PULMONARY ENDOTHELIAL CELLS

Bruce R. Pitt

Department of Pharmacology
University of Pittsburgh School of Medicine
Pittsburgh, Pennsylvania 15261 U.S.A.

We (*Riggs et al., 1989*) and others (*Gardaz et al., 1988*) have noted that phorbol myristate acetate (PMA) reduces serotonin (5-HT) extraction in intact lungs. This effect can be produced in perfused rabbit lungs without the presence of granulocytes (*Myers and Pitt, 1988; Merker and Gillis, 1988*) suggesting a direct effect of PMA on lung tissue. The effect does not seem to be merely a function of loss of vascular surface area due to pulmonary vasoconstriction since PMA-induced depression of 5-HT occurs without a significant change in the maximal velocity of pulmonary angiotensin-converting enzyme in the presence of: a) catalase in the intact lung (*Riggs et al., 1989*); or b) papaverine in the perfused lung (*Myers and Pitt, 1988*). Furthermore, PMA inhibits 5-HT uptake in the perfused lung at a time when surface area is unchanged as assessed by minimal changes in propranolol uptake (*Merker and Gillis, 1988*). Although the mechanism underlying PMA-induced decrease in 5-HT transport is unclear, we recently reported that PMA inhibits 5-HT uptake in bovine pulmonary artery endothelial cells in culture and that this effect is associated with translocation of protein kinase C (*Myers et al., 1989*). We also noted that staurosporine, a putative inhibitor of PKC, blocked the PMA-induced depression in 5-HT uptake in cultured pulmonary artery endothelial cells (*Myers et al., 1989*) as well as perfused lungs (*Weng and Pitt, 1990*). Other activators of PKC, including diterpene derivatives (e.g. phorbol dibutyrate) and non-phorbol esters (e.g. mezerein), also inhibited 5-HT uptake in both cultured cells (*Myers et al., 1989*) and perfused lungs (*Weng and Pitt, 1990*). These data are consistent with a negative modulatory role of activation of intracellular PKC on endothelial cell transport of 5-HT. We hypothesize that such activation may underlie other observations regarding decreased serotonin transport and acute lung injury (*Pitt et al., in press*).

REFERENCES

Gardaz, J.P., P. Py, P.M. Suter and A.F. Junod: Effects of oleic acid, alpha naphthylthiourea and phorbol myristate acetate-induced microvascular damage on indexes of pulmonary endothelial function in anesthetized dogs. *Am. Rev. Resp. Dis.* 137:1350, 1988.

Merker, M. and C.N. Gillis: Propranolol and serotonin removal in lung injury. *J. Appl. Physiol.* 65:2579, 1988.

Myers, C.L., J.S. Lazo and B.R. Pitt: Translocation of protein kinase C is associated with inhibition of 5-HT uptake by cultured endothelial cells. *Am. J. Physiol.* 257 (*Lung Cell Mole. Physiol.* 1):L253, 1989.

Myers, C.L. and B.R. Pitt: Selective effect of phorbol ester on kinetics of serotonin removal and ACE activity of in situ perfused rabbit lungs. *J. Appl. Physiol.* 65:377, 1988.

Pitt, B.R., C.L. Myers and C.N. Gillis: The biochemical pharmacology of pulmonary amine disposition. In: *Pulmonary Biology of the Normal Lung*. Ed: N. Voelkel. Teleford Press, In Press

Riggs, D., A.M. Havill, B.R. Pitt and C.N Gillis: Pulmonary angiotensin converting enzyme kinetics following acute lung injury in the anesthetized rabbit. *J. Appl. Physiol.* 64:2508, 1989.

Weng, W. and B.R. Pitt: Activation of protein kinase C inhibits extraction of serotonin by perfused rat lungs, *in situ*. *Am. J. Physiol.* 258:(*Lung Cell. Mole. Physiol.* 2:L289, 1990.

N-HYDROXYLATION AND VASCULAR RELAXATION:
AN ALTERNATE EDRF HYPOTHESIS

G. Thomas and Peter W. Ramwell

Georgetown University Medical Center
Washington, D.C. U.S.A.

The concept of endothelium-dependent relaxation is now ten years old, but there is still controversy as to the chemical nature of the endothelium dependent relaxing factor (EDRF) or factors. The similarities in the pharmacological properties of EDRF with nitrovasodilator drugs and the demonstration that endothelial cells release nitric oxide (NO) has lead to the suggestion that EDRF is an endogenous nitrovasodilator (*Furchgott, 1988; Palmer et al., 1988*).

The generation of EDRF is believed to be associated with the oxidation of the basic amino acid L-arginine (*Palmer et al., 1988*). However, L-arginine when added to endothelium-intact vessels has no vasodilatory properties except in millimolar concentrations (*Thomas et al., 1986*). In contrast, several substituted arginine compounds such as Na-benzoyl L-arginine ethyl ester (BAEE) are potent vasodilators both *in vitro* and *in vivo* and generate the corresponding citrulline, the second product formed during EDRF formation (*Thomas et al., 1988*). In addition, the dipeptide form of arginine is also a potent vasodilator when compared to L-arginine and also generates citrulline (*Thomas et al., 1986*). The relaxation elicited by the substituted arginine compounds is inhibited by the EDRF inhibitors L-monomethyl L-arginine, methylene blue, hemoglobin and superoxide anion (*Thomas et al., 1988; Thomas and Ramwell, 1989*)

Several arginine peptides and substituted guanidine compounds are present in endothelial cells but little is known of their biological properties. The weak vasodilatory effect of L-arginine in comparison to the potent effect of substituted arginine compounds indicate that the precursor of EDRF is an arginine containing moiety or a substituted guanidine.

The identification of NO as EDRF, however, depends on acid activation in the presence of potassium iodide (*Palmer et al., 1988*). Several endogenous compounds are converted to NO by such a chemical treatment. These include nitosoamines, hydroxylamines and oximes (*Thomas and Ramwell, 1989*). Like EDRF they are powerful vasodilators and increase cyclic GMP formation (*Thomas and Ramwell, 1989*). Many of these compounds, particularly hydroxylamines (R-NHOH) and oximes (R=NOH), directly bind to heme-containing compounds. Hence, they can activate guanylate cyclase, a heme-containing enzyme, without being converted to nitric oxide. The hydroxylamines and oximes generate nitric oxide on interaction with superoxide anion and hydrogen peroxide, and this is associated with increased biological activity. Thus, compounds such as hydroxylamines and oximes may generate NO depending on oxygen tension, temperature and pH.

From these results we suggest the following scheme for EDRF generation from an arginine compound or a substituted guanidine (R).

R --- R-NHOH ---- R-NOH ----R-NO ----NO

The nature of the R group determines the stability and ease of conversion to NO. Several substituted hydroxylamines and oximes exhibit vasodilatory properties without the generation of NO (*Thomas and Ramwell, 1989*). The vasodilator effects of the hydroxylamines and oximes are inhibited by both methylene blue and hemoglobin (*Thomas and Ramwell, 1989*). These results clearly support the notion that EDRF is a hydroxylamine or an oxime with NO being an active oxidation product derived from them. This will explain the discrepancies in the literature regarding the identity of EDRF as NO. Both the hydroxylamines and oximes avidly bind to proteins and cell membrane components. Hence, it is unlikely that they will be present in the effluent from the endothelial cells, whereas, gaseous lipophilic compounds such as NO readily permeate cell membrane and hence could be detected in the extracellular space. Some of the NO may be still attached to the R group and released as R-NO which rapidly undergoes decomposition to generate free NO. Such a NO carrier molecule (R-NO) may have the same biological activities as free NO. We have shown previously that NO-carrying molecules like streptozotocin are vasodilators and activate guanylate cyclase (*Thomas and Ramwell, 1989*). However, the amount of the NO-carrying molecule released from the endothelium may be insignificant since very little citrulline is detected outside intact tissues.

In conclusion, we suggest that endothelial cells generate hydroxylamines and oximes from an arginine moiety or a guanidino compound. These hydroxylamines and oximes directly activate guanylate cyclase without necessarily being converted to NO. Thus, NO or a NO-carrying molecule may be post-event products from the endothelial cells. (*This work was supported by National Institutes of Health Grants HL 36802 and HL 40069*).

REFERENCES

Furchgott, R.F. In: *Vasodilation: Vascular smooth muscle, peptides, autonomic nerves and endothelium*, Vanhoutte, P.M., ed., Raven Press New York, pp. 401-414, 1988.

Palmer, R.M.J., Ashton, D. S. and Moncada, S. *Nature (London)* 333:664-666, 1988.

Thomas, G., Hecker, M. and Ramwell, P.W. *Biochem. Biophys. Res. Commun.* 158:177-180, 1989.

Thomas, G., Mosthaghim, R. and Ramwell, P.W. *Biochem. Biophys. Res. Commun.* 141: 446-451, 1986.

Thomas, G. and Ramwell, P.W. *Biochem. Biophys. Res. Commun.* 154:332-338, 1988.

Thomas, G. and Ramwell, P.W. *Biochem. Biophys Res. Commun.* 164:889-893, 1989.

Thomas, G. and Ramwell, P.W. *European J. Pharmacol.* 161:279-280, 1989.

INVESTIGATION OF ANF RECEPTOR FUNCTION IN CULTURED BOVINE PULMONARY ARTERY ENDOTHELIAL CELLS

Eileen M. Redmond and Alan K. Keenan

Department of Pharmacology
University College
Dublin 4, Ireland

Atrial Natriuretic Factor (ANF) is a peptide hormone which exhibits a number of hormonal effects that influence cardiovascular homeostasis, including diuresis, natriuresis and relaxation of vascular smooth muscle. The lung is one of the first organs to receive the peptide on its release from the right atrium and recent experimental studies, using isolated arterial and venous preparations, have demonstrated that ANF can induce relaxation of the pulmonary vasculature (*Ignarro et al., J. Appl. Physiol. 60:1128, 1986; Brink et al., Eur. J. Pharm. 150:397, 1988*). Alterations of ANF release could therefore be of physiopathologic importance in patients with pulmonary artery hypertension. Specific ANF binding sites have previously been located on endothelial cells, which can be regarded as the first tissue encountered by the blood born hormone; the functional consequences of activation of these receptors remains to be clarified. The purpose of this study was to characterize ANF receptor binding and coupling to particulate guanylate cyclase (pGC) in a primary culture of bovine pulmonary artery endothelial cells (BPAEC). Furthermore, a preliminary investigation of ANF's possible role as a cytoprotectant was undertaken.

Cells, (obtained from the University of Miami), were maintained and subcultured in M199 containing 5% fetal calf serum, 5% Nu-serum, 50U/ml penicillin/streptomycin, 2.5µg/ml fungizone and 2mM glutamine. The cells were grown with 95% air-5% C_2 in a humidified incubator maintained at 37^0C and seeded in 24-well plates (Nunclon) at a density of approximately 100,000 cells/well for subsequent experiments. For binding analyses, cell monolayers were incubated with (a) various concentrations (0.05-3.0nM) of [^{125}I]rANF(99-126) (saturation analysis) or (b) 0.5nM [^{125}I]rANF(99-126) alone or in the presence of increasing concentrations of unlabelled rANF(99-126) or one of its analogs (competitive displacement analysis) for 30 min at 37^0C. Non-specific binding was determined in the presence of 100 fold molar excess of unlabelled rANF(99-126). Following washing, cell monolayers were solubilized with 1M NaOH to determine cell-bound radioactivity. For internalization studies, surface-bound radioactivity was removed by "acid-stripping" with 0.2M acetic acid pH 2.5, containing 0.5M NaCl for 6 min at 4^0C. Intracellular cyclic GMP, (cGMP) accumulation, stimulated by ANF and its analogs, in the presence of 0.5mM IBMX, was determined by radioimmunoassay after TCA solubilization of the cells. "Extruded"/extracellular cGMP present in the medium bathing the cells was also measured following extraction with 1.2M $HCLO_4$ and neutralization with KOH before acetylation and radioimmunoassay. To visualize actin, BPAEC monolayers grown on circular coverslips were washed twice in PBS, fixed in 3.7% formaldehyde in PBS at 20^0C for 10 min, and washed twice in PBS before permeabilization for 3-5 min in -20^0C acetone. The coverslips were air-dried and incubated for 20 min at 20^0C in 200µl of 0.165µM rhodamine-conjugated phalloidin, washed twice in PBS and mounted on slides with PBS and glycerol for epifluorescence microscopy.

Saturation binding of [^{125}I]rANF(99-126) to BPAEC indicated a single high affinity site with K_D 0.62 ± 0.08 nM and B_{max} 102 ± 20 fmol/10^6 cells (n=4) corresponding to 61,500 sites/cell. Binding of [^{125}I]rANF(99-126) was displaced by unlabelled rANF(99-l26), rANF(l03-125), rANF(l03-123), des[Cys^{105}, cys^{121}]rANF(l04-126) (#SC-46313), des[Phe^{106}, Gly^{107}, Ala^{115}, Gln^{116}rANF(104-126)(#SC-46542) and des[Gln^{116}, Ser^{117} Gly^{118}, Leu^{119}, Gly^{120}] (C-ANF) in a dose - dependent manner with IC$_{50}$ values being comparable (0.4-4.3nM). However, in contrast to rANF(99-126) and rANF(l03-125), #SC-46313, #SC-46542, C-ANF and rANF(l03-123) failed to completely displace [^{125}I]rANF(99-126) binding even at a concentration of 1μM, leaving 5%, 1%, 7.8% and 6.5% respectively, undisplaced. These latter four analogs, therefore apparently, fail to recognize a small proportion of the ANF receptor pool. In the presence of 1μM unlabelled #SC-46313 or #SC-46542, rANF(99-126) completely displaced "residual" [^{125}I]rANF(99-126) binding. The IC$_{50}$ for rANF(99- 126) was similar in the absence or presence of 1μM unlabelled #SC-46313 or #SC-46542, (0.4 compared with 0.63 and 0.2nM). A time- and temperature-dependent internalization of [^{125}I]rANF(99-126) was demonstrated. ANF produced a time- and dose-dependent increase in cellular cGMP. Maximal fold increases over basal varied between 5 and 12 for 0.1μM ANF. The order of potency for cGMP production was rANF(99-126) > rANF(l0l-125) > rANF(l03-125) >> C-ANF, with #SC-46313, #SC-46542 and rANF(l03-123) giving no increase over basal. rANF(99-126)- and rANF(l03-125)-stimulated production of cGMP was followed by a time-dependent release of the nucleotide into the incubation medium; this release was negligible at short incubation times (when cellular cGMP content peaked), but thereafter increased steadily for up to 2hr. In response to 0.1nM PMA (a potent activator of protein kinase C), BPAEC F-actin stress fiber pattern and distribution was greatly disrupted, compared to untreated monolayers, an effect which appeared to be attenuated in the presence of rANF(99-126).

These experiments show that specific high affinity ANF receptors are present on BPAEC. There appears to be at least 2 ANF receptor subtypes present in this system; one receptor (ANF-B), linked to pGC is present as the minor component, while the so-called ANF-C or clearance receptor forms the major part of the total receptor pool. With time a significant proportion of the peptide receptor complex is internalized at 37^{0}C. Stimulation of the ANF-B receptors induces increased cellular cGMP production followed by subsequent extrusion to the extracellular space. Finally ANF diminished the PMA-induced changes in the actin cytoskeleton supporting a cytoprotective role for this peptide. [*This work was supported by EOLAS (The Irish Science and Technology Agency) and by the Irish Heart Foundation.*]

POTENTIATION OF ENDOTHELIN-1 INDUCED PULMONARY VASOCONSTRICTION BY ACIDOSIS

Pnina G. Rosenkranz, S. Rimar and C.N. Gillis

Departments of Anesthesiology, Pharmacology and Pediatrics
Yale University
New Haven, Connecticut, U.S.A.

Acidosis blunts the effect of many pulmonary vasoconstrictors (*Porcelli, et al., 1971*). However, previous studies have demonstrated that acidosis markedly increases the binding of ^{125}I Endothelin-1 (ET-1) to rabbit lung membranes (*Rosenkranz and Gillis, 1990*). We therefore sought to assess the effect of acidosis on ET-1 induced pulmonary vasoconstriction.

Rabbit lungs were perfused (30 ml/min) *in situ* by recirculating Krebs-3% albumin solution (pH=7.4) aerated with 95% O_2, 5% CO_2 at 37°C. The pH of the medium was decreased to 7.0 with 2N HCl for 5 minutes with no significant effect on pulmonary artery pressure (n=20). ET-1 16 nM was then recirculated at pH 7.4 for 10 minutes. When the pH of the perfusate was again decreased to 7.0, the pulmonary artery pressure increased by 30 ± 3% ($p < 0.01$) from a baseline of 14.9 ± 7.4 mm Hg (n=9). Similar results were obtained 5 and 45 minutes after ET-1 was added. There was no evidence of pulmonary edema; the wet:dry weight ratio was 5.4 ± 0.7 (n=4). Specificity of this effect was demonstrated by the fact that acidosis did not increase pulmonary artery pressure after preconstriction with either U46619 (n=6) or histamine (n=3).

The fact that ET-1 induced pulmonary vasoconstriction is potentiated by acidosis, as well as the apparent specificity of this effect, may prove important in understanding its role in pathophysiology as well as its mechanism of smooth muscle contraction. (Supported by USPHS Grants HL 13315, RR05358, HL 40863 and HL 07272).

REFERENCES

Porcelli, R.J. and Bergofsky, E.H.   *J. Appl. Physiol.* 31:679-685, 1971.
Rosenkranz, P.G. and Gillis, C.N. *Am. Rev. Respir. Dis.* 141:A481, 1990.

THE EFFECT OF IN VIVO AND IN VITRO CAPSAICIN TREATMENT ON ENDOTHELIUM DEPENDENT RELAXATION

Thomas M. Scott and Karen Drodge

Faculty of Medicine
Memorial University of Newfoundland
St. John's
Newfoundland, Canada

The ability of vasoactive agents to induce endothelium-dependent vascular relaxation appears to depend on a number of factors. We have examined the contribution of specific peptidergic perivascular nerve fibers to endothelium dependent vascular relaxation elicited by acetylcholine. Adult rats were treated with capsaicin (50mg/kg on two consecutive days) under ether anesthesia. An equal number of rats was treated in the same way with capsaicin vehicle (10% Tween 80, 10% ethanol in normal saline). One week after treatment of adult rats, treated and age-matched untreated rats were anesthetized (sodium pentobarbitone, 35mg/kg i.p.) and the superior mesenteric arterial bed removed. The superior mesenteric artery was cannulated and perfused at a constant rate of flow (3ml/min) with Krebs' solution. Methoxamine ($0.4 - 1 \times 10^{-5}$M, gift from Burroughs Wellcome) was added to the perfusing solution to produce a constriction of the arterial bed sufficient to raise the pressure above 100mmHg. Capsaicin (10^{-5}M), acetylcholine (10^{-8}M, 5×10^{-7}M), A23187 (10^{-8}M, 5×10^{-8}M) and sodium nitroprusside (10^{-4}M) were tested for their ability to produce relaxation. Following application of acetylcholine or A23187, 1-arginine (10^{-4}M) was added to the perfusing solution once a stable pressure had been achieved. This protocol was also carried out on vessel beds from treated and untreated rats following perfusion for ten minutes with Krebs' solution containing capsaicin (10^{-5}M). At the conclusion of each experiment, the vessel beds were processed for immunohistochemical demonstration of substance P and calcitonin gene-related peptide immunoreactive perivascular innervation.

It was determined that both *in vivo* and *in vitro* capsaicin treatment resulted in increased sensitivity to methoxamine. Capsaicin treatment reduced the ability of the tissue to relax to acetylcholine by up to 45% following *in vivo* treatment and to acetylcholine and A23187 by up to 95% following *in vitro* treatment. *In vivo* capsaicin treatment resulted in the loss of the ability to demonstrate substance P or calcitonin gene-related peptide immunoreactive perivascular nerve fibers. Addition of L-arginine did not restore the ability of the agents to induce relaxation.

Although capsaicin has been shown to inhibit acetylcholine-elicited vascular relaxation following *in vivo* and *in vitro* treatment, there is only circumstantial evidence that peptidergic perivascular innervation is involved, and that the mechanism underlying the phenomenon is common to both treatments reported.

KAWASAKI DISEASE

Stanford T. Shulman

Northwestern University Medical School
2300 Children's Plaza
Chicago, Illinois 60614, U.S.A.

Kawasaki Disease (KD) is an acute inflammatory arteritis that preferentially affects coronary arteries and primarily occurs in young children. The acute illness is characterized by fever for at least five days, rash, conjunctival injection, inflamed oral mucosa, red and swollen hands and feet, and cervical lymphadenopathy. Epidemiologic data indicate a median age of 2 years, a male:female ratio of 1.5:1, world-wide distribution with highest rates in Japanese and Korean children, epidemics at 2 1/2 to 3 1/2 year intervals, and lack of apparent person-person contact. More than 100,000 cases have been registered in Japan, and approximately 3000 cases/year occur in the U.S. Coronary artery abnormalities (aneurysms, stenosis, myocardial infarction) develop in 20-25% of patients with KD, and in developed countries KD now rivals rheumatic fever as the leading cause of acquired heart disease in children. Late coronary artery histologic changes include striking myo-intimal proliferation and fragmentation of the internal elastic lamina. The etiology is unknown, but clinical and epidemiologic features strongly support an infectious etiology. The pathogenesis of KD is likely to involve immune mechanisms in that 1) profound immunoregulatory aberrations are present, 2) serum antibodies cytotoxic for cytokine-stimulated endothelial cells have been demonstrated, and 3) activated monocytes and CD4-positive lymphocytes that express class II antigens dominate the vasculitic inflammatory infiltrate. Recently, coronary endothelial cells expressing class II antigens were demonstrated in affected but not in unaffected vessels of an autopsied patient. Despite incomplete data regarding the etiology and pathogenesis of KD, remarkably effective therapy has been developed recently. High dose intravenous gammaglobulin (IVGG) with aspirin is approximately 85% effective in preventing the development of coronary abnormalities, reducing their prevalence to approximately 3%. The mechanism of action of IVGG in KD is unknown, but the prognosis of patients with KD has very significantly improved with this treatment. The precise role of vascular endothelial cells in KD requires future studies.

DIFFERENT MECHANISMS OF HOMOLOGOUS AND HETEROLOGOUS DESENSITIZATION OF THROMBIN INDUCED ENDOTHELIAL PROSTACYCLIN PRODUCTION

H. Halldorsson and *Gudmundur Thorgeirsson

Department of Pharmacology
University of Iceland
Iceland

*Department of Medicine
Landspitalinn
University Hospital of Iceland
Reykjavik, Iceland

Endothelial cells respond to a number of agonists by a burst of prostacyclin production. Several workers have described desensitization of this response, but conflicting evidence has been published regarding the mechanism of desensitization; whether it is homologous (agonist specific) or heterologous and whether inactivation of cyclooxygenase is involved. Previously, we have shown concurrent homologous desensitization of inositol phosphate production and prostacyclin production, and that neither desensitization is mediated by protein kinase C. The purpose of the present study was to determine the relation between the intensity of a first thrombin stimulus and a subsequent response to a repeat thrombin, histamine or ionophore A23187 stimulation. We further sought to determine the target of desensitization by measuring the generation of inositol phosphates and release of arachidonic acid in addition to production of prostacyclin.

Following thrombin stimulation of cultured human umbilical vein endothelial cells, only homologous desensitization of inositol phosphate production was observed but both homologous and heterologous desensitization of arachidonic acid release and prostacyclin production occurred. For any given dose of the first stimulant there was much greater effect on the homologous response than on the heterologous response. Pretreatment with only 0.08 U of thrombin, a dose causing minimal prostacyclin production, was sufficient to virtually abolish the response to a subsequent thrombin stimulation, whereas pretreatment with 1U of thrombin was required to diminish the response to a subsequent stimulation with histamine or A23187 by 50%. These differences suggest different mechanisms. The homologous desensitization probably involves the receptor, whereas the present results suggest that the target of heterologous desensitization is distal to calcium mobilization in the signal transduction pathway. The possibilities include decreased activity of Phospholipase A_2 or decreased pool of accessible arachidonic acid.

PRO OXIDANT EFFECTS OF NORMOBARIC HYPEROXIA IN RAT TISSUES

M. Ahotupa, E. Mantyla, V. Peltola,
A. Puntala and Hannu Toivonen*

Department of Physiology
University of Turku
Turkey

*Department of Anesthesia
Helsinki University Central Hospital
Helsinki, Finland

Rats were exposed to 100% oxygen atmosphere for 12, 36 or 48 hours. Their lungs, brain, liver and kidney were studied for signs of free radical oxidant damage.

Macroscopically, the brains started gaining weight at 12 hours followed by the lungs at 36 hours. At 48 hours the brain/body weight ratio was increased by 58% with no change in tissue protein content, whereas, in lungs 33% weight gain was due to non-protein edema, and lung wet/dry weight ratio rose from 2.5 to 6.9. Liver water content, on the contrary, decreased as indicated by the 24% decrease in liver/body weight ratio together with corresponding increase in liver protein concentration. Kidney weight and protein content remained unaltered.

Free radicals generate a fluorescent chromolipid via peroxidation of tissue lipids (*Esterbauer et al., Biochem. J. 238:405-409, 1988*). The concentration of this chromolipid increased in the kidneys from the control of 6560 ± 460 (arbitrary units) to 8850 ± 110 already after 12 hours of oxygen exposure and remained at this level after 36 or 48 hours exposure (8680 ± 360 and 9050 ± 290, respectively). Similar increase was detected after 36 and 48 hours in brains (from 392 ± 79 to 781 ± 115 and 933 + 109, respectively) and lungs (from 730 ± 30 to 930 ± 60 and 1110 ± 110, respectively), whereas no increase could be detected in the liver. The early increase of this chromolipid in kidneys was explained by its urinary excretion as it was detectable in the urine already after 6 hours of oxygen exposure and the amount excreted correlated with the length of hyperoxic period.

No signs of lipid peroxidation could be detected by the measurement of diene conjugation double bonds in the same tissues. Neither was there any increase in thiobarbituric acid reactive lipid peroxidation products in lungs, liver or serum of the same animals.

The antioxidant enzyme levels were measured in lungs and brain. Brain Cu/Zn superoxide dismutase (SOD) increased from the control of 588 ± 41 µg/g tissue to 760 ± 37 and 791 ± 30 µg/g after 36 or 48 hours exposure, respectively, whereas, catalase activity remained under the detection limit of our assay. Lung catalase activity decreased from 1.02 ± 0.16 mg/g tissue to 0.52 ± 0.04 and 0.56 ± 0.03 mg/g after 36 or 48 hours exposure, respectively. Lung SOD activity could not be quantified because of the formation of interfering epinephrine oxidizing compound in lungs during oxygen exposure.

In conclusion, decreased catalase activity during oxygen exposure could create hydrogen peroxide accumulation during oxidant attack on tissues. After hyperoxic challenge, fluorescent chromolipid formation seems to be more sensitive than diene conjugation or thiobarbituric acid reactive compounds in detecting oxidant-caused lipid peroxidation.

COMPARATIVE STUDY OF PATHOLOGIC CARDIOVASCULAR BIOMINERALIZATION

Branko B. Tomazic, C. Siew and W.E. Brown

ADAHF Paffenbarger Research Center
National Institute of Standards and Technology
Gaithersburg, Maryland 20899, U.S.A.

One of the most severe consequences of atherosclerosis is the irreversible formation of calcific deposits in different sites of the cardiovascular system. The pathogenesis appears to be related to the initial damage of the endothelium (*Ross and Glomset, 1976*), which increases endothelial permeability. When a lesion forms, the natural vasodilation/vasoconstriction performance of endothelium and underlying cellular layers is impaired. This condition can result in progressive atherogenesis, leading to pathologic mineralization. Dystrophic calcification is a major problem associated with prolonged use of bioprosthetic heart valves (*Schoen, 1989*). The biomineralization and deterioration of bioprostheses appears to be a fast process, compared to degeneration of natural heart valves. This may be due to the absence of the vital protective endothelium layer, which in healthy tissues prevents active transport of atherosclerotic plaque ingredients. Wickham et al. (1988) documented an increase of intracellular calcium in infected endothelial cells. This increase may be a valid messenger of inorganic ion diffusion from extracellular fluid into intracellular calcification sites.

In the present study, we investigated the process of mineralization of glutaraldehyde-pretreated bovine pericardium, BP, the material used for fabrication of artificial heart valves, under *in vitro* and *in vivo* conditions as follows: *in vitro*: BP segments were immersed in a metastable calcium phosphate solution and mineralization was followed at constant pH = 7.40. Mineralization took place only on the surface of the BP *in vivo*: BP segments were implanted subcutaneously in rats and retrieved at selected times. Histologic (van Kossa) and microscopic (SEM, EDX) examination of 10 μm cross sections of implants proved that only intrinsic dystrophic calcification took place. After implantation, the tissue became inflamed and implants were swollen. These abnormalities probably increased BP permeability and vasodilation, allowing active diffusion of inorganic ions through the inactive endothelium into subsurface calcification sites. Analysis of BP calcific deposits and thermodynamic considerations are in accord with previous findings on the composition of atherosclerotic and bioprosthetic calcific deposits (*Tomazic et al., 1989*). Conclusions of this study support a mechanism for calcification that involves octacalcium phosphate as a precursor to cardiovascular apatitic biomineral. (*Supported in part by NHLBI grant HL30035*).

REFERENCES

Ross, R. and Glomset, J.A. The pathogenesis of atherosclerosis. *N. Engl. J. Med.* 295:420, 1976.

Schoen, F.J. Interventional and Surgical Cardiovascular Pathology. *Clinical Correlations and Basic Principles.* W.B. Saunders Company, Philadelphia p.137, 1989.

Tomazic, B.B., Brown, W.E., Queral, L.A. and Sadovnik, M. Physicochemical characterization of cardiovascular calcified deposits, *Atherosclerosis* 69:5-19, 1989.

Wickham, N.W.R., Vercellotti, G.M., Moldow, C.F., Visser, M.R. and Jacob, H.S. Measurement of intracellular calcium concentration in intact monolayers of human endothelial cells. *J. Lab. Clin. Med.* 12:157-167, 1988.

INHIBITION OF EDRF SYNTHESIS REDUCES THE PENTAGASTRIN-INDUCED HYPERAEMIA OF THE RAT GASTRIC MUCOSA

Claire E. Walder, C. Thiemermann and J.R. Vane

The William Harvey Research Institute
St.Bartholomew's Hospital Medical College
Charterhouse Square
London, United Kingdom

The release of local vasodilator mediators plays a pivotal role for the regulation of gastric mucosal blood flow. In the rat gastric microcirculation *in vivo*, endothelium-dependent increases in blood flow were induced by acetylcholine and vagal stimulation (*Kitagawa et al., 1987*). Furthermore, inhibition of EDRF (NO) synthesis dose-dependently reduces rat gastric blood flow, indicating that endogenous NO plays a role in the modulation of basal blood vessel tone in the gastric vasculature (*Pique et al., 1989*). The present study was designed to investigate the role of endothelium-derived relaxing factor (EDRF) as a potential mediator of the pentagastrin-induced hyperaemia in the rat gastric mucosa *in vivo*, under conditions where prostacyclin mediation was eliminated.

Male Wistar rats (215-310g; starved for 20-26 h) were anaesthetized with inactin (120 mg kg^{-1} i.p.) and treated with indomethacin (5 mg kg^{-1} i.v.). The stomach was exposed via a mid-line laparotomy and opened along the greater curvature. Gastric mucosal blood flow (GMBF) was continuously monitored by using a standard laser Doppler flow probe placed on the mucosal surface of the main body of the stomach and expressed as perfusion units (PU).

Infusion of pentagastrin (750 ng kg^{-1} min^{-1} i.v. for 10 min) significantly enhanced GMBF throughout the infusion period. N^G-monomethyl-L-arginine (L-MA; 1 mg kg^{-1} min^{-1} i.v.; n=6) or Nx-nitro-L-arginine methyl ester (L-NA; 0.1 mg kg^{-1} min^{-1} i.v.; n=3) two inhibitors of nitric oxide synthesis (*Palmer et al., 1988* and *Moore et al., 1989*) infused 10 min prior to and throughout pentagastrin administration, significantly increased mean arterial blood pressure (MAP) by 27% and 18%, respectively. Although neither inhibitor had any effect on GMBF, L-MA and L-NA attenuated the pentagastrin-induced hyperaemia by 82% and 84%, respectively (P<0.05). L-arginine (3 mg kg^{-1} min^{-1} i.v.; n=6) significantly reversed the increase in MAP and the inhibition of pentagastrin-induced hyperaemia brought about by L-MA alone.

These results show that in the absence of prostacyclin, EDRF is a mediator of the pentagastrin-induced hyperaemia in the rat gastric mucosa. (*The William Harvey Research Institute is supported by a grant from the Glaxo Group Research Limited. C.T. is a research fellow of the "Thyssen-Stiftung" Koln, F.R.G.*)

REFERENCES

Kitagawa, H. et al., *Eur. J. Pharmacol.* 133:57-63, 1987.
Moore, P.K. et al., *Br. J. Pharmacol.* 98:905P, 1989.
Palmer, R.M.J. et al., *B.B.R.C.* 153:1251-1256, 1988.
Pique, J.M. et al., *Eur. J. Pharmacol.* 174:293-296, 1989.

PROPERTIES OF A REPERFUSIBLE MODEL OF PHOTOCHEMICALLY INDUCED THROMBOTIC STROKE IN THE RAT MIDDLE CEREBRAL ARTERY TERRITORY

Brant D. Watson, H. Kanemitsu, R. Prado and W.D. Dietrich

Department of Neurology
Cerebral Vascular Disease Research Center
University of Miami
Miami, Florida 33101, U.S.A.

Exploration of therapies for cerebrovascular accidents, or stroke, may be constrained by the use of animal models in which cerebral ischemia is induced by temporary or permanent mechanical occlusion of brain arteries in conjunction with common carotid artery occlusion. With such models, reproducible infarcts may be produced, but at the intentional cost of inhibition of natural processes which may be protective, such as activation of the collateral circulation. In order to more realistically simulate clinical stroke, we have developed a model of thrombotic stroke in the middle cerebral artery (MCA) territory of the rat. The occlusions are formed *in situ* in response to argon laser irradiation of the MCA following intravenous injection of a photosensitizing dye, either rose bengal (20 mg/kg, with irradiation at 514.5 nm) or flavin mononucleotide (FMN) (37 mg/kg, with irradiation at 457.9 nm). The resultant photochemical reaction generates thrombi with different structures and stabilities toward thrombolysis. With rose bengal, the occlusions appear white under an operating microscope, and consist exclusively of a tight meshwork of agglutinated platelets which are extremely resistant to lysis by tissue plasminogen activator (rt-PA). Positive fluorescent antibody staining for Factor VIII and fibrinogen suggested that fibrin is not present, an impression confirmed by scanning (SEM) and transmission (TEM) electron microscopy. Intraplatelet binding may be facilitated by thrombospondin-stabilized fibrinogen bridges among glycoprotein IIb-IIIa receptors. In contrast, occlusions mediated by FMN photochemistry appear reddish and are lysed by rt-PA. Fluorescent antibody staining for Factor VIII and fibrinogen is negative, while TEM reveals platelet thrombi intercalated with short strands of a material which may be fibrin in the process of formation. A structureless, apparently proteinaceous secretion is also commonly present at the endothelial surface and throughout the occlusions. The efficiency of formation, and stability to either spontaneous recanalization or to that induced by rt-PA, appear to be inversely related to the blood pressure, although thrombus composition appears (by TEM) to be unrelated to it. Currently, the MCA is irradiated with three beams derived from the primary laser beam by means of a Ronchi ruling. The independently positionable beams are focussed with cylindrical lenses and steered by wedge prisms. With this approach, the thicknesses of the formed occlusions can be minimized by sharp focussing, thus minimizing the amount of rt-PA required to effect complete recanalization of the arterial segment whose total length (proximal to the rhinal branch and extending just distal to the inferior cerebral vein, ca. 1.3 mm) is much longer than the sum of individual occlusion thicknesses (100 μm each). Mechanical occlusion of this segment was reported previously to generate a large, reproducible infarct in the MCA territory (*Bederson et al., 1986*).

FMN-mediated photochemical occlusion of the MCA in Sprague-Dawley rats yields inconsistent infarct volumes, however. Infarct inconsistency may be related to local (as well as distal) activation of collateral circulation. The undamaged arterial segments between the occlusions are observed to dilate (coincident with blood pressure surges) and fill with blood despite the apparent lack of retrograde flow from MCA branches (e.g., rhinal), inasmuch as the branches and proximal trunk are occluded simultaneously by means of specific beam placement. This segmental infiltration of blood also destabilizes the occlusions and often leads to spontaneous thrombolysis. Under these circumstances, in the blood pressure range of 90-110 mmHg (but not at higher pressures), striatal infarct volume was found to be proportional to time of occlusion until spontaneous recanalization (cortical infarcts were not seen). Because infarct susceptibility depends on arterial blood pressure and collateralization, the rheological and histopathological characteristics of this minimally invasive and recanalizable model of arterial thrombotic stroke indicate that pharmacological manipulation of the collateral circulation may be necessary clinically.

ENDOTHELIAL CELL ACTIVATION BY NEUTROPHIL RELEASE PRODUCTS

Nicholas W.R. Wickham*, C.F. Moldow, S.P. Severson, H.S. Jacob, and G.M. Vercellotti

University of Minnesota
Minneapolis, Minnesota U.S.A.

The vascular endothelium comprises an extremely active endocrine and autocrine organ. We have shown previously that measurement of intracellular calcium (Ca_i), using the calcium-responsive fluorescent probe FURA 2 in endothelial cell (EC) monolayers grown on glass cover slips, is a highly sensitive method of detecting cell activation due to soluble inflammatory mediators, such as thrombin, histamine and bradykinin. Indeed, 0.005 u/ml thrombin is enough to cause a rise of Ca_i 15% above base-line, whereas, an increase in phosphatidyl-inositol (PI) metabolism, to which Ca_i changes are usually (but not inextricably) linked, could not be detected until cells were exposed to 0.1-0.5 u/ml thrombin. We have also shown that endothelial cell surface-expressed platelet activating factor (PAF) is increased in a dose-related fashion with exposure to similar minute amounts of thrombin. Furthermore, the PAF on thrombin-treated EC is capable of enhancing stimulated-PMN responses. In contrast, we now wished to establish whether inflammatory cells such as PMN and/or their release products, which are known to disrupt EC during severe acute inflammation, were capable of activating EC in the sub-acute situation, i.e. when PMN are present in suspension in physiological concentration and then activated, in this case by exposure to FMLP. FURA 2-loaded EC monolayers were exposed to PMN at 10^7/ml held in suspension in a magnetically stirred cuvette placed in a thermostatically controlled cuvette holder in a Spex Fluorolog II scanning spectrofluorometer. Unstimulated PMNs had no effect on basal Ca_i, but stimulation with FMLP at 10^{-7} M, produced an increase in Ca_i of 30% over 1 minute, with a fall to near basal concentration (~80nM) over 2 minutes. Pseudo-stimulation of PMN with HBSS produced no effect, and FMLP had no direct effect on EC alone. PMN release products obtained by collecting the supernatant from PMNs maximally stimulated with FMLP/CB produced a similar response at equivalent PMN concentrations. Although reagent hydrogen peroxide (100µM - 10mM) produced a dose-dependent increase in Ca_i, oxygen radical scavengers did not abolish the response to PMN release products. As proteolytic enzymes (thrombin, trypsin) cause increased Ca_i and PI turnover, it is interesting that purified PMN elastase (100-300 µg/ml) had no effect, and the potent elastase inhibitor - alpha 1 protease inhibitor only partially blocked the PMN-induced Ca_i rise. Another PMN lysosomal product, Cathepsin G, a strongly cationic protease, was a powerful stimulus to EC, producing a 200% increase in Ca_i, that was inhibited in the presence of heparin, a strongly anionic molecule. Supernatants from PMN sonicates, a more efficient (up to 10-fold) method of producing PMN lysosomal release products, were similarly potent stimuli to EC Ca_i and were also inhibited in the presence of heparin. Furthermore, we were also able to demonstrate that PI turnover and EC-PAF expression in response to sonicates increased in a dose-dependent fashion, whereas no increase was demonstrable with the weaker FMLP/CB stimulated PMN release-products. Thus, we have demonstrated that stimulated PMN are able to signal to EC in a graded fashion, that may allow non-disruptive functional changes in EC activity at low-level stimuli, but may equally provoke more profound changes in EC exposed to stronger inflammatory PMN stimuli. These changes may enhance the inflammatory process by a positive feed-back mechanism by virtue of increased EC-PAF expression and subsequent PMN priming. Furthermore, PMN cationic proteins such as Cathepsin G, would seem to make a major contribution to this process. (*present address: St. Helier Hospital Carshalton, United Kingdom)

BLOCKADE OF CD18-DEPENDENT NEUTROPHIL ADHESION LIMITS MYOCARDIAL INFARCT SIZE FOLLOWING CORONARY ARTERY OCCLUSION IN CYNOMOLGUS MONKEYS

Raymond J. Winquist, P. Frei, M. McFarland, P. Harrison, G. Letts, R. Rothlein and *T. Hintze

Boehringer Ingelheim Pharmaceuticals Incorporated
Ridgefield, Connecticut 06877, U.S.A.

*Department of Physiology
New York Medical College
Valhalla, New York 10595, U.S.A.

Neutrophil accumulation into ischemic myocardium is believed to underlie many of the deleterious sequelae observed with coronary artery occlusion and reperfusion. Excessive margination of and release of reactive and cytotoxic substances from activated neutrophils have been associated with the increased coronary vascular resistance, depressed ventricular wall function and myocardial necrosis resulting from ischemic and reperfusion injury. Neutrophil deposition in ischemic tissue is initiated by the attachment of these leukocytes to the vascular endothelium via various plasmalemmal adhesion macromolecules. These molecules include the lymphocyte function-associated antigen (LFA-1) family of glycoproteins located on the surface of neutrophils. We have initiated studies which attempt to elucidate the importance of the LFA-1 family of glycoproteins in mediating the possible deleterious effects of neutrophil infiltration in a primate model of myocardial ischemic and reperfusion injury. Male cynomolgus monkeys (5.0 to 6.7 kg) were anesthetized with Nembutal and implanted with catheters for monitoring hemodynamics (BP, HR, LVP, dP/dt, ECG) and administering radioactive microspheres for determining left ventricular area at risk (AAR) of ischemic injury. The left anterior descending coronary artery was occluded for 90 minutes followed by 4 hours of reperfusion. Animals received either saline or R15.7 (1.0 mg/kg i.v.), a monoclonal antibody (MAB) directed against the (primate) beta subunit (CD10) of LFA-1. Hearts were quickly removed following reperfusion, sectioned, and stained with triphenyletetrazolium chloride to image infarct size. Animals in both the saline (49 ± 3%, n=4) and R15.7 (49 ± 3%, n=5) treated groups had similar AAR (% of left ventricle). There were no appreciable differences in the hemodynamics monitored between the two groups. One saline-treated animal died from ventricular fibrillation during the occlusion period. Infarct size (% of AAR) was significantly ($p<0,05$) less in animals receiving R15.7 (11 ± 3%) compared to animals receiving saline (28 ± 7%). In summary, blockade of CD18-dependent neutrophil adhesion pathways limits the development of tissue necrosis in this primate model of ischemic and reperfusion injury. Possibly consistent with these findings is the observation that ICAm-1, a ligand for the LFA-1 family of glycoproteins, is located on the coronary endothelium.

INFLUENCE OF HYPERTHERMIA ON THE FIBRINOLYTIC POTENTIAL OF HUMAN ENDOTHELIAL CELLS

Johann Wojta, M. Holzer, G. Christ, R.L. Hoover, B.R. Binder

Institut for Medizinische Physiologie
Der Universitat Wien
Vienna, Austria

The effect of exposure to hyperthermia on the fibrinolytic potential of human umbilical vein endothelial cells (HUVEC) in culture was studied. HUVEC responded to exposure to 42^0C with a time dependent increase in plasminogen activator inhibitor type 1 (PAI-1) activity and antigen and a decrease in tissue-type plasminogen activator (t-PA) antigen. Hyperthermic treatment did not influence cell viability as judged by ^{51}Cr release. The effect of short term exposure to hyperthermia on PAI-1 activity and antigen could not be reversed by reexposure of the cells to 37^0C for 24 hours as evidenced by continuously increased amounts of PAI-1 released into the conditioned media. t-PA release, however, increased significantly over control in the 24 hours at 37^0C following short term exposure to hyperthermia. No difference in PAI-1 antigen present in the extracellular matrix of heat treated HUVEC as compared to HUVEC kept at 37^0C could be found. Furthermore it could be demonstrated by Northern blotting techniques that hyperthermia induced the increase in PAI-1 by increased synthesis as evident by the presence of elevated m-RNA levels. Our data supports the idea that hyperthermia is one stress factor influencing the fibrinolytic potential of endothelial cells.

THE RELEASE OF VASOACTIVE MEDIATORS FROM ARTERIAL AND VENOUS CULTURED ENDOTHELIAL CELLS

Elizabeth G. Wood, J.A. Mitchell, P. D'Orleans-Juste and J.R. Vane

The William Harvey Research Institute
St. Bartholomew's Hospital Medical College
Charterhouse Square
London EC1M 6BQ, United Kingdom

Cultured endothelial cells (EC) retain their ability to release endothelium derived relaxing factor (EDRF) (*Cocks et al., 1985*), prostacyclin (PGI$_2$) (*Gryglewski et al., 1986*) and endothelin (ET) (*Yanagasawa et al., 1988*). However, the comparative release of vasoactive mediators by EC cultured from different vascular beds has not been studied. Hence, in this study bovine EC from arterial and venous vessels are compared for their ability to release EDRF, PGI$_2$ and ET.

Endothelial cells were harvested from bovine aortae and vena cava, grown to confluence and then seeded on to microcarrier beads or in 10 cm petri dishes. The EC on beads were packed into a jacketed chromatography column and perfused with warmed (37°C), oxygenated (95% O$_2$/5% CO$_2$) Krebs' solution containing superoxide dismutase (10 U/ml). To assess EDRF release, the effluent from the columns superfused a bioassay cascade of 4 spirally cut rabbit aortic strips which were denuded of the endothelium. The effluent from the columns was collected and the release of PGI$_2$ determined by using a specific radioimmunoassay (RIA) for 6-keto-PGF$_{1a}$. EC grown to confluence in 10 cm petri dishes and incubated for up to 4 hours with or without thrombin (0.2 IU/ml) were used to study the release of ET (*Emori et al., 1989*). ET-like immunoreactivity was measured by a sandwich antibody RIA for ET-1 and ET-2.

Venous EC released EDRF in response to bradykinin (BK) at a threshold concentration of 0.05 pmol, whereas the threshold concentration for arterial EC was 1 pmol (n=4), indicating that there are more functional BK receptors on the cell surface of the venous EC. Arterial and venous EC when stimulated maximally by BK or ADP released similar amounts of EDRF. Infusions of N^G-monomethyl-L-arginine (30 mM) inhibited the basal release of EDRF along with that induced by BK. Infusion of L-arginine (100 mM), but not D-arginine, reduced this inhibition. Arachidonic acid (10-30 pmol) dose dependently induced a similar release of PGI$_2$ from both cell types. Basal levels of ET, measured after 1 min and 4 h, were of 0.08 ± 0.04 and 0.28 ± 0.05 ng/ml for venous EC and 0.16 ± 0.05 and 0.80 ± 0.14 ng/ml for arterial EC. Thrombin enhanced the detectable levels of ET by 70% after 4 h incubation in both venous and arterial EC, leading to levels of 0.43 ± 0.08 ng/ml and 1.37 ± 0.09 ng/ml respectively ($p<0.05$, n=6).

Thus, venous and arterial EC release vasodilators, such as EDRF and PGI$_2$ to a similar extent, whereas arterial EC release more ET. These findings are of particular interest as veins contract more readily to ET than arteries (*D'Orleans-Juste et al., 1988*). They may also indicate a functional difference between the two cell types. (*The William Harvey Research Institute is supported by a grant from Glaxo Group Research Ltd. PDJ is a Scholar of the Canadian Heart Foundation.*)

REFERENCES

Cocks et al., *J. Cell Physiol.* 123:310-320, 1985.
D'Orleans-Juste et al., *Br. J. Pharmacol.* 95:809, 1988.
Emori et al., *Biochem. Biophys. Res. Comm.* 160:93-100, 1989.
Gryglewski et al., *Br. J. Pharmacol.* 87:685-694, 1986.
Yanagasawa et al., *Nature* 332:411-415, 1988.

X.   LIST OF PARTICIPANTS

AHMAD, Mushtaq, Ph.D.	Department of Ophthalmology, Emory University, 1327 Clifton Road, N.E., Atlanta, GA, U.S.A.
ASBERT, Monica Alsina	Hormonal Laboratory, Hospital Clinic i Provincial, Villarroel 170, 08036 Barcelona, SPAIN
BAKKALOGLU, Aysin, M.D.	Department of Pediatrics/Nephrology, Hacettepe University, Ankara, TURKEY
BRANCO, Nuno, M.D.	Office of Gerais De Material Aeronautico, Portuguese Airforce, 2615 Alverca, PORTUGAL
BRIGHAM, Kenneth L., M.D.	Department of Medicine, Vanderbilt University, Nashville, TN, U.S.A.
BURKE-GAFFNEY, Anne,	Department of Pharmacology, University College, Foster Avenue, Belfast, Dublin 4, IRELAND
BUSSE, Rudi, Ph.D.	Department of Angewandt Physiologie, Albert Ludwigs Universitat, Hermann Herder strasse 7, 7800 Freiburg, GERMANY
CALLOW, Allan D., M.D., Ph.D.	Department of Surgery, Washington University Medical Center, 4960 Audubon Avenue, St. Louis, MO 63110, U.S.A.
CATRAVAS, John D., Ph.D.	Department of Pharmacology & Toxicology, Medical College of Georgia, Augusta, GA 30912-2300, U.S.A.
CARVALHO, Dulce, M.D.	Medicine I Service, Santa Maria Hospital, Professor Egas Roniz Avenue, 1600 Lisbon, PORTUGAL
CECCHELLI, Romeo, Ph.D.	Service de Recherche, Institut Pasteur, 20, bd Louis XIV, Lille, FRANCE
CEREDA, Roberta	Research and Development, Italfarmaco, via dei Lavoratori, 54, Cinisello Balsami (MI) 20092, ITALY
CHAMPENEY, Richard	Department of Neurosurgery, Level A2, Addenbrook's Hospital, Hill's Road, Cambridge, CB2 200, ENGLAND

CHEN, Xilin, M.D. — Department of Pharmacology & Toxicology, Medical College of Georgia, Augusta, GA 30912-2300, U.S.A.

CHESTER, Adrian, M.D. — Thoracic and Cardiac Surgical Unit, Harefield Hospital, Harefield, Middlesex UB9 6JH, UNITED KINGDOM

CONNOLLY, Daniel, T., Ph.D. — Department of Cell Culture & Biochemistry, Monsanto Company, 800 N. Lindbergh Boulevard, St. Louis, MO 63167, U.S.A.

COONEY, Deirdre, Ph.D. — Children's Research Centre, Our Lady's Hospital for Sick Children, Dublin 12, IRELAND

CUNHA-RIBEIRO, Luis, M., M.D. — Department of Physiology, Hospital S. Joao, 4200 Porto, PORTUGAL

DALAVANGA, Yotanna, M.D. — Department of Anatomy & Histology, Ioannina Medical School, Ioannina, GREECE

DAFFONCHIO, Luisa, M.D. — Department of Pharmacy, Instituto di Science Farmacologiche, 20133 Milano, ITALY

DASHWOOD, Michael, Ph.D. — Thoracic & Cardiac Surgery Unit, Harefield Hospital, Harefield, Middlesex UB9 6JH, UNITED KINGDOM

DAVIS, Larry E., M.D. — Neurology Service, VA Medical Center, 2100 Ridgecrest Drive, Albuquerque, NM 87108, U.S.A.

DEHOUCK, M.P., Ph.D. — Service De Recherche, Institut Pasteur, 20, bd Louis XIV, Lille, FRANCE

DEJANA, Elisabetta, M.D. — Instituto Mario Negri, 62 Via Eretria, Milano, ITALY

DEMIRCIOGLU, F. Ferkan, M.D. — Faculty of Medicine, Gazi University, GMK bul 18/5, Demirtepe 06440, Ankara, TURKEY

DIGLIO, Clement A., Ph.D. — Department of Pathology, Wayne State University School of Medicine, 9374 Scott Hall, 540 East Canfield, Detroit, MI 48201, U.S.A.

FEHR, Jorg, M.D. — Department of Medicine, A Hof 149, University Hospital, CH-8091 Zurich, SWITZERLAND

FOEGH, Marie, M.D. — Department of Surgery, Georgetown University, Reservoir Road, Washington, D.C., U.S.A.

GILLIS, C. Norman, Ph.D. — Department of Anesthesiology, Yale University, 333 Cedar Street, New Haven, CT 06511, U.S.A.

GIULIANI, Paola	Department of Research & Development, Italfarmazo, via dei Lavoratori, 54, Cinisello Balsami (MI) 20092, ITALY
GORDON, John, Ph.D.	Department of Research & Development Division, British Bio-technology Limited, Brook House, Watlington Road, Cowley, Oxford OX4 5LY, UNITED KINGDOM
GRAMMAS, Paula, Ph.D.	Department of Pathology, Wayne State University, Gordon H. Scott Hall, 540 East Canfield Avenue, Detroit, MI 48201, U.S.A.
GREISLER, Howard, P., M.D.	Department of Surgery, Loyola University Medical Center, Stritch School of Medicine, 2160 South First Avenue, Maywood, IL 60153, U.S.A.
GURTNER, Gail, M.D.	Division of Pulmonary Medicine, New York Medical College, Valhalla, NY 10595, U.S.A.
HAWORTH, Sheila, M.D.	Department of Pediatric Cardiology, Institute of Child Health, 30 Guilford Street, London WC2N 1EH, England, UNITED KINGDOM
HAYES, Brendan, Ph.D.	Department of Surgery, Washington University, 4960 Audubon Avenue, St. Louis, MO 63100, U.S.A.
HEISTAD, Donald D., M.D.	Department of Internal Medicine & Pharmacology, The University of Iowa College of Medicine, Iowa City, IA 52242, U.S.A.
HOLDRIGHT, Diana R.	2 Malm Close, Rickmansworth, Hertfordshire, England WD3 1NR, UNITED KINGDOM
HUBER, Kurt, M.D.	Department of Cardiology, University of Vienna, Garnisongasse 13, A-1090 Vienna, AUSTRIA
HUNT, Beverley J.	Thoracic & Cardiac Surgical Unit, Harefield Hospital, Harefield, Uxbridge Middlesex UB9 6JH, Harefield 3737, UNITED KINGDOM
HUTCHESON, Iain Robert	Department of Diagnostic Radiology, University of Wales College of Medicine, Heath Park, Cardiff CF4 4XN, Wales, UNITED KINGDOM
INNERARITY, Thomas L., Ph.D.	Gladstone Foundation Laboratory, University of California, 2550 23rd Street, PO Box 40608, San Francisco, CA 94140, U.S.A.

KARA, Bilgin, M.D.	Zanburi mah, Gilihisarlila, Selcuk University, Keskesler ap. No:7/6, Konya, TURKEY
KEENAN, Alan K., Ph.D.	Department of Pharmacology, University College, Foster Avenue, Belfast, Dublin 4, IRELAND
KEFALIDES, Nicholas A., M.D.	Connective Tissue Research Section, University of Pennsylvania, 3624 Market Street, Philadelphia, PA 19104, U.S.A.
KONTOS, Hermes A., M.D., Ph.D.	Department of Medicine, Medical College of Virginia, Richmond, Virginia 23298, U.S.A.
KORKMAZ, Seher, M.D.	Department of Pharmacology, Ankara University, 06339 Ankara, TURKEY
KUHN, Michaela, M.D.	Department of Clinical Pharmacology, Hannover Medical School, Konstanty Gutschow Street 8, 3000 Hannover 61, GERMANY
LANGLEBEN, David, M.D.	Department of Medicine-Division of Cardiology, The Sir Mortimer B. Davis-Jewish General Hospital, 3755 Chemin De La Cote Ste-Catherine, Montreal, Quebec H3T 1E2, CANADA
LARSEN, Glenn, Ph.D.	Genetics Institute, 87 CambridgePark Drive, Cambridge, MA 02140, U.S.A.
LIBBY, Peter, M.D.	Vascular Medicine and Atherosclerosis Unit, Cardiovascular Division Department of Medicine, Harvard Medical School and Brigham and Women's Hospital, 75 Francis Street, Boston, MA 02115, U.S.A.
LOSKUTOFF, David J., Ph.D.	Research Institute of Scripps Clinic, La Jolla, CA 92037, U.S.A.
LUSCHER, Thomas F., M.D.	Department of Medicine-Cardiology Division, University Hospital, CH-4031 Basel, SWITZERLAND
MANTOVANI, Alberto, M.D.	Instituto Mario Negri, Via Eretria 62, 20157 Milan, ITALY
MARAGOUDAKIS, Michael, Ph.D.	Department of Pharmacology, University of Patras, Patras, GREECE
MARCZIN, Nandor, M.D.	Department of Pharmacology & Toxicology, Medical College of Georgia, Augusta, GA 30912-2300, U.S.A.

MERONI, Pier Luigi, M.D.	Institute of Internal Medicine, Padiglione Granelli-Ospedale Policlinico, Via Francesco Sforza, 35, 20122-Milan, ITALY
MORRISSEY, Gary J.	Victoria Hospital, 375 South Street, London, Ontario, CANADA
MOSER, Rene, M.D.	University Hospital, Division of Hematology/Laboratory, Schonleistr. 10 Ch-8032, Zurich, SWITZERLAND
MULLER, Thomas, H., Ph.D.	Department of Pharmacology, Gymnasium str. 16, Box 17 55, D7950 Biberach, GERMANY
OBRIG, Tom G., Ph.D.	23 Burhans Place, Delmar, NY, U.S.A.
OKTEM, Huseyin, A.,	Institute of Biochemistry/Biology Research Center, Hungarian Academy of Sciences, Box 521, 6701 Szeged, HUNGARY
O'NEIL, Greg, Ph.D.	Thoracic and Cardiac Surgical Unit, Harefield Hospital, Harefield, Middlesex UB9 6JH, UNITED KINGDOM
ORFANOS, Stylianos, M.D., Ph.D.	Department of Medicine, Ioannina General Hospital, Ioannina, GREECE
OWNBY, Charlotte L., Ph.D.	Department of Physiological Sciences, Oklahoma State University, College of Veterinary Medicine, Stillwater, OK 74078-0353, U.S.A.
PAGE, Christopher, Ph.D.	Thoracic and Cardiac Surgical Unit, Harefield Hospital, Harefield, Middlesex UB9 6JH, UNITED KINGDOM
PAPADIMITRIOU, Evangelia	Department of Pharmacology, University of Patras, Patras, GREECE
PAPAIOANNOU, Stamatis E., Ph.D.	Department of Pharmacy, University of Patras, School of Health Sciences, Patras, GREECE
PEARSON, Jeremy, Ph.D.	Biology Division, Clinical Research Center, Harrow, Middlesex, UNITED KINGDOM
PHAN, Sem, H., Ph.D., M.D.	Department of Pathology, The University of Michigan Medical School, 1301 Catherine Road, Ann Arbor, MI 48109-0602, U.S.A.
PITT, Bruce R., Ph.D.	Department of Pharmacology, University of Pittsburgh School of Medicine, 518 Scaife Hall, Pittsburgh, PA 15261, U.S.A.

RAMWELL, Peter W., Ph.D.	Department of Physiology and Biophysics, School of Medicine, 3900 Reservoir Road N.W., Washington D.C. 20007, U.S.A.
REDMOND, Eileen	Department of Pharmacology, University College, Foster Avenue, Belfast, Dublin 4, IRELAND
REES, Daryl, Ph.D.	The Wellcome Research Laboratory, Langley Court, South Eden Park Road, Beckenham, Kent BR3 3BS, UNITED KINGDOM
REMUZZI, Giuseppe, M.D.	Instituto Di Ricerche Farmacologiche, Mario Negri, 24100 Bergamo, ITALY
ROSENKRANZ, Pnina G., M.D.	Department of Pediatrics, Yale University, 333 Cedar Street, New Haven, CT 06510, U.S.A.
RYAN, Una S., Ph.D.	Director of Health Sciences, Monsanto Company, Mail Zone R1B, 800 N. Lindbergh Boulevard, St. Louis, MO 63167, U.S.A.
SCHONHARTING, Martin, Ph.D.	Werk Kalle-Albert Klinische Forschung, Hoechst Aktiengesellschaft, Post Fach 35 40, 6200 Wiesbaden 1, GERMANY
SCOTT, T.M., Ph.D.	Memorial University of New Foundland, St. John's, Newfoundland, CANADA
SEIFERT, Paul, Ph.D.	Department of Immunology, Hospital Broussais, 96 rue Didot, Paris 75014, FRANCE
SHULMAN, Stanford T., M.D.	Division of Infectious Diseases, Northwestern University Medical School, 2300 Children's Plaza, Chicago, IL 60614, U.S.A.
SOYDAN, Inan, M.D.	Department of Internal Medicine, Ege University, Bornova/Izmir, TURKEY
SOYDAN, Saliha, M.D.	Department of Pathology, Ege University, Bornova/Izmir, TURKEY
THORGEIRSSON, Gudmundur, M.D., Ph.D.	Department of Medicine, National University Hospital, Box 10, 121 Reykjavik, ICELAND
TOIVONEN, Hannu, M.D., Ph.D.	Department of Anesthesiology, Helsinki University Hospital, Helsinki, FINLAND
TOMAZIC, Branko B., Ph.D.	Research Associate, Paffenbarger Research Center, National Institute of Standards & Technology, Gaithersburg, MD 20899, U.S.A.

TZANELLA, Marinella, M.D., Ph.D.	Department of Endocrinology, Evaggelismos General Hospital, Athens, GREECE
VAN MOURIK, Jan, Ph.D.	Blood Coagulation, Center Laboratory of Netherlands, Plesmanlaan 125, 1006 Amsterdam, NETHERLANDS
VARANI, James, Ph.D.	Department of Pathology, University of Michigan Medical School, Ann Arbor, MI 48109, U.S.A.
WALDER, Claire E.	The William Harvey Research Institute, St. Bartholomew's Hospital Medical College, Charterhouse Square, London ED1M 6BQ, UNITED KINGDOM
WATSON, Brant D., Ph.D.	Department of Neurology (D4-5), University of Miami, Box 016960, Miami, FL 33101, U.S.A.
WICKHAM, Nicholas,	Flat H510 Du Cane Court, Balham High Road, London SW17 7JT, UNITED KINGDOM
WILLIAMS, Timothy J., Ph.D.	Department of Applied Pharmacology, Cardiothoracic Institute, Dovehouse Street, London SW3 6LY, UNITED KINGDOM
WINQUIST, Raymond J., Ph.D.	Department of Pharmacology, Boehringer Ingelheim Pharmaceuticals, Inc., 90 East Ridge, Box 368, Ridgefield, CT 06877, U.S.A.
WOJTA, J., Ph.D.	Institut fur Medizinische Physiologie, Der Universitat Wien, Vienna, AUSTRIA
WOOD, Elizabeth	St. Bartholomew's Hospital Medical College, Charterhouse Square, London ED1M 6BQ, UNITED KINGDOM
YACOUB, Magdi H., M.D.	Harefield Hospital, Harefield, Middleesex UB9 6JH, UNITED KINGDOM

1. Soydan, I.	21. Argyropoulos, L.	41. Thorgeirsson, G.	61. Winquist, R.
2. Soydon, S.	22. Rosenquist, D.	42. Brigham, K.	62. Libby, P.
3. Papioannou, S.	23. Orphanos, S.	43. Carvalho, D.	63. Kuhn, M.
4. Bakkaloglu, A.	24. Tzanella, M.	44. Branco, N.	64. Watson, B.
5. Demircioglu, F.	25. Ryan, U.	45. Chen, X.	65. Cunha-Ribeiro, L.
6. Callow, A.	26. Shulman, S.	46. Tomazic, B.	66. Mantovani, A.
7. Alsina, M.	27. Varani, J.	47. Rees, D.	67. Mueller, T.
8. Gillis, N.	28. Daffonchio, L.	48. Luscher, T.	68. Seifert, P.
9. Catravas, J.	29. Giuliani, P.	49. Hunt, B.	69. Heistad, D.
10. Ahmad, M.	30. Dashwood, M.	50. Busse, R.	70. Wickham, N.
11. Innerarity, T.	31. Redmond, E.	51. Cooney, D.	71. Dejana, E.
12. Keenan, A.	32. Obrig, T.	52. Pearson, J.	72. O'Neil, G.
13. Burke-Gaffney, A.	33. Wood, E.	53. Greisler, H.	73. Meroni, P.
14. Maragoudakis, M.	34. Huber, K.	54. van Mourik, J.	74. Chester, A.
15. Walder, C.	35. Cecchelli, R.	55. Toivonen, H.	75. Loskutoff, D.
16. Dehouck, M.	36. Diglio, C.	56. Williams, T.	76. Page, C.
17. Dalavanga, Y.	37. Ownby, C.	57. Wojta, J.	77. Remuzzi, G.
18. Korkmaz, S.	38. Gurtner, G.	58. Cereda, R.	78. Marczin, N.
19. Papadimitriou, E.	39. Kontos, H.	59. Haworth, S.	
20. Grammas, P.	40. Morrissey, G.	60. Hayes, B.	

XI. INDEX

INDEX

ACE, *see* angiotensin converting enzyme
Acetylcholine, 34, 44, 48, 51, 52, 57, 65, 142, 144-146, 149-151, 153-155, 170, 175-180
ACh, *see* acetylcholine
Adenosine uptake, 112
Adult respiratory distress syndrome, 3, 11, 55, 56, 64, 66, 91, 233
AECA, 117-122
  in transplant rejection, 117, 118
Aging, 167, 171, 172
Alloantigens, 121, 124, 226-229
AMP, 102, 112-115, 132
  angiogenesis, 71, 74, 75, 200, 233
Angiotensin Converting Enzyme, 56, 62-65, 234
  inhibitors, 62, 63, 64
Anti-endothelial cell, 117, 122-125
Anti-oxidant, 63
Antibodies, 15, 16-18, 25, 69, 84, 87, 88, 89, 90, 92, 99, 107, 117, 118, 119, 121-125, 128, 129, 134, 159, 191, 195, 196, 225, 227, 228
Apolipoprotein B, 24, 25, 183, 192, 194--196
Apolipoprotein E, 183, 194-196
Arachidonate metabolites, 158
ARDS, *see* adult respiratory distress syndrome
Arteriolopathy, 159-162
Atherosclerosis, 13, 14, 17, 18, 22-28, 33, 34, 36-45, 50, 51, 82, 87, 89, 94, 123, 124, 151, 152, 154, 156, 167, 172, 183, 184, 186-191, 193-196, 203, 233
ATP, 111-116, 132, 133, 147

Blood-brain barrier, 48-51, 167-173, 180

cAMP, 106, 199, 201
Cellular immunity, 122
Cerebral blood vessels, 167, 169
cGMP, 145, 156
Chemoattractant, 73, 75, 83
Chemotaxis, 23, 73, 82
CyA, *see* cyclosporine

Cyclosporine, 157, 162-166
  chronic nephrotoxicity, 159, 161
Cytokines, 66, 79-84, 89-94, 97, 101, 118--121, 124, 129, 134, 198
Cytotoxicity, 119, 121-125, 129, 132, 133, 135, 128
  cell-mediated, 119

Electrolysis, 59, 63, 64, 66
Endothelial cell, 8, 11, 15, 16, 23-25, 27, 28, 32, 34, 36, 38, 41, 43, 44, 47-49, 50, 55, 57, 59, 63, 64, 66, 71-76, 79, 80, 82-84, 88, 90-95, 97-101, 104-106, 111-117, 122-125, 127-135, 141, 149-151, 154-159, 161-164, 166, 175, 178, 194-197, 200, 201, 203-210, 225-229, 234-236
EDRF, *see* endothelium derived relaxing factor
cell growth, 71, 74, 75
endothelium-dependent responses, 8, 11, 15, 16, 18, 23, 25, 38, 41, 44, 45, 47-51, 55-57, 59, 61, 64-67, 79, 83, 84, 87-89, 93, 94, 101, 103, 104, 106, 111-117, 122-125, 127, 129, 130, 133, 134, 139, 141, 142, 144, 145, 147, 149-164, 167-172, 175-180, 194, 202-204, 225, 233--236
endothelium-derived relaxing factor, 49, 50, 57, 59, 61, 63, 65, 66, 112, 115, 142, 145-147, 149, 150, 153-157, 171, 175-178, 180, 236
pulmonary endothelium, 11, 65
Endothelin, 55, 56, 66, 67, 113, 116, 142, 146, 154-156, 159-165, 236
Endotoxins, 91, 157
Epidemiology, 27

Fibrinolysis, 69, 80, 92, 97, 103-105, 122
Fibroblasts, 71, 83, 84, 93, 94, 102, 105, 117-119, 191, 200, 227, 229
Fibrosis, 90, 100, 104, 159

Genetic disorders, 183, 189, 195

315

Glanzmann's thrombasthenia, 228, 229
Glomerular filtration rate, 158, 165
Glomerulosclerosis, 160
Glucose transport, 72, 74, 75

Hemolytic-uremic syndrome, 161, 209
Histamine, 11, 87, 88, 94, 104, 115, 142, 144, 146, 153, 157
Hydrogen peroxide, 47-49, 59, 127, 130, 134, 135, 176-178
Hydroxyl radical, 48, 49, 130, 134, 176, 177, 179
Hypersensitivity, 36, 87-90, 94
Hypertension, 29, 31, 39, 44, 47-52, 64, 115, 150, 152, 153, 155, 158, 167-173, 177-180, 233, 234

IFN, *see* interferon
Inflammation, 8, 44, 52, 69, 72, 73, 75, 79, 80, 82, 83, 87-89, 92, 94, 95, 134, 233
Integrins, 121, 197-201, 225-229
Interferon, 34, 89, 90, 119, 129
Interleukin, 80-84, 89, 90, 92-95, 101, 103, 106, 119, 124, 129, 134, 157
Intracellular, 13, 61, 100, 102, 111-115, 130, 145, 197, 199, 201, 210
  signalling, 113

Kawasaki disease, 117-119, 121, 123, 124

LDL, 13, 19, 20, 150, 183-191, 193, 194, 235, 236
Leukocytes, 14, 25, 49, 50, 79-82, 84, 87-89, 91-93, 101, 121, 125, 129, 141, 225, 226
Lipoprotein, 32, 150, 155-157, 183, 184, 186, 187, 189, 193-196
Lymphocytes, 27, 28, 34, 45, 69, 74, 75, 81-84, 88-91, 119, 121, 124, 128
Lymphokines, 34, 45, 75, 89

Mitomycin, 157, 163, 164
Monocytes, 13, 16, 18, 24, 69, 73-75, 82, 84, 93, 118-121, 123, 124, 133, 128, 141, 151, 194

Neovascularization, 37, 42, 43, 69
Neutrophils, 8, 11, 23, 25, 51, 71, 74, 82, 83, 88, 93, 99, 121, 123, 127, 129-135, 128
NO (nitric oxide), 4, 10, 15, 31, 33, 38-41, 49, 50, 57, 58, 61, 63, 64,

Oxygen radical, 59, 133, 176, 177

PGI$_2$, *see* prostacyclin
Plasminogen activation, 97
Platelet, 27, 32, 38, 43, 48-52, 55, 66, 70, 75, 76, 79, 82, 83, 84, 88, 89, 92-94, 98, 99, 101, 103, 105, 106, 107,

Platelet (continued) 111, 113, 120, 121-123, 124, 127, 141, 142, 147, 149, 151, 152-157, 160, 161, 162, 171, 172, 175, 179, 198-200, 201, 203, 204, 207-210, 225-229
Platelet derived-growth factor (PDGF), 70
Prostacyclin, 8, 25, 32, 62, 66, 79, 84, 112, 114, 115, 120, 122, 123, 125, 142, 144, 149, 150, 153, 156, 162, 165, 166, 209
  synthesis, 84, 165, 209
  stimulating factor, 162
Prostaglandins, 90, 158
Purines, 111, 112, 114
  extracellular purines, 111, 114
  P$_2$ receptors, 112-114

Renal plasma flow, 158
Renal vasoconstriction, 158-162, 165
Renin-angiotensin system, 158

Serotonin, 48, 51, 55, 56, 61, 65, 66, 141, 145, 149, 152, 155, 157, 168, 172
Shock, 6, 66, 91-95
Smooth muscle cells, 13, 22, 27, 28, 34, 36, 39, 42-44, 71, 81, 82, 84, 87, 89-95, 104, 111, 112, 114, 115, 118, 119, 141, 145, 151, 152, 155, 160, 162, 193-195, 198, 227-229
Superoxide anion, 50, 51, 113, 116, 130, 178, 179
Systemic lupus erythematous, 117
Systemic sclerosis, 117-123, 164

T-lymphocytes, 27, 28, 34
TF, *see* tissue factor
Therapy, 8-11, 39, 101, 153, 157- 162, 164, 165, 233, 234, 236
Thromboxane A$_2$, 23, 24, 57, 113, 146, 149, 151, 155, 157-159, 162
Tissue Factor, 120
TN, *see* tissue necrosis factor
Transplant, 39, 42-44, 91, 117, 118, 158, 159, 161-166
Tumor necrosis factor, 71-73, 75, 76, 79, 80, 82-84, 89, 91, 95, 101, 103, 106, 119, 123, 129, 130, 134, 201
TXA$_2$, *see* thromboxane A$_2$
Type 1 plasminogen activator inhibitor, 97, 105, 106

Vascular endothelial growth factor, 69-71
Vascular injury, 44, 45, 47, 48, 50, 51, 88, 92, 101, 161-163, 203, 208
Vascular permeability factor, 69-76
VEGF, *see* vascular endothelial growth factor
Vessel wall, 48-50, 82, 90-92, 97, 100, 101, 111, 114, 116, 141, 147, 152, 160, 162, 194, 203, 204, 225

von Willebrand factor, 120, 123, 197-200, 203, 204, 206, 209, 210, 228
von Willebrand's disease, 206, 207, 209, 210
VPF, *see* vascular permeability factor

Wegener's granulomatosis, 117-119, 123, 124
Weibel-Palade bodies, 94, 204, 206, 209

Xanthine oxidase, 47, 61, 64, 65, 132-134
Xanthine/hypoxanthine, 132

GPSR Compliance

The European Union's (EU) General Product Safety Regulation (GPSR) is a set of rules that requires consumer products to be safe and our obligations to ensure this.

If you have any concerns about our products, you can contact us on

ProductSafety@springernature.com

In case Publisher is established outside the EU, the EU authorized representative is:

Springer Nature Customer Service Center GmbH
Europaplatz 3
69115 Heidelberg, Germany

www.ingramcontent.com/pod-product-compliance
Lightning Source LLC
Chambersburg PA
CBHW082022250426
43749CB00045B/858